Frontier Discoveries and Innovations in Interdisciplinary Microbiology

Pratyoosh Shukla
Editor

Frontier Discoveries and Innovations in Interdisciplinary Microbiology

Springer

Editor
Pratyoosh Shukla
Enzyme Technology and Protein
Bioinformatics Laboratory
Department of Microbiology
Maharshi Dayanand University
Rohtak, Haryana, India

ISBN 978-81-322-2609-3 ISBN 978-81-322-2610-9 (eBook)
DOI 10.1007/978-81-322-2610-9

Library of Congress Control Number: 2015955557

Springer New Delhi Heidelberg New York Dordrecht London

Printed on acid-free paper

Springer (India) Pvt. Ltd. is part of Springer Science+Business Media (www.springer.com)

Foreword

Newcastle University

This exciting new book on the latest microbiological techniques is an outstanding collection of contributions from notable authors in various disciplines of microbiology, highlighting cutting-edge discoveries and innovations. It covers a broad diversity of contemporary and interdisciplinary microbiological topics, such as algal biotechnology, biodiversity, biofuel and bioenergy, bioinformatics and metagenomics, environmental microbiology, enzyme technology, marine biology, food and medical microbiology as well as agricultural and soil microbiology.

An important feature of this book is that it presents the latest areas of technological progress and directs them towards collaborative and interdisciplinary research. It also highlights the roles of rewarding interactions in academic research that are essential for development of the most progressive biotechnologies. Most of the chapters in this book are dedicated to industrially important aspects in microbiology ranging from microbial production of antioxidant-glutathione, enzyme engineering, probiotics, and functional aspects of xylanases to some very concerned topics like potential health hazards in recreational water environments, fullerene nanocomposites, etc. In addition, the book also highlights essential topics involving new discoveries and their impact on microbiology and biotechnology.

While the contributions presented through the 13 chapters review the current state of the art in various research areas they also describe the latest trends in the development of core biotechnologies for the immediate future.

Therefore, it is fully justified to congratulate the Editor of the book, Prof. Pratyoosh Shukla, for the successful completion and publication, which would not have happened without his enormous efforts and dedicated work. I very much appreciate his sincere efforts in bringing a book together, which in a true sense emphasizes frontier discoveries and innovations in microbiology.

School of Biology
Newcastle University
United Kingdom

Hans-Peter Klenk

Contents

About the Editor

Prof. Pratyoosh Shukla, Ph.D. is working as Head, Department of Microbiology, at Maharshi Dayanand University, Rohtak, India. Prof. Shukla is also working as General Secretary of the 75-year-old Association of Microbiologists of India (AMI), one of the reputed scientific organizations in the world.

He has more than 14 years of research and teaching experience in reputed universities of India and post-doctoral experience at Durban University of Technology, Durban, South Africa, and worked as Research Professor at the University of Cincinnati under the Indo-US Science and Technology Forum (IUSSTF) and the American Society of Microbiology (ASM) program. He has written 4 books, 15 book chapters, and one patent; edited 3 biotech journals (Springer); and published more than 65 peer-reviewed papers in reputed national and international journals. He has received several awards, including the Prof. S.B. Saksena, F.N.A., Award in Life Sciences (1999), the Best Presentation Award (Senior Category-2006) by NCSTC, India, the NRF-DUT PDF Award in Enzyme Biotechnology (2008), the Danisco India Award in Probiotics and Enzyme Technology (2010), and the Indo-US Professorship Award (2014) by Indo-US Science and Technology Forum (IUSSTF) and the American Society of Microbiology (ASM) and has been selected as Scientist in Southern Ocean Expedition (2011), and the American Society of Microbiology (ASM).

He is serving as Editor of the *Indian Journal of Microbiology* (Springer) and as editorial board member and reviewer in many peer-reviewed SCI journals. He also served as Editor-in-Chief of the *Internet Journal of Microbiology*, Internet Scientific Publishers, USA, during 2008–2009. His current research interests are in the areas of Enzyme Technology, Protein Bioinformatics, and Systems Biology.

Studies on Biosynthetic Production of Antioxidant Glutathione Using Microbial Cultures

1

Poonam Singh Nigam
and Richard Owusu-Apenten

Abstract

Glutathione (γ-L-glutamyl-L-cysteinyl-glycine) is a well-known major antioxidant produced in living cells at the concentrations of 1–10 mM. It protects cells from nutritional, environmental, and oxidative stresses. Structurally, it is a nonprotein thiol compound made up of three amino acids – glutamate, cysteine, and glycine with an unusual γ-peptide bond. To meet the increasing commercial demand of GSH as a functional food supplement, various methods for its production methods have been tried. Forms of glutathione (GSH and GSSG) could be produced by enzymatic methods, which are expensive and unprofitable. The microbial biosynthesis could be achieved by fermentation method using natural or engineered microorganisms such as some strains of yeasts *Saccharomyces cerevisiae* and *Candida utilis* and some strains of bacteria *Escherichia coli* and *Lactococcus lactis*. However, the former approach is not commercially viable because of its high production cost as compared to the latter one, which is more practical and also cost-effective. Certain microorganisms such as *Saccharomyces cerevisiae* and *Candida utilis* have been studied as potential microorganisms for GSH production. For enhanced production of glutathione, using a bifunctional enzyme encoded by gshF from *Streptococcus thermophilus* has been expressed in *Escherichia coli*.

Keywords

Antioxidant • Glutathione • Fermentation • *Saccharomyces cerevisiae* • Oxidative stress • Osmotic stress • Esterification

P.S. Nigam (✉) • R. Owusu-Apenten
School of Biomedical Sciences, Faculty of Life and Health Sciences, University of Ulster,
Coleraine BT52 1SA, Northern Ireland, UK
e-mail: p.singh@ulster.ac.uk

P. Shukla (ed.), *Frontier Discoveries and Innovations in Interdisciplinary Microbiology*,
DOI 10.1007/978-81-322-2610-9_1

1.1 Significance of Glutathione

Glutathione (GSH) possesses life-sustaining functions as it is an important antioxidant synthesized by mammals. The deficiency of GSH has been reported to cause several diseases (Ballatori et al. 2009; Lu 2009) due to the oxidative stress (Valencia et al. 2001). GSH a nonprotein thiol compound is found in cells at the concentration of 1–10 mM. GSH is found to be the most important antioxidant, because its anti-oxidizing capability is 100 times more than simple antioxidants. Therefore, there is a high demand and research on production of synthetic antioxidant (Nutracam 2007; Xiong et al. 2009; Das et al. 2011).

The antioxidants and their antioxidant activities are the subject of interest for the nutritionists (Nutrition Advisor 2007) and health professionals (World IPO 2007) and so its increasing demand in pharmaceutical and food industries. Antioxidants are the factors that protect cells against free radicals by preventing the oxidation of other molecules. The common examples of antioxidants are vitamin C, flavonoids, lipoic acid, vitamin E, and glutathione. Natural antioxidants including vitamin C, flavonoids, and coumarins are widely found in plants, many have been extracted from various plants using ultrasonic methods, and their antioxidant capacities have been studied in detail (Adam et al. 2009). Antioxidants are those compounds which prevent the natural formation of reactive oxygen species (free radicals) inside the living cells (Wu et al. 2004).

Glutathione (g-L-glutamyl-L-cysteinyl-glycine), a tripeptide, is a low molecular weight nonprotein thiol compound (Li et al. 2004). It is made up of three amino acids, namely, cysteine, glutamic acid, and glutamine. There is a γ-peptide bonding between glutamate and cysteine which makes this molecule very resistant toward its hydrolysis by most peptidases (Anderson 1998). GSH exists in two forms:

1. Reduced form of glutathione with molecular weight 307.3 g/mol – referred as *GSH*
2. Oxidized form of glutathione with molecular weight 612.6 g/mol – referred as *GSSG*

These forms of GSH can be measured using a practical HPLC method (Yilmaz et al. 2009).

1.2 Properties of Glutathione

GSH is associated with in various vital functions such as in the maintenance of cells, as radical scavenger, in detoxification and restoration of redox potential (Santos et al. 2007; Noctor et al. 1998). Glutathione was first identified in baker's yeast cells extracted with ethanol as early as in 1888. Its molecular structure was established in 1921, and it was named as glutathione. Reduced form of glutathione is γ-glutamylcysteinylglycine, a nonprotein thiol compound, which has been reported (Meister and Tate 1976) to be present in most aerobic organisms.

Glutathione is one of the major antioxidant, which has been detected at the concentration of up to 10 mM in living organisms and mammalian cells. GSH reacts with toxic compounds to form GSH conjugates in enzymatically or nonenzymatically catalyzed reactions.

The significant functions of GSH at cellular level have been placed in three main categories:

(a) *Antioxidant:* This is the most important function of GSH. It protects cells against oxidative damage caused by the reactive oxygen species. It is also found to have an antiaging effect on the body.
(b) *Detoxifier:* It detoxifies the thiol group of foreign compounds by conjugating with the exogenous electrophiles and diverse xenobiotics.
(c) *Immunity booster:* GSH plays an important role in white blood cells production.

GSH has its role in the regulation of cellular events like protein synthesis, gene expression, and protein glutathionylation and in the regeneration of vitamin C and vitamin E to its unoxidized state, conversion of prostaglandin H_2 into prostaglandins D2 and E2 by endoperoxide isomerase. GSH converts methylglyoxal to D-lactate in a glyoxalase microbial pathway, and it also actively contributes in spermatogenesis and sperm maturation. Commercially, GSH has established its application in several cosmetic products, where it is used as an emulsifier and moisturizing agent, to enhance the whitening of skin and as an antiaging agent in sunprotection products (Wu et al. 2004; Bachhawat et al. 2009). If the cells become deficient of GSH, this condition may lead to certain diseases such as Alzheimer's, Parkinson's, cystic fibrosis, sickle cell anemia, cancer, and HIV (Wu et al. 2004). Wang et al. (2008) have reported that GSH can be used in brewing industries to maintain the high cell viability and to enhance the flavor of beer.

1.3 Occurrence of Glutathione

Natural glutathione is found in several fruits and vegetables, like orange, peach, squash, asparagus, potato, spinach, tomato, grapefruit, etc. The consumption of some vegetables, such as broccoli, cabbage, Brussels sprouts, cauliflower, kale, and parsley, in human diets contributes in two ways – firstly, providing their naturally occurring GSH and secondly also stimulating the human body to self-produce this powerful antioxidant.

Though there are many natural sources to extract GSH, but economically, yeast cells are proved to be the most suitable resource to extract GSH. The yeast can be easily and practically grown on larger scale in shorter time period and have a low-cost maintenance and storage compared to vegetables and fruits. Yeasts have exceptional capability to accumulate GSH in its cells, and low operational costs are involved in its extraction (Vitamin Stuff 2013). Research work has been focused on some specific yeast strains, e.g., *Saccharomyces*, *Candida*, *Kluyveromyces*, *Pichia*, *Rhodotorula*, *Hanserzula*, *Debaryomyces*, *Torulopsis*, or the fission yeast genus *Schizosaccharomyces*, which produced glutathione in simple cultivation medium. For the large-scale production of GSH, two yeasts – *Saccharomyces cerevisiae* and *Candida utilis* – are widely used (Cha et al. 2004; Liang et al. 2008a, b).

1.4 Biosynthesis of Glutathione

Several researches were actively carried out on fermentative and enzymatic production of GSH (Yamada et al. 1984; Murata 1989). Thornalley (1991) studied the esterification of glutathione for its bioavailability. Anderson (1998) overviewed the biosynthesis and modulation of glutathione. The optimization of biosynthesis of GSH using yeast cells of *Saccharomyces cerevisiae* has been studied by many workers (Liu et al. 1999; Chalil et al. 2010; Nigam and Owusu-Apenten 2011).

Glutathione can be enzymatically produced in the presence of ATP and its precursor amino acids, namely, L-glutamic acid, L-cysteine, and glycine (Lu 2013). But the commercialization of GSH was obtained by fermentative production process and not by the method of enzymatic production, because of its high production cost (Li et al. 2004). Later on the large-scale production of GSH was carried out by fermentation methods using *Candida utilis* and *Saccharomyces cerevisiae* (Wei et al. 2003a, b).

It is synthesized from its constituent amino acids by two ATP-dependent reaction steps which are catalyzed by two enzymes γ-glutamylcysteine synthetase (GSH1) and glutathione synthetase (GSH2). In the first step, γ-glutamylcysteine is formed from glutamate and cysteine, catalyzed by the enzyme γ-glutamyl synthetase. In the second step, glycine is added to the C-terminal of γGC by the enzyme glutathione synthetase (Li et al. 2004). The strong electron-donating factor of glutathione is generally due to the presence of a sulfhydryl group (SH) on the cystenial portion. It is mainly present in the body in reduced form; GSH is oxidized to GSSG by reacting with free radicals catalyzed by enzymes glutathione peroxidase or glutathione *S*-transferase. Regeneration of GSH is an NADPH dependent and is catalyzed by the enzyme glutathione reductase (Grant 2001). GSH reacts enzymatically or nonenzymatically with toxic products to form GSH conjugates (Anderson 1998). In humans, the metabolism takes place in the liver by absorbing amino acids to increase GSH level for detoxifying chemicals (Valencia et al. 2001).

Due to its vast functions and significance, the bioavailability of GSH to humans was of great concern. Many techniques were undertaken to make glutathione available on large scale but at a lower cost. Chemical method was one of the earliest method in which modified version of benzyl carbonate method was used to make synthetic glutathione. But the GSH synthesized in this process was optically inactive mixture of D- and L-isomers; therefore, the optical resolution was required to separate the L-form from the D-isomer which hindered this method (Bachhawat et al. 2009).

Enzymatic method has tried to avoid the problems and cost involved in chemical synthesis of GSH. The immobilized microbial cells and enzymes have also been studied in the bioreactor systems. The enzymatic method required the participation of ATP and coenzymes; however, the regeneration of ATP and the low activity of GSH1 and GSH2 became the limiting factors in GSH biosynthesis. This method was also expensive and time consuming because of the usage of three precursor amino acids. The biosynthetic approach via fermentative production method is the most economical one, which involves the sugar material as the fermentation raw substrate, and microorganisms are employed to carry out the fermentative-biosynthetic process (Li et al. 2004). Yeast strains are the most suitable microorganism for this purpose, yeasts like *Saccharomyces cerevisiae* and *Candida utilis* have been found to possess higher intracellular content of GSH. The biggest advantage using yeasts is that these yeasts can be easily grown on cheaper sources of sugar substrates, and the other major advantages in using yeast are that they do not cause any endotoxic reactions in humans. In research to increase yield in biosynthesis of GSH, recombinant techniques have been used to overexpress gshp1 and gshp2 in *S. cerevisiae* and *E. coli*, but any increased level of glutathione yield could not be achieved (Bachhawat et al. 2009). Research has been conducted by Li et al. (2011) on the production of glutathione using a bifunctional enzyme encoded by gshF from *Streptococcus thermophilus* which was expressed in *Escherichia coli*. The gene coding for a novel bifunctional enzyme that is catalytically involved in the reaction for glutathione synthesis, gshF, was cloned from *Streptococcus thermophilus* SIIM B218 and expressed in *Escherichia coli* JM109. The induced *E. coli* JM109 (pTrc99A-gshF) showed the accumulation of 10.3 mM GSH in 5 h in the presence of the precursor amino acids and ATP. Addition of higher concentrations of the precursor amino acids and ATP induced *E. coli* JM109 (pTrc99A-gshF) to produce up to 36 mM GSH. The molar yield of GSH based on added cysteine was 0.9 mol/mol and based on added ATP yield was found 0.45 mol/mol. Li et al. (2011) found that when ATP was replaced with glucose, *E. coli* JM109 (pTrc99A-gshF) could produce 7 mM GSH in 3 h. ATP was generated by *Saccharomyces cerevisiae* for GSH production. *E. coli* JM109 (pTrc99A-gshF) produced 33.9 mM GSH in 12 h with a yield of 0.85 mol/mol based on added L-cysteine in the presence of glucose and the pmr1 mutant of *S. cerevisiae* BY4742.

1.5 Parameters Control in GSH Biosynthesis

A number of researches were carried out to study the cellular responses of different microorganisms in order to maintain metabolic viability and vitality of strains for use in industrial fermentation. Thus, a number of practical approaches were carried out to enhance the fermentative glutathione production, and this could be performed either by increasing the cell biomass by high cell density cultivation or by optimizing the fermentation process parameters (Liang et al. 2008a, b, c). The overall production of GSH has been focused in three approaches: (a) production methods, (b) extraction methods, and (c) detection methods.

Initially glutathione was produced extracellularly by *Candida tropicalis* pk233. In this method, the filamentous form of organism was grown in the presence of ethanol. Initially at higher concentration of ethanol, the production of extracellular glutathione was low, but it increased rapidly along growth time, and the maximum glutathione was obtained after 96 h of cultivation at 5 % (v/v) concentration of ethanol (Yamada et al. 1984). Optimization of medium was one of the basic approaches for higher GSH production. Liu et al. (1999) optimized the growth medium by using peptone, magnesium sulfate, and glucose for enhanced synthesis of GSH by *S. cerevisiae* and obtained a yield of 124 mg/L. Later on Cha et al. (2004) improved the yield up to 204 mg/L by using a different medium composition consisting of glucose, yeast extract, KH_2PO_4, and cysteine.

Amino acids like cysteine, glutamic acid, and glycine are required to form glutathione. Hence, the enhanced GSH production was studied using amino acid supplementation using *S. cerevisiae* (Wen et al. 2004) and by amino acid modulation (Wen et al. 2006). Cysteine had a major impact, so a two-step strategy was undertaken where cysteine was added at the beginning of culturing, and then other amino acids were added after sometime which resulted in 2.67 times increase in GSH yield. Single shot of cysteine addition was shown to give higher yield as compared to continuous addition, because more amount of cysteine ceased the cell growth. Later studies used a strategy of combining amino acids with other parameters, where a feedback control strategy was used and the GSH yield and productivity could be increased by 25 and 70 % from *S. cerevisiae*. Santos et al. (2007) optimized culture conditions of *S. cerevisiae*, and GSH yield achieved was 154.5 mg/L after 72 h incubation. Subsequent studies in 2008 were designed on control of dissolved oxygen (DO) along with cysteine addition in *C. utilis*. It was undertaken by adapting a two-stage strategy where the cysteine was added initially; DO was initially controlled at 5 % v/v for the first 3 h, and then it was gradually increased to 20 % in the following 12 h, such strategy resulted in 1767 mg/L GSH yield (Liang et al. 2008a). In the same year, again, *C. utilis* was used by adapting precursor amino acids with ATP which was added after 15 h of amino acid addition resulting in GSH yield of 2043 mg/L after 72 h of incubation (Liang et al. 2008b).

Since yeast cells consist of stress molecule that is synthesized and stored under high stress conditions, therefore, other parameters like salinity, high temperature, and osmotic pressure were undertaken to obtain maximum GSH without inhibiting the cell growth. Earlier in 2003, effect of temperature was studied on *C. utilis*, and a two-stage temperature control strategy was chosen for increasing the cell density as well as higher GSH production by growing cells at 30 °C for 8 h and glutathione production at 26 °C ranging 235–385 mg/L (Wei et al. 2003a). With regard to pH studies in *C. utilis*, a low pH strategy was applied initially in 2005 and then later on combined low pH with cysteine addition (Liang et al. 2008c). The pH less than 1.5 with single and double shift addition led to 673 and 558 mg/L GSH excretion into the medium (Liang et al. 2008b, c). Dong et al. (2007) studied the effect of high pressure on glutathione accumulation in *S. cerevisiae*, in this approach the yield of GSH obtained was 103 g/mg GSH at 1.0 MPa.

Glutathione in reduced form is electron donor for GPx reaction (Izawa et al. 1995). Having this knowledge that exposure of cells to osmotic and oxidative stress results in inhibition of cell growth, the strategy of multiple oxidative stress using H_2O_2 and osmotic stress using NaCl was applied conditions on *C. utilis*. The maximum GSH yield obtained was 218 mg/L and 238 mg/L, respectively, under these two conditions. Increased level of intracellular cysteine and γ-glutamyl synthetase and GSH reductase were noticed indicating the involvement in GSH accumulation against the condition of stress imposed (Liang et al. 2009a, b).

The downstream processing for the extraction of GSH after the fermentation process is the most important step in GSH recovery. The main objective is the selective separation of compound from the fermented medium. Despite the fact that extraction of GSH is the most important step, only a little work has been carried out on the extraction methods, and hence the detailed information is not available. Earlier glutathione was extracted by using aqueous acetone in Pirie's method which extracted around 60 % of the GSH. In 1988, glutathione extraction was carried out from *S. cerevisiae* by selenite. Selenite was added into the medium, and intracellular GSH was leaked out turning the yeast cell culture red due to the presence of elemental selenium. In this reaction glutathione reductase was involved in the reduction of glutathione selenotrisulfide to glutathione selenopersulfide or to GSH and elemental selenium (Iizuka et al. 1988). In one work copper precipitation method and ion exchange chromatography were used for GSH extraction (Bachhawat et al. 2009). In 2005, a comparison of various techniques was made based on factors like extraction time and temperature. The techniques used were ultrasonic methods, homogenizing method, and autolysis. The autolysis method was found to be an efficient technique among others (Salleh et al. 2005).

The solvent ethanol was used for the extraction in cells of *S. cerevisiae* without disrupting the yeast cells by Xiong et al. (2009), thus having advantage over other techniques as it can maintain cells under appropriate condition with intact plasma membrane. This method required less time, and the solvent ethanol used for the extraction could be recycled, which reduced the cost of the overall production (Xiong et al. 2009). Surfactants were used to optimize the extracellular accumulation of GSH in a fermentation production process using *S. cerevisiae* by Wei et al. (2003b). GSH production was found mainly affected by the factors like processing temperature, extraction time, the concentration of the yeast cells, and the solvent used (Salleh et al. 2005); however, the extraction using ethanol has proven beneficial to extract GSH for a higher yield. Another important parameter was glutathione detection in yeast cells which was determined by the popular method known as Tietze method. This method includes the use of DTNB also known as Ellman's reagent which reacts with GSH and produces an oxidized form GSSG and 2-nitro-5-thiobenzoic acid, which is yellow in color (Biomax Co. Ltd 2007). The formation of color can be determined by OD measurement at 412 nm using a spectrophotometer. The advantage of using this reagent was its convenience and short detection time, and DTNB produced can be recycled.

Many conventional techniques have been used like encapsulating the glutathione in the liposome or in combination with other pharmacological products like vitamins or statin, a cholesterol-lowering product (World Intellectual Property Organization 2007). In some research work, glutathione precursor is made available which induces glutathione formation inside the body (Nutrition Advisor 2007). The pure form of GSH is presently supplemented in the form of oral pills or tablets. However, the major drawback of this supplementation is that glutathione is broken down into its three constituent amino acids in the digestive system by the enzymes γ-glutamyl transpeptidases, and thus the glutathione doesn't reach the end target (Nutra Cam 2007).

The bioavailability of GSH can be improved by esterifying the GSH with alcohol and using HCl as a catalyst. The monoester form of GSH can be regarded as major bioavailable form of glutathione (Thornalley 1991), as it can be easily taken up by the cells.

1.6 Future Prospects

The biosynthesis of GSH using yeast cells, particularly *S. cerevisiae*, has been studied by several researchers, because of its capability of regenerating sufficient ATP for GSH biosynthesis and due to simplified glycolytic pathway. Yeast cells can be used in large-scale biosynthesis of GSH due to the fact that these compared to other microorganisms are more capable of thriving under adverse growth conditions (Murata 1989). The potential ability of organism to produce higher amount of glutathione under osmotic stress condition using NaCl, and under oxidative stress using H_2O_2, could be used by applying the strategy of varying the doses and periods of stress factors. The advantages of use of ethanol in extraction studies and the comparison of different alcohol solvents should be explored for the effective extraction of glutathione to obtain maximum yield under high cell density achieved in fermentation process.

A modified HPLC method may be used to determine the oxidized and reduced form of GSH in animal tissues which would be faster and more sensitive and would not require derivatization process unlike normal method (Yilmaz et al. 2009). The bioavailability on larger scale can be enhanced by esterification of glutathione using the background information from previous studies (Thornalley 1991).

The future prospects for the microbial biosynthesis of GSH should be focused on enhancing the yield of glutathione through the recovery of dense yeast cell mass and making it bioavailable using optimized method and suitable chemicals on a large scale. The large-scale fermentative production of glutathione can be done producing active yeast cell cultures in bioreactors under optimized conditions mainly DO and some stress agents. The formulation of GSH, according to its required applications, can be devised in the form of tablets, powder, or cream. Esterified GSH can be tested on animals for its bioavailability and functions and for the pharmaceutical applications.

References

Adam M, Dobias P, Eisner A, Ventura K (2009) Extraction of antioxidants from plants using ultrasonic methods and their antioxidant capacity. J Sep Sci 32:288–294

Anderson M (1998) Glutathione: an overview of biosynthesis and modulation. Chem Biol Interact 11:1–14

Bachhawat A, Ganguli D, Kaur J, Kasturia N, Thakur A, Kaur H, Kumar A, Yadav A (2009) Glutathione production in yeast. In: Satyanarayan T, Gunje G (eds) Yeast biotechnology diversity and applications. Springer, New York, pp 259–280

Ballatori N, Krance SM, Notenboom S, Shi S, Tieu K, Hammond CL (2009) Glutathione dysregulation and the etiology and progression of human diseases. Biol Chem 39:191–214

Biomax. Co. Ltd (2007) http://www.biomaxkorea.com/product/dojindo/dpo-appli/stress/stress.htm

Cha J, Park J, Jeon B, Lee Y, Cho Y (2004) Optimal fermentation conditions for enhanced glutathione production by *Saccharomyces cerevisiae* FF-8. J Microbiol 42:51–55

Chalil S, Das AJ, Nigam P, Owusu-Apenten R (2010) Novel approaches of production and downstream processing of glutathione from microbial resources, In: International Congress in Bioprocesses in the Food Industries (ICBF), University of Parana, Curitiba, Abstract book p 3

Das A, Chalil S, Nigam P, Magee P, Janneh O, Owusu-Apenten R (2011) Glutathione transferase-P1-1 binding with naturally occurring ligands: assessment by docking simulations. J Biophys Chem 2:401–407

Dong Y, Yang Q, Jia S, Qiao C (2007) Effects of high pressure on the accumulation of trehalose and glutathione in the *Saccharomyces cerevisiae* cells. Biochem Eng J 37:226–230

Grant CM (2001) Role of glutathione/glutaredoxin and thioredoxin systems in yeast growth and response to stress conditions. Mol Microbiol 39:533–541

Iizuka M, Murata K, Kimura A (1988) Induction of glutathione leakage from *Saccharomyces cerevisiae* cells by selenite. Agric Biol Chem 52:613–614

Izawa S, Yoshiharu I, Kimura A (1995) Oxidative stress response in yeast: effect of glutathione on adaptation to hydrogen peroxide stress in *Saccharomyces cerevisiae*. FEBS Lett 368:73–76

Li Y, Wei G, Chen J (2004) Glutathione: a review on biotechnological production. Appl Microbiol Biotechnol 66:233–242

Li W, Li Z, Yang J, Ye Q (2011) Production of glutathione using a bifunctional enzyme encoded by gshF from *Streptococcus thermophilus* expressed in *Escherichia coli*. J Biotechnol 154:261–268

Liang G, Cheng D, Chen J (2008a) A novel strategy of enhanced glutathione production in high cell density cultivation of *Candida utilis* – cysteine addition combined with dissolved oxygen controlling. Enzyme Microb Technol 42:284–289

Liang G, Liao X, Du G (2008b) Elevated glutathione production by adding precursor amino acids coupled with ATP in high cell density cultivation of *Candida utilis*. J Appl Microbiol 105:1432–1440

Liang G, Du G, Chen J (2008c) Enhanced glutathione production by using low-pH stress coupled with cysteine addition in the treatment of high cell density culture in *Candida utilis*. Lett Appl Microbiol 46:507–512

Liang G, Guocheng D, Chen J (2009a) Salt-induced osmotic stress for glutathione production in *Candida utilis*. Enzyme Microb Technol 45:324–329

Liang G, Liao X, Du G (2009b) A new strategy to enhance oxidative stresses in *Candida utilis*. Bioresour Technol 100:350–355

Liu C, Hwang C, Liao C (1999) Medium optimization for glutathione production by *Saccharomyces cerevisiae*. Process Biochem 34:17–23

Lu SC (2009) Regulation of glutathione synthesis. Mol Aspects Med 30:42–59

Lu SC (2013) Glutathione synthesis. Biochim Biophys Acta (BBA) Gen Subj 1830:3143–3153

Meister A, Tate SS (1976) Glutathione and related γ-glutamyl compounds: biosynthesis and utilization. Annu Rev Biochem 45:559–604

Murata K (1989) Immobilization, chemical and industrial application and biosynthetic preparation of glutathione and related compounds. Coenzymes Cofactors 3:187–242

Nigam P, Owusu-Apenten R (2011) Production and downstream processing of glutathione from microbial sources. In: International conference on new horizons in biotechnology, Trivandrum, Abstract book p 52

Noctor G, Arisi MA, Jouanin L, Kunert K, Rennenberg H, Foyer C (1998) Glutathione: biosynthesis, metabolism and relationship to stress tolerance explored in transformed plants. J Exp Bot 49:623–647

NutraCam (2007) Glutathione supplement, antioxidant glutathione chewing gum. http://www.thqueen.com/

Nutrition Advisor (2007) Glutathione, GSH and whey proteins. Information for physicians. http://www.nutritionadvisor.com/immunocal_balch.php

Salleh MB, Farouk AE, Jamal P (2005) Comparison studies among the methods used in isolating the GSH from baker yeast. J Appl Sci 5:151–154

Santos LO, Gonzales TA, Úbeda BT (2007) Influence of culture conditions on glutathione production by *Saccharomyces cerevisiae*. Appl Microbiol Biotechnol 77:763–769

Thornalley PK (1991) Esterification of reduced glutathione. Biochem J 275:535–539

Valencia E, Marin A, Hardy G (2001) Glutathione – nutritional and pharmacological viewpoints: part III. Nutrition 17:696–697

Vitamin Stuff (2013) Antioxidant benefits, information on supplements. http://www.vitaminstuff.com/glutathione.html

Wang Z, He X, Lium N, Zhang B (2008) Construction of self-cloning industrial brewing yeast with high glutathione and low-diacetyl production. Int J Food Sci Technol 43:989–994

Wei G, Li Y, Du G, Chen J (2003a) Application of a two-stage temperature control strategy for enhanced glutathione production in the batch fermentation by *Candida utilis*. Biotechnol Lett 25:887–890

Wei G, Li Y, Dua G, Chen J (2003b) Effect of surfactants on extracellular accumulation of glutathione by *Saccharomyces cerevisiae*. Process Biochem 38:1133–1138

Wen S, Zhang T, Tan T (2004) Utilization of amino acids to enhance glutathione production in *Saccharomyces cerevisiae*. Enzyme Microb Technol 35:501–507

Wen S, Zhang T, Tan T (2006) Maximizing production of glutathione by amino acid modulation and high-cell-density fed-batch culture of *Saccharomyces cerevisiae*. Process Biochem 46:2424–2428

World Intellectual Property Organization (2007) Liposomally encapsulated reduced glutathione, including with other pharmacologic preparation, capable of administration as an oral, topical, intraoral or transmucosal preparation, for reversal and prevention of oxidation of cholesterol and of low density lipoprotein, vol. no. 053810

Wu G, Fang Y, Yang S, Lupton J (2004) Glutathione metabolism and its implications for health. J Nutr 43:489–492

Xiong Z, Guo M, Guo Y (2009) Efficient extraction of intracellular reduced glutathione from fermentation broth of *Saccharomyces cerevisiae* by ethanol. Bioresour Technol 100:1011–1014

Yamada Y, Tan Y, Kamihara T (1984) Production of extracellular glutathione by *Candida tropicalis* Pk 233. J Gen Microbiol 130:3275–3278

Yilmaz O, Keser S, Tuzcu M, Guvenc M, Cetintas B, Irtegun S, Tastan H, Sahin K (2009) A practical HPLC method to measure reduced (GSH) and oxidized (GSSG) glutathione concentrations in animal tissues. Anim Vet Adv 8:343–347

Enzyme Engineering for Oligosaccharide Biosynthesis

2

David Talens-Perales, Julio Polaina, and Julia Marín-Navarro

Abstract

Oligosaccharides are compounds of great interest in the food and pharmaceutical industry, mainly due to their prebiotic action and other health-promoting effects, in addition to other properties, acting as low-caloric sweeteners, antioxidants, or antimicrobial agents. Oligosaccharide functionality depends on their chemical structure, which relies on the nature of the building sugar residues, the linkage type, and the degree of polymerization. Their production is based either in the partial hydrolysis of polysaccharides or in the synthesis from other sugars with lower degree of polymerization. This methodology is described along the first part of this chapter, making special emphasis on the enzymatic-based technology, which shows clear advantages over chemical methods. Glycoside hydrolases and glycosyltransferases are the two enzyme classes involved in oligosaccharide production. They operate as highly specialized molecular tools for the generation of the different types of oligosaccharides. In the second part of the chapter, the strategies employed for improving the properties of these enzymes are explained. Enzyme engineering, through site-directed mutagenesis or directed evolution, has succeeded in delivering enzymatic variants with enhanced oligosaccharide yields, a different product profile or a wider substrate specificity. Moreover, immobilization of enzymes on solid supports through different methods has allowed their reutilization during repeated batches or their implementation in continuous systems, and, in some instances, it was also accompanied by an increased activity or stability.

D. Talens-Perales • J. Polaina • J. Marín-Navarro (✉)
Instituto de Agroquímica y Tecnología de Alimentos, CSIC, Av. Agustín Escardino 7, Paterna (Valencia) E46980, Spain
e-mail: jumana@iata.csic.es

P. Shukla (ed.), *Frontier Discoveries and Innovations in Interdisciplinary Microbiology*, DOI 10.1007/978-81-322-2610-9_2

Overall, this chapter provides an overview about the facts and potentials of oligosaccharide production methodologies.

Keywords
Prebiotic oligosaccharide • Glycoside hydrolase • Glycosyltransferase • Glycosynthase • Transglycosylation • Site-directed mutagenesis • Directed evolution • Protein immobilization

2.1 Introduction

Oligosaccharides are low molecular weight carbohydrates with a degree of polymerization (DP) between 2 and 10 that may be linear or branched (Weijers et al. 2008). They are crucial for the development, growth, and function of organisms, being present in essential biomolecules as glycolipids and glycoproteins (Raman et al. 2005). Oligosaccharides show a broad structural diversity. In eukaryotes, they are synthesized in the endoplasmic reticulum and Golgi apparatus by several glycosyltransferases. In prokaryotes, N-linked oligosaccharides are synthetized in the periplasmic space, whereas *O*-linked oligosaccharides are produced in the cytoplasm and surface appendages, such as pili and flagella (Szymanski and Wren 2005; Weijers et al. 2008). The structural stability of these compounds depends on the nature of the sugar residues, their degree of polymerization and the linkage type.

Oligosaccharides have been commercialized since the 1980s as low-calorie bulking agents. More recently they have received much interest in food and pharmaceutical industry because of their prebiotic properties and the increased consumer interest in health-promoting products (Patel and Goyal 2011). They are also used in food industry as jelly agents, antioxidants, and humectants and in the pharmaceutical industry for drug delivery. The chemical diversity of different oligosaccharides and their applications is shown in Table 2.1.

2.2 Oligosaccharide Production

The two major methods for the production of oligosaccharides are depolymerization of polysaccharides and the synthesis from precursors of lower DP. Both procedures can be carried out by either chemical or enzymatic technologies (Courtois 2009). Chemical hydrolysis of polysaccharides uses diluted or concentrated solutions of strong acids (up to 2 M), such as hydrochloric, sulfuric, trifluoroacetic, formic, or nitrous acid, at temperatures between 50 °C and 90 °C. Oligosaccharides from fucans, pectins, and galactans are produced by this method (Courtois 2009). The reaction time of the process has been significantly reduced using microwave-induced acid hydrolysis, obtaining gluco-oligosaccharides from branched glucan of *Leuconostoc mesenteroides* (Patel and Goyal 2011). An alternative method that employs higher temperatures (150–200 °C) for hydrolysis in the absence of acid has been used for the production of XOS (Aachary and Prapulla 2011; Carvalheiro et al. 2004; Chen et al. 2014; Samala et al. 2015). Chemical synthesis of oligosaccharides involves the union of at least two sugar molecules by glycosidic bonds. This bond is formed by a nucleophilic displacement of a leaving group attached to the anomeric carbon of a sugar moiety (Fig. 2.1a). The reaction is enhanced in the presence of an activator, generally a Lewis acid, which assists the departure of the leaving group (Weijers et al. 2008). Traditional strategies for chemical oligosaccharide synthesis require extensive protecting/deprotecting steps and/or leaving group manipulation between each glycosylation stage, making the whole process cumbersome and decreasing the efficiency and yield (Fig. 2.1b). An alternative method of chemical synthesis emerged with the development of the armed-disarmed approach by Fraser-Reid and coworkers, based on the chemoselectivity principle (Fraser-Reid et al. 1990). With this method the

Table 2.1 Chemical description of different oligosaccharides, applications, and properties

Substrate	Oligosaccharide type	Chemical description	Applications/properties	References
Lactose	Galactooligosaccharides (GOS)	(D-Galactose)$_n$ connected by ß-(1,3), ß-(1,4), or ß-(1,6) linkages and bonded in most cases to a terminal glucose moiety at the reducing end	Functional food (immune stimulation, prebiotic, anti-adhesive effect against pathogens, calcium absorption improvement)	Crittenden and Playne (1996), Hernot et al. (2009), Macfarlane et al. (2008), Moreno et al. (2014), Whisner et al. (2013)
	Lactulose and lactulose-derived GOS (Lu-GOS)	Lactulose: 4-O-β-D-galactopyranosyl-D-fructose Lu-GOS: (D-galactose)$_n$-fructose	Pharmaceutical (anti-constipation agent, treatment for hepatic encephalopathy) Functional food (prebiotic effect)	Cardelle-Cobas et al. (2011), Crittenden and Playne (1996), Hernandez-Hernandez et al. (2012), Prasad et al. (2007), Schuster-Wolff-Bühring et al. (2010)
	Lactosucrose	O-ß-D-Galactopyranosyl-(1,4)-*O*-*α*-*D*-glucopyranosyl-(1,2)-ß-D-fructofuranosidase (lactosyl fructoside)	Functional food (prebiotic effect and increased mineral absorption)	Crittenden and Playne (1996), Mu et al. (2013)
Sucrose/inulin	Palatinose/isomaltulose	[6-*O*-α-D-Glucopyranosyl-D-fructose]$_n$	Functional food (prebiotic effect) Food additive (substitute sweetener)	Crittenden and Playne (1996)
	Fructooligosaccharides (FOS)	(D-Fructose)$_n$ linked in most cases to a terminal sucrose at the reducing end ^{1}F-FOS: D-fructose units interconnected through ß-(1,2) linkages (inulin type) ^{6}F-FOS: D-fructose units interconnected through ß-(2,6) linkages (levan type) ^{6}G-FOS: D-fructose units connected to the C6 of the terminal glucose (neoseries)	Functional food (prebiotic effect, relieve constipation) Industrial uses (processed food adjuvant, cosmetics)	Closa-Monasterolo et al. (2013), Crittenden and Playne (1996), Monsan and Ouarné (2009), Sabater-Molina et al. (2009)

(continued)

Table 2.1 (continued)

Substrate	Oligosaccharide type	Chemical description	Applications/properties	References
Starch	Maltooligosaccharides (MOS)	(D-glucose)$_n$ connected by α-(1,4) linkages	Industrial sweetener	Moon and Cho (1997)
	Isomaltooligosaccharides (IMOS)	(D-glucose)$_n$ connected by α-(1,6) linkages	Functional food (prebiotic effects, anti-cariogenic effects) Industrial applications (food additives)	Goffin et al. (2011)
	Gentiooligosaccharides (GeOS)	(D-Glucose)$_n$ connected by ß-(1,6) linkages	Bifidogenic activity	Rycroft et al. (2001)
	Cyclodextrins (CDs)	Cyclic maltooligosaccharides formed by 6–12 glucose units α-CDs: 6 units ß-CDs: 7 units γ-CDs: 8 units	Biotechnological applications in food, pharmaceuticals, and chromatographic techniques, as carrier and stabilizer of liposoluble vitamins, polyunsaturated fatty acids, and drugs, and for emulsion of fat oils	Del Valle (2004), Li et al. (2007b), Singh et al. (2002)
	ß-glucan oligosaccharides	(D-Glucose)$_n$ connected by ß-(1,3) linkages	Stimulatation of immunity and decreased infectious complications in humans and animals	Otaka (2006)
Others	Human milk oligosaccharides (HMOs)	Oligosaccharides formed by D-glucose, D-galactose, *N*-acetylglucosamine, L-fucose, and sialic acid as basic building blocks	Prebiotic effect	Bode (2009)
	Chitooligosaccharides (COS)	(D-Glucosamine)$_n$ connected by β-(1,4) linkages	Antimicrobial, antiviral, antioxidant, and immunostimulant	Kim and Rajapakse (2005)
	Xylooligosaccharides (XOS)	(D-Xylose)$_n$ connected by ß-(1,4) linkages	Immunomodulatory properties, antimicrobial activity, and cosmetic effects	Aachary and Prapulla (2011)
	Feruloyl xylooligosaccharides (FeXOS)	Xylooligosaccharides esterified with trans-ferulic acid at position *O*-4 from some α-D-xylopyranosyl residues	Pharmaceutical (protection against oxidative damage of low-density lipoprotein and erythrocytes)	Ishii and Tobita (1993), Katapodis and Christakopoulos (2008)

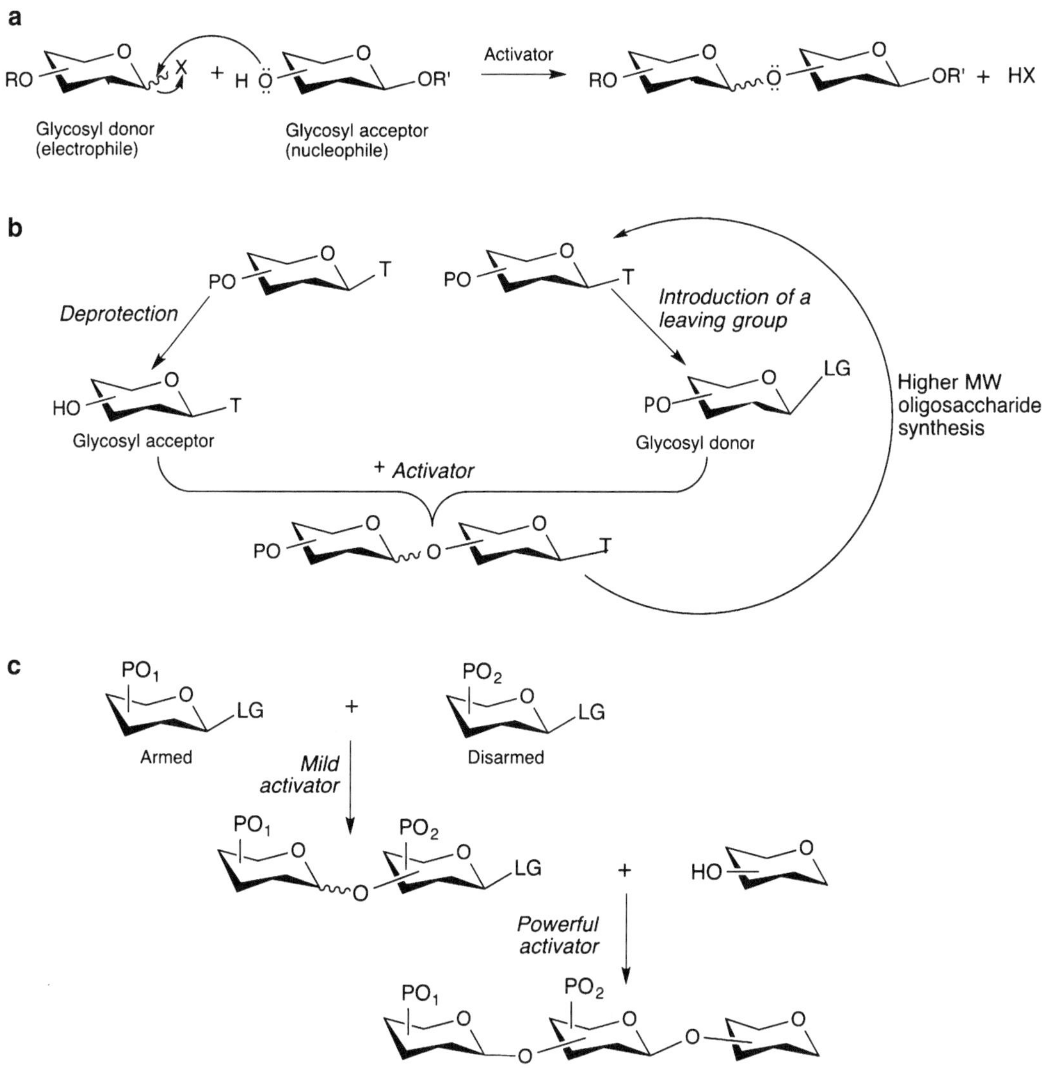

Fig. 2.1 Strategies for chemical synthesis of oligosaccharide. (**a**) Nucleophilic substitution between the glycosyl donor (electrophile) and the glycosyl acceptor (nucleophile). The leaving group is represented as X. (**b**) Traditional strategy for chemical oligosaccharide synthesis, involving protecting/deprotecting steps and the introduction of leaving groups. *PO* protecting group, *T* temporary stable anomeric substituent, *LG* leaving group. (**c**) Armed-disarmed method for oligosaccharide synthesis. The protecting groups (PO_1 and PO_2) modulate the reactivity of the glycosyl molecules. The departure of the leaving group (LG) from the "armed" sugar is promoted by a mild activator, whereas for the "disarmed" sugar, a powerful activator is required

glycosyl donor reactivity is modulated entirely through the choice of protecting group, and the resulting disaccharide can then be used directly in a subsequent glycosylation (Fig. 2.1c). This procedure has been applied in the solid-phase synthesis in which the assembly of the different glycosyl molecules occurs in donor building blocks attached to a solid support, reducing the protection steps and facilitating product isolation (Weishaupt et al. 2010). This solid-phase method is used for making carbohydrate microarrays, with applications in biological and medical research (Park et al. 2013).

The methods of chemical synthesis are expensive and their yields are low. Their use is therefore restricted to obtain high-added-value oligosaccharides for analytical or medical purposes. For large-scale industrial oligosaccharide production, the use of enzymes offers considerable advantages, as will be discussed in the following sections. The enzymes used are glycosylhydrolases (GHs) and glycosyltransferases (GTs), which are classified in different families, based on sequence homology (Lombard et al. 2014). GHs hydrolyze glycosidic bonds, but in some instances they also catalyze transglycosylating reactions. Both possibilities involve the cleavage of a glycosidic bond. Within the active site of GHs and GTs, different residues form distinct pockets, or subsites, which dock the glycosyl residues framing the glycosidic bond that is going to be cleaved. These subsites determine substrate specificity. According to the nomenclature established by Davies et al., the subsites are designated with integer numbers from $-n$ to $+n$, numbered from the nonreducing to the reducing end of the substrate, with the cleavage occurring between subsites −1 and +1 (Davies et al. 1997).

2.2.1 Enzymatic Hydrolysis of Polysaccharides

Enzymatic hydrolysis is more attractive than chemical hydrolysis because it is environmentally friendly, it does not use hazardous chemical products, it allows an easy control of depolymerization extent, and it does not produce undesired side products (Chapla et al. 2010; da Silva et al. 2014). Oligosaccharides such as xylooligosaccharides (XOS), feruloyl xylooligosaccharides (FeXOS), maltooligosaccharides (MOS), gentiooligosaccharides (GeOS), fructo-oligosaccharides (FOS), chitooligosaccharides (COS), and cyclodextrins (CDs) can be obtained by enzymatic hydrolysis from polysaccharides.

XOS are released from xylan, a widely available polysaccharide, by endo-ß-(1,4)-xylanase (EC 3.2.1.8) and exo-ß-(1,4)-xylanase (EC 3.2.1.37) (Maalej-Achouri et al. 2009). Xylanases from different microorganisms are used, although fungi are generally considered more efficient producers than bacteria or yeasts (Ahmed et al. 2009). Wheat bran arabinoxylan has been used as a substrate to obtain XOS by combining the action of endo- and exo-xylanases and α-l-arabinofuranosidase (EC 3.2.1.55), yielding 0.3 g of XOS per g of soluble polysaccharides (Manisseri and Gudipati 2010). Commercial xylanase from *Trichoderma viride* released xylobiose, xylotriose, and xylose from the grass *Sehima nervosum* (Samanta et al. 2012). Xylan from alkali-treated corncobs was hydrolyzed by a xylanase from *Aspergillus foetidus* to obtain a mix of XOS with DPs from 2 to 5, with a yield of 6.7 mg XOS/g of raw corncobs (Chapla et al. 2010; Chapla et al. 2012). Thermophilic xylanases (Xyn A/B/C and Xyn W belonging to family GH10) from *Humicola insolens* yielded xylose, xylobiose, and xylotriose using brewing industry waste as a source of xylan (Du et al. 2013). FeXOS result from the hydrolysis of xylan esterified with trans-ferulic acid, which is present in a high proportion of vegetal waste material (Katapodis and Christakopoulos 2008). Endoxylanases from diverse sources show different substrate binding efficiencies to partially feruloyl-substituted xylan. *Thermoascus aurantiacus* GH10 endoxylanase produced FeXOS using as substrate wheat bran dietary fiber previously treated with *Bacillus subtilis* GH11 endoxylanase (Katapodis and Christakopoulos 2008).

Cyclodextrin glycosyltransferase (CGTase, EC 2.4.1.19), belonging to the GH13 family (Lombard et al. 2014), is employed to produce α-, ß-, or γ-CDs from starch or starch derivatives (Li et al. 2007b). In the first step the enzyme cleaves an α-1,4 glycosidic bond at position n from the nonreducing end of an amylose chain (n value being between 9 and about 100) (Terada et al. 1997). The resulting oligosaccharide of n units forms a covalent intermediate with the enzyme through the C-1 of the residue at the reducing end. Subsequently, this C-1 undergoes an intramolecular nucleophilic attack by the 4-hydroxyl group of the nonreducing end, releasing a cyclodextrin of n glucosyl units (Li et al. 2007b). Archaea and bacteria may secrete CGTases in order to take over the starch present in the medium since cyclodextrins cannot be metabolized by competing microorganisms

(Leemhuis et al. 2010). Most CGTases characterized so far yield a mixture of α- and ß-CDs as major products. A specific production of ß-CD was achieved with a CGTase from *Paenibacillus illinoisensis* (Lee et al. 2013). Interesting properties of γ-CDs, such as higher water solubility and bioavailability, have fostered the search of CGTases that synthetize this product (Li et al. 2007b). For instance, CGTases from *Bacillus sp.*, *Bacillus thuringiensis, Paenibacillus macerans*, and *Bacillus clarkia*, yield γ-CDs with a high specificity (Goo et al. 2014; Lee et al. 2013; Li et al. 2007b; Wu et al. 2012).

FOS can be obtained from inulin or levan by enzymatic hydrolysis with either inulinases or levanases, respectively (both types of enzymes belonging to the GH32 family). Like other carbohydrate hydrolases, inulinases may have endo or exo action. Endoinulinases (2,1-ß-D-fructan fructanohydrolases; EC 3.2.1.7) specifically hydrolyze internal ß-(2,1) glycosidic bonds releasing inulotriose, inulotetraose, and inulopentaose (Mutanda et al. 2014). Exoinulinases (ß-D-fructohydrolases; EC 3.2.1.80) cleave the terminal fructosyl residue from inulin and are used to obtain high-fructose syrup (Apolinário et al. 2014). Many yeast and bacterial strains produce inulinase, for instance, species belonging to *Pichia*, *Cryptococcus*, *Candida*, *Yarrowia*, *Debaryomyces*, *Xanthomonas*, and *Pseudomonas*. However, molds like *Aspergillus* and *Penicillium* and the yeast *Kluyveromyces* are more frequently used (Paixão et al. 2013). *Bacillus licheniformis* endolevanase is able to produce FOS containing ß-(2,6) linkages from levan, with a yield of 97 %, based on total sugar content (Porras-Domínguez et al. 2014).

Chitinases are GH18 and GH19 glycoside hydrolases (EC 3.2.1.14.) (Lombard et al. 2014) which hydrolyze ß-1,4 linkages between N-acetylglucosamine units from chitin yielding COS (Kim and Rajapakse 2005). Hexa-N-acetylchitohexaose was produced by chitin hydrolysis using the chitinase ASCHI6 isolated from *Aeromonas schubertii* (Liu et al. 2014). Chitinases from *Bacillus cereus* produced a mix of COS with DPs between 2 and 5 (Wang et al. 2012).

2.2.2 Enzymatic Synthesis of Oligosaccharides

The enzymatic synthesis of oligosaccharides can be carried out by using either glycosidases with transferase capacity or glycosyltransferases. In both cases, the catalytic mechanism implies the transfer of a glycosyl residue from one of the two sugar substrates involved in the reaction (glycosyl donor) to the other substrate (glycosyl acceptor).

Glycosidases (EC 3.2.1) are enzymes with the capability to hydrolyze glycosidic bonds. Depending on the stereochemistry of the product compared to that of the glycosyl donor, glycosidases are classified into retaining or inverting enzymes. The inverting mechanism occurs in a single step involving two critical residues: One acts as general base, activating the water molecule that carries out the nucleophilic attack on the anomeric carbon. The other acts as general acid and protonates the leaving sugar. Instead, retaining glycosidases proceed through a double nucleophilic substitution mechanism. In the first step, a nucleophilic residue in the active site attacks the anomeric carbon of the substrate, and the leaving group is protonated by an acid/base catalyst. This results in a covalent intermediate between the enzyme and the glycosyl donor, which is subsequently released by the nucleophilic attack of a water molecule activated by the acid/base catalyst. The retaining mechanism confers glycosidases the potential to catalyze transglycosylation reactions, which occur when a second sugar molecule enters the catalytic site of the covalent intermediate and acts as glycosyl acceptor (Fig. 2.2) (Davies and Henrissat 1995).

Glycosyltransferases (EC 2.4) catalyze the transfer of sugar moieties from donor to acceptor molecules thereby forming glycosidic bonds. The acceptor substrate may be as simple as a monosaccharide or as complex as a heteropolysaccharide, oligosaccharide, protein, nucleic acid, or lipid (Lairson et al. 2008; Weijers et al. 2008). These enzymes are responsible for the synthesis of most cell-surface glycoconjugates. Like glycosidases, glycosyltransferases can be retaining or inverting, although their detailed molecular mechanism is

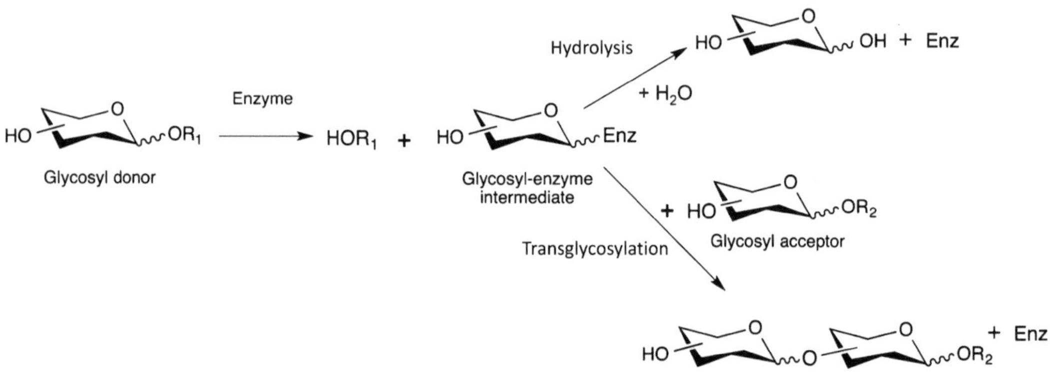

Fig. 2.2 Synthesis of oligosaccharides with nonactivated donors can be carried out by retaining glycosidases. The enzyme (*Enz*) forms a covalent intermediate with a glycosyl group from the glycosyl donor. This group can be transferred to a water molecule (hydrolysis) or to a glycosyl acceptor (transglycosylation), releasing the free enzyme to start a new cycle of catalysis

not completely understood. Inverting glycosyltransferases utilize a direct displacement S_N-2-like mechanism assisted by an enzymatic base catalyst. Retaining glycosyltransferases operate through a double-displacement reaction involving a covalently bound glycosyl-enzyme intermediate, except for members from families GH4 and GH109 which use an elimination/addition mechanism with a transient oxidation, and retaining ß-hexosaminidases that use an intramolecular nucleophile (Lairson et al. 2008; Mark et al. 2001).

Enzymes with transglycosylating activity have been classically classified as *non-Leloir* or *Leloir* transferases depending on the nature of the glycosyl donor. Enzymes that use activated donor carbohydrates bound to nucleosides mono- or diphosphate (CMP, UDP, GDP, or TDP) are classified as *Leloir*. Enzymes that use nonactivated carbohydrate donors or monophosphate-substituted sugars are classified as non-*Leloir*. In this chapter, however, we have employed a more practical classification into two groups of enzymes, based on their capability to synthesize oligosaccharides using either activated or nonactivated donors.

2.2.3 Synthesis of Oligosaccharides with Nonactivated Donors

Lactose is the substrate for the synthesis of galactooligosaccharides (GOS) and related compounds like lactosucrose and lactulose, which represent about 40 % of annual oligosaccharide production (Gänzle 2012). β-Galactosidases transfer the galactosyl moiety from lactose to an acceptor. Lactose can also act as galactosyl, glucosyl, or fructosyl acceptor through the action of ß-galactosidases, glucansucrases, and fructansucrases, respectively.

GOS synthesis is catalyzed by β-galactosidases, with lactose acting as both glycosyl donor and acceptor. The transglycosylation/hydrolysis ratio depends on the structural conformation of the active site of the enzyme, but it is also influenced by other factors such as lactose concentration (Bruins et al. 2003; Ji et al. 2005), water activity, and, to a minor extent, reaction temperature (Millqvist-Fureby et al. 1998). The chemical structure and DP of the resulting oligosaccharide also rely on structural features of the active site that have not been fully explained yet. The most common products are the trisaccharides 6′-galactosyllactose, 3′-galactosyllactose, and 4′-galactosyllactose. GOS yields obtained with different enzymes vary between 2 % and 70 %, based on the total sugar content (Park and Oh 2010; Torres et al. 2010). Fungi, bacteria, and archaea have been used as enzyme sources for GOS production. Currently, commercial enzymes are obtained from from *Kluyveromyces lactis*, *Streptococcus thermophilus*, *Bacillus circulans*, *Aspergillus oryzae*, and *Bifidobacterium bifidum*. The three former belong to the GH2 family, and synthesize GOS with yields up to 44 %, 25 %, and 39 %, respectively (Rodriguez-Colinas et al. 2011, Torres et al. 2010, Song et al. 2011). Different β-galactosidases from *Bifidobacterium bifidum* produce GOS with yields ranging between 15–47 % (Goulas et al. 2009). They are used for the production of ß-(1,3) GOS (*B. bifidum*), ß-(1,4) GOS (*B. circulans*), and ß-(1,6) GOS (*Aspergillus oryzae*, *Streptococcus thermophilus* and *Kluyveromyces lactis*) (Goulas et al. 2009; Torres et al. 2010; Rodriguez-Colinas et al. 2011). GH35 ß-galactosidases from *Aspergillus* and *Penicillium* species have rendered 30–45 % GOS yields (Torres et al. 2010), whereas GH42 ß-galactosidase from *Bacillus longum* synthesized GOS with lower (10 %) yields (Torres et al. 2010). Since high lactose concentrations have a strong positive influence in GOS yield, the use of high temperatures is desirable in order to increase lactose solubility (Roos 2009). This can be accomplished with ß-galactosidases that are stable at temperatures between 70 °C and 90 °C (Park and Oh 2010). Moreover, these enzymes usually show high structural resistance, convenient for industrial applications. Therefore, different thermostable β-galactosidases have been studied. GH1 enzymes from *Sulfolobus solfataricus*, *Pyrococcus furiosus*, and *Thermus caldophilus* produced GOS with 53 %, 40 %, and 77 % yield, respectively (Torres et al. 2010). *Thermotoga maritima* β-galactosidase from GH2 family yielded 18 % GOS (Ji et al. 2005), whereas that from *Geobacillus stearothermophilus*, belonging to the GH42 family, yielded only 2 % GOS (Placier et al. 2009). The enzymes from *Sulfolobus solfataricus*, and *Pyrococcus furiosus* produce ß-(1,3) and ß-(1,6) GOS (Torres et al. 2010), whereas those from *Thermus sp.*, *Thermotoga maritima*, and *Geobacillus stearothermophilus* show a higher preference for the synthesis of ß-(1,3) GOS (Akiyama et al. 2001; Marín-Navarro et al. 2014; Placier et al. 2009).

Lactulose is synthesized by transgalactosylation when the enzymatic hydrolysis of lactose occurs in the presence of fructose. In this case the galactosyl moiety from lactose is linked to the C-4 atom of fructose. The yield depends on the enzyme and the ratio of concentrations of lactose and fructose. Lactulose is produced in the early stage of the reaction and can be consumed in the final stages. Because of this, it is necessary to stop the reaction or separate the enzyme from the product (Panesar and Kumari 2011). Some ß-galactosidases and ß-glycosidases have been used for lactulose production. GH1 ß-glycosidases from *Aspergillus oryzae, Pyrococcus furiosus*, and *Sulfolobus solfataricus* yielded 30, 44, and 12.5 % lactulose relative to lactose, respectively (Panesar and Kumari 2011; Kim et al. 2006; Wang et al. 2013). Lactulose has also been synthesized from lactose in the absence of fructose by combining a ß-galactosidase from *Kluyveromyces lactis* and a glucose isomerase from *Streptomyces murinus*, with a conversion yield of 19 % (Wang et al. 2013). Another method for lactulose production using only lactose as substrate has made use of a thermostable cellobiose-2-epimerase with isomerase activity, from *Caldicellulosiruptor saccharolyticus*, yielding 58 % lactulose (Kim and Oh 2012).

Lactosucrose has been obtained with levansucrase from *Zymomonas mobilis* using a 1:1 mass ratio of lactose and sucrose as substrate, achieving a conversion efficiency of 28 %. Since this enzyme is inhibited by the glucose released in the reaction, the former yield could be increased up to 43 % by combining the action of levansucrase with glucose oxidase thus transforming the produced glucose into D-gluconolactone, which does not exert inhibitory effect (Han et al. 2009). Trisaccharides derived from lactulose (Lu-GOS) have been synthesized by transglycosylation using the ß-galactosidase from *K. lactis*, with a yield of 14 % from total carbohydrate content (Martinez-Villaluenga et al. 2008). The galactosyl unit from a lactulose molecule was transferred to another lactulose at two different positions releasing two types of Lu-GOS.

Fructosyltransferases (FTase, EC 2.4.1) and ß-fructofuranosidases (FFase, EC 3.2.1.26) with high transfructosylating activity, belonging to GH32 and GH68 families, are the enzymes involved in the synthesis of FOS from sucrose, which acts both as donor and acceptor of the fructosyl group (Dominguez et al. 2014; Ganaie et al. 2013). Fructosyl units can be connected through ß-(2,1) linkages (inulin-type FOS or ^{1F}FOS, as 1-kestose) or ß-(2,6) linkages (levan-type FOS or

6F FOS, as 6-kestose). In neoseries FOS or 6G FOS, a ß-(2,6) link connects a fructosyl unit to the glucosyl moiety of sucrose, as in neokestose (Linde et al. 2012; Porras-Domínguez et al. 2014). Commercial FOS have been produced by enzymes from fungi and bacteria, with different yields and product specificity. Several fungal strains, mostly from the genus *Aspergillus* and *Penicillium*, produce intracellular and extracellular FTases that synthesize mainly inulin-type FOS, with yields ranging between 25 % and 70 % from total sugar content (Dominguez et al. 2014). FOS mixtures include 1-kestose, 1-nystose, and 1-fructofuranosyl nystose as predominant oligosaccharides (Ganaie et al. 2013). There are fewer reports on the production of levan-type FOS as a major product of enzymatic transfructosylation. GH68 levansucrase from *Zymomonas mobilis* synthesizes a mixture of 6-kestose, 1-kestose, and neokestose with different ratios depending on the reaction conditions and a 6-kestose maximum specificity of 49.1 % of total FOS (Vigants et al. 2013). In contrast, *Schwanniomyces occidentalis* GH32 FTase produces FOS with a 16.4 % yield, 6-kestose representing 75 % of total FOS (Álvaro-Benito et al. 2007). Production of neoseries FOS has been reported by FTases from *Xanthophyllomyces dendrorhous*, *Penicillium oxalicum*, and *Penicillium citrinum* with 57 %, 19 %, and 11 % conversion yields to neokestose, respectively (Ning et al. 2010; Xu et al. 2015; Hayashi et al. 2000).

IMOS can be produced by dextransucrases (EC 2.4.1.5) belonging to the family GH70 (Lombard et al. 2014). These enzymes catalyze the transfer of the glucose moiety from a sucrose molecule to a glucose acting as acceptor. *Leuconostoc mesenteroides* dextransucrase is able to produce a mixture of IMOS with DPs between 4 and 9. Maximum conversion yields (between 70–90 %, in terms of glucose equivalents) were achieved with a sucrose/maltose ratio of 2:1 (Lee et al. 2008). α-Glucosidases (EC 3.2.1.20) are exo-glycosidases grouped in GH13 and GH31 families that catalyze the release of glucose from the nonreducing end of short oligosaccharides. Some α-glucosidases with transglycosylating activity, such as the enzymes from *Aspergillus niger*, *S. cerevisiae*, or *Xanthophyllomyces dendrorhous*, synthetize IMOS (DPs between 2 and 4) from maltose with conversion yields between 30 % and 60 %, from total sugar content (Duan et al. 1995; Fernández-Arrojo et al. 2007).

In the case of GeOS synthesis, a few examples have been reported. A mixture of GeOS with different polymerization degrees (between 3 and 9) has been achieved with a combination of ß-glucosidase and ß-(1,6)-glucanase (GH5) from *Penicillium multicolor*. In this case, ß-glucosidase synthesizes gentiotriose using gentiobiose as substrate, and this trisaccharide is elongated by ß-(1,6)-glucanase yielding the final GeOS mix (Fujimoto et al. 2009). GH70 alternansucrase (EC 2.4.1.140) from *Leuconostoc mesenteroides* is capable of synthesizing GeOS with DPs between 3 and 5 in presence of gentibiose (Côté 2009).

Fucosyl-*N*-acetylglucosamine, one of the components of HMOs, can be synthesized using fucosidases with transglycosylating capability such α-L-fucosidases AlfB and AlfC from *Lactobacillus casei* that yield fucosyl-α-1,3-N-acetylglucosamine and fucosyl-α-1,6-N-acetylglucosamine, respectively, using *p*-nitrophenyl-α-L-fucopyranoside (pNP-fuc) as fucosyl donor and *N*-acetylglucosamine (GlcNAc) as acceptor (Rodriguez-Diaz et al. 2011).

2.2.4 Synthesis of Oligosaccharides with Activated Donors

Synthesis of oligosaccharides with activated donors is carried out by two main groups of enzymes (Fig. 2.3). Phosphorylases can act as glycosyltransferases in their reverse reaction using phosphorylated glycosyl donors, whereas Leloir glycosyltransferases use glycosyl donors activated with nucleotide mono- or diphosphate (CMP, UDP, GDP, or GTP).

In their forward reaction, phosphorylases catalyze the phosphorolysis of glycosyl linkages at the nonreducing end of carbohydrates, releasing a monosaccharide 1-phosphate in the presence of phosphate. In the reverse reaction, these enzymes catalyze the transfer of the activated monosaccharide to a carbohydrate acceptor. In one particular case, the enzyme α-1,4-glucan-maltose-1-P

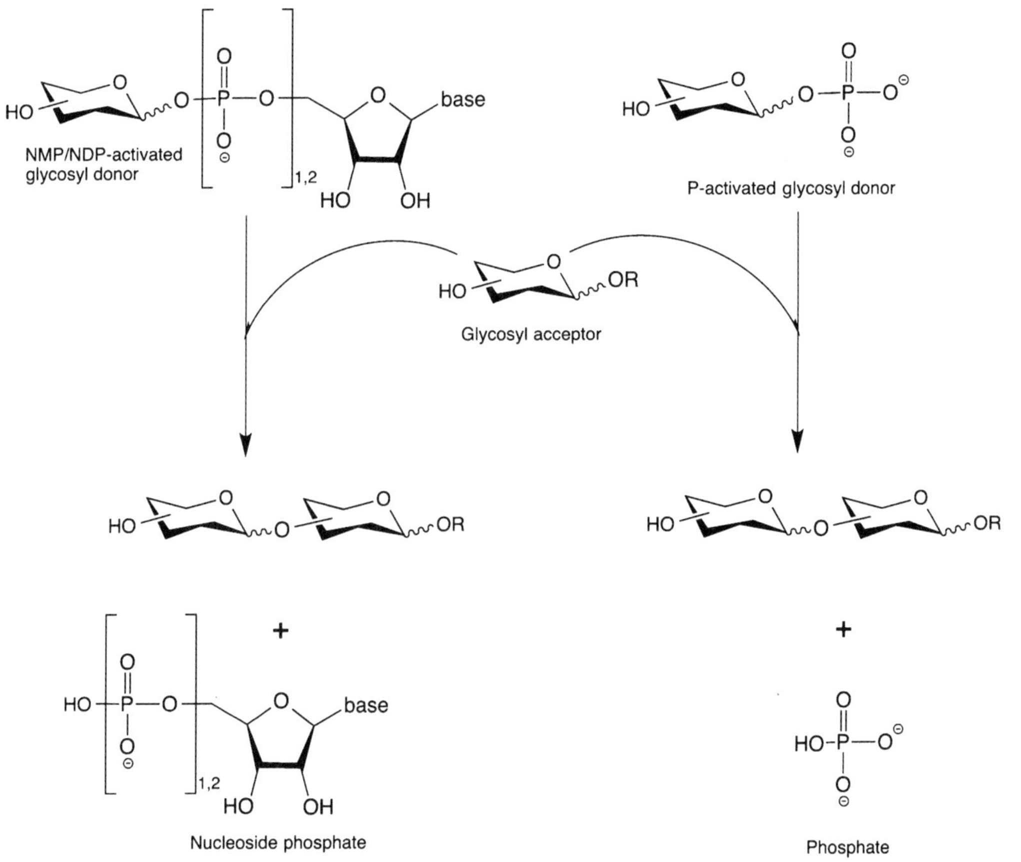

Fig. 2.3 Synthesis of oligosaccharides with activated donors is carried out by phosphorylases (right side) and Leloir glycosyltransferases (left side). The activated glycosyl donor is a phosphorylated sugar or a nucleotide mono- or diphosphate (CMP, UDP, GDP, or GTP), respectively

maltosyltransferase (EC 2.4.99.16) has been described to release maltose 1-phosphate from glycogen (Nakai et al. 2013), which would allow the transglycosylation of a disaccharide in a single step in the reverse reaction. Reversible phosphorolysis of α-glucosides is catalyzed by phosphorylases belonging to glycoside hydrolase families GH13 and GH65 and to glycoside transferase families GT4 and GT35. ß-Glucosides, ß-galactosides, and ß-manosides are substrates of phosphorylases belonging to families GH94, GH112, and GH130, respectively. Family GH94 also includes a chitobiose phosphorylase capable of transferring N-acetyl-glucosamine moieties (Honda et al. 2004). Despite the wide diversity of potential natural sugars able to act as glycosyl acceptors, little use of phosphorylases to produce oligosaccharides has been made due to the reduced number of enzymes characterized so far (Nakai et al. 2013).

GH65 kojibiose phosphorylase (EC 2.4.1.230) from *Thermoanaerobacter brockii* catalyzes the synthesis of kojibiose (α-D-glucopyranosyl-(1,2)-D-glucopyranose) by transglycosylation using ß-D-glucose-1-phosphate as glycosyl donor and D-glucose as acceptor, yielding inorganic phosphate. This enzyme is also able to use other glycosyl acceptors such as L-sorbose, maltose, and glucosyl-maltose to yield glucosyl-l-sorbose, glucosyl-sucrose, and glucosyl-maltose, respectively, as novel oligosaccharides (Chaen et al. 2001). Maltose phosphorylase (EC 2.4.1.8) from *Propionibacterium freudenreichii* catalyzes the transglycosylation of ß-D-glucose from ß-D-glucose-1-phosphate to different acceptors such as D-glucose, D-glucosamine, *N*-acetylglucosamine, mannose, 2-deoxy-D-glucose, and 2-deoxy-D-xylose (Aisaka et al. 1996), making this enzyme suitable for the synthesis of oligosaccharides with a broad chemical diversity.

Leloir enzymes require the previous synthesis of an activated substrate, which is a nucleotide mono- or diphosphate glycoside. Such compounds are synthesized by the consecutive action of a kinase, which phosphorylates the sugar, and a nucleoside transferase/pyrophosphorylase (Weijers et al. 2008). Therefore, the use of this methodology requires the implementation of in situ coupled reactions either in vitro or in vivo (Weijers et al. 2008). Leloir transferases are classified in two different superfamilies depending on their structure: GT-A and GT-B, and both of them include retaining and inverting enzymes.

The requirement of a nucleotide-activated donor makes *Leloir* glycosyltransferases less attractive than those using nonactivated sugars for industrial production of low-added-value oligosaccharides. However, these enzymes show a high promiscuity, accepting a rather wide range of substrates which leads to the synthesis of a diversity of products. Indeed, *Leloir* glycosyltransferases are able to synthesize most human milk oligosaccharides (HMOs), such as trisaccharides (sialyllactose and fucosyllactose) or other complex oligosaccharides (lacto-*N*-biose derivatives). These compounds are very interesting to be used as functional ingredients in commercial infant formula because of their prebiotic and immunogenic effects (Bode 2009). The synthesis of these oligosaccharides requires the combined action of several enzymes and cofactors. This has been carried out both by in vitro and whole-cell approaches, being the latter more convenient for industrial purposes, due to the high costs associated to enzyme purification and cofactor synthesis (Table 2.2).

2.3 Enzyme Engineering

As it has been described in previous sections, glycosidases and GTs have proven to be useful tools for oligosaccharide synthesis in vitro. However, sometimes native enzymes do not provide good performance, specificity, or stability. In this case, one can either look for new enzymes in nature or change the properties of an existing enzyme by protein engineering. The procedures used to improve enzyme performance are mutagenesis, either random or site directed (Bornscheuer 2013), and immobilization (Homaei et al. 2013).

2.3.1 Directed Evolution

Directed evolution attempts to mimic the natural selection process by recursive rounds of mutation, recombination, and selection (Kittl and Withers 2010) Random mutagenesis carried out by error-prone PCR is generally used to obtain the initial library of mutants. The major limitations of this approach are the inherent bias of existing mutagenesis methods (Wong et al. 2006) and the requirement in many instances of technically complex high-throughput screening methods. However, this approach is a powerful tool for molecular engineering of enzymes, and it does not require any previous information about the structural-function relationships of the enzyme of interest. Directed evolution has successfully been used to increase the transglycosylating activity of a ß-glycosidase (GH1) from *Thermus thermophilus*. Mutant F401S/N282T, using either maltose or cellobiose as acceptors and *o*-nitrophenyl-β-D-galactopyranoside as donor, showed molar yields of trisaccharides up to 60 % and 75 %, respectively, whereas the wild-type yield was only 8 %. Galactosyl transfer to maltose ocurred with a major β-(1,3) regioselectivity, whereas with cellobiose both β-(1,3) and β-(1,6) isomers were synthesized. These mutations may reorient the glycone at the −1 subsite allowing a better transfer to the acceptors (Feng et al. 2005). A similar approach was used to obtain α-L-arabinofuranosidase (GH51) from *Thermobacillus xylanilyticus* with better transglycosylation and lower hydrolytic activity with mutants N344P and N344Y (Arab-Jaziri et al. 2013). α-L-Fucosidase (GH29) from *Thermotoga maritima* was converted into α-L-transfucosidase. The change of activity required mutations at only three positions (T264A, Y267F, L322P). This was carried out by a first step of random

Table 2.2 Oligosaccharide production with *Leloir* glycosyltransferases

Final product	Method	Involved enzymes/microorganisms[f]	Reaction scheme	References
Sialyllactose	Whole cells	(1) *Corynebacterium ammoniagenes* for UTP synthesis (2) CTP synthetase* (3) CMP-NeuAc synthetase* (4) α-2,3-sialyltransferase* *Recombinantly produced by two *E. coli* strains	UTP[1] + ATP + Gln →[2] CTP + ADP + Pi + Glu CTP + NeuAc →[3] CMP-NeuAc + PPi CMP-NeuAc + lactose →[4] sialyllactose + CMP	a
NeuAc (sialyllactose precursor)	Whole cells	(1) GlcNAc epimerase* (2) NeuAc synthetase* *Recombinantly produced by one *E. coli* strain	GlcNac →[1] ManNAc ManNAc + PEP →[2] NeuAc + Pi	b
Sialyllactose	Enzymatic	(1) GlcNAc 2-epimerase (2) NeuAc aldolase (3) CMP-NeuAc synthetase (4) α-2,3-sialyltransferase (5) CMP kinase (6) Acetate kinase (7) Polyphosphate kinase	GlcNac →[1] ManNAc ManNAc + pyruvate →[2] NeuAc NeuAc + CTP →[3] CMP-NeuAc + PPi CMP-NeuAc + lactose →[4] sialyllactose + CMP + *two CTP regeneration systems*: (I) CMP + ATP →[5] CDP + ADP CDP + acetylPi →[6] CTP + Acetate ADP + acetylPi →[6] ATP + acetate (II) CMP + polyPn →[7] CDP + polyPn-1 CDP + polyPn-1 →[7] CTP + polyPn-2	c
Fucosyllactose	Enzymatic/whole cells	(1) GDP-D-mannose-4,6-dehydratase* (2) GDP-L-fucose synthase* (3) Fucosyltransferase *Purified enzymes or recombinantly produced by *S. cerevisiae* or *E. coli*	GDP-mannose →[1] GDP-4-keto-deoxymannose GDP-4-keto-deoxymannose + NADPH →[2] GDP-L-fucose + $NADP^+$ GDP-L-fucose + lactose →[3] fucosyllactose	d
Lacto-*N*-biose I	Enzymatic	(1) Sucrose phosphorylase (2) UDP-glucose-hexose-1-phosphate urydyltransferase (3) Lactobiose *N*-phosphorylase (4) UDP-glucose-epimerase	Sucrose + Pi →[1] Glc1P + Frc Glc1P + UDP-Gal →[2] UDP-Glc + Gal1P Gal1P + GlcNAc →[3] LNB + *UDP-Gal regeneration system:* UDP-Glc →[4] UDP-Gal	e

a Endo et al. 2000; b Ishikawa and Koizumi 2010; Koizumi et al. 2000; c Han et al. 2012; d Albermann et al. 2001; e Nishimoto and Kitaoka 2007; f Numbers in brackets indicate the corresponding step in the "Reaction scheme" column. *Glu* glucose, *Gln* glucosamine, *Gal* galactose, *Frc* fructose, *NeuAc* *N*-acetylneuraminic acid, *PEP* phosphoenol pyruvate, *GlcNAc* *N*-acetyl-D-glucosamine, *ManNAc* *N*-acetyl-D-mannosamine, *PolyP* polyphosphate, *Pi* phosphate, *PPi* pyrophosphate, *Glc1P* glucose 1-phosphate, *Gal1P* galactose 1-phosphate, *LNB* lacto-*N*-biose, *NADPH* reduced nicotinamide dinucleotide phosphate, $NADP^+$ oxidized nicotinamide dinucleotide phosphate mutagenesis after which selected mutations were combined to yield a 32-fold increase in the transferase/hydrolytic ratio while keeping 60 % of the overall wild-type enzyme activity (Osanjo et al. 2007). In another instance, aiming to prevent product hydrolysis, cyclodextrin glucanotransferase (GH13) from *Thermoanaerobacterium thermosulfurigenes* was subjected to error-prone PCR, searching for mutant versions of the enzyme with lower hydrolytic rate. Two mutations were identified, located at the outer region of the active site (W232R and S77P). These mutations lowered the hydrolytic activity up to 15-fold while keeping the cyclization activity (Arnold et al. 2009). Significant increase of GOS production was achieved by mutagenesis of ß-galactosidase (GH42) from *Geobacillus stearothermophilus*. A random mutagenesis approach led to the isolation of mutant R109K with a sixfold higher yield than that obtained with the wild-type enzyme. Saturation mutagenesis at this position allowed the characterization of mutations R109V and R109W, which caused an even higher (around 11-fold) yield increment, up to 23 % from total carbohydrate content (Placier et al. 2009). Random mutagenesis of ß-fructofuranosidase from *Schwanniomyces occidentalis* increased 6-kestose production around twofold over the parent type, with the mutations Q78L and F523V. While Gln78 is located close to the active site, Phe523 resides out of the catalytic domain and probably exerts a long-distance effect (de Abreu et al. 2011). Kojibiose phosphorylase from *Thermoanaerobacter brockii* was also improved by random mutagenesis. Two mutants, S676N and N687I, showed higher productivity of koji-oligosaccharides than the wild type (Yamamoto et al. 2006).

2.3.2 Site-Directed Mutagenesis

Knowing the structure and reaction mechanism of a given protein allows designing specific mutations to improve its activity or change its substrate specificity. Work done with glycosyl hydrolases has had as a major objective to increase the transglycosylation/hydrolysis ratio. Site-directed mutagenesis of ß-glucosidase (GH1) from *Pyrococcus furiosus* at residues within the catalytic pocket revealed that the F426Y mutant had an increased GOS yield (up to 45 % from total carbohydrate content) compared to the wild type (40 %). Moreover, the double mutant M424K/F426Y had better transglycosylating properties at low substrate (lactose) concentrations compared to the wild type and the corresponding single mutants (Hansson et al. 2001). In another GH1 enzyme, the ß-galactosidase from *Sulfolobus solfataricus*, wild-type GOS production (50.9 %) was increased up to 58 % and 62 % as a consequence of mutations F359Q and F441Y, respectively (Wu et al. 2013). This F441 is homologous to the aforementioned F426 and F401 from the GH1 β-glycosidases from *P. furiosus* and *T. thermophilus*, respectively (Feng et al. 2005; Hansson et al. 2001), all resulting in a higher GOS yield when mutated. The observed effect may be caused by a reorientation of the glycone at the −1 subsite to a position more favorable for transglycosylation. On the other hand, the F359Q mutation may increase the affinity of the +1 subsite for the lactose molecule acting as glycosyl acceptor (Wu et al. 2013).

Within invertases belonging to the GH32 family, different studies with plant and yeast enzymes showed that substitution of one or the two asparagines within the so-called β-fructofuranosidase motif (WMNDPGN), harboring the catalytic nucleophile, resulted in increased FOS synthesis (Alvaro-Benito et al. 2010; Lafraya et al. 2011; Ritsema et al. 2006; Schroeven et al. 2008). One of the most dramatic changes was observed with the invertase from *Saccharomyces cerevisiae*, where N21S and N24S amino acid replacements caused a sixfold increase of 6-kestose synthesis. The combination of the aforementioned changes with the W19Y substitution (also within the β-fructofuranosidase motif) further increased the synthesis of 6-kestose up to tenfold, compared to the wild type. Another conserved motif within hydrolytic GH32 enzymes is the "ECP" motif harboring the acid/base catalyst. The substitution of Pro by Val in this motif also increased 6-kestose synthesis by the invertase from *S. cerevisiae* and the β-fructofuranosidase from *S. occidentalis* by sixfold and twofold, respectively (Alvaro-Benito et al. 2010; Lafraya et al. 2011). GH32 enzymes from yeasts synthesize mainly 6-kestose, whereas their plant homologues produce mainly 1-kestose. Although the −1 subsite is rather conserved, the +1 subsite is more divergent between the two groups. The second Trp in the "WGW" motif in hydrolytic plant enzymes has been proposed to be part of the sucrose acceptor subsite, and its replacement by Tyr in the invertase from *Arabidopsis thaliana* doubled the transferase capacity of the enzyme (Altenbach et al. 2009). Not only the transglycosylation yield but also substrate and product specificity can be modulated by site-directed mutagenesis. Donor and acceptor substrate selectivity has been altered in a plant GH32 enzyme, transforming a fructan-fructan fructosyltransferase into a sucrose-sucrose fructosyltransferase (Lasseur et al. 2009; Van den Ende et al. 2009). On the other hand, the introduction of the N228A substitution changed the FOS product profile of the N21S mutant of *S. cerevisiae* invertase, from a highly specific 6-kestose synthesis to an equimolar production of 6-kestose and 1-kestose. This Asn 228, together with Trp 291, was proposed to build the acceptor sucrose-binding platform in yeast enzymes (Lafraya et al. 2011). Mutations at an equivalent residue to this Asn 228 in the β-fructofuranosidase from *S. occidentalis* also resulted in a broader product spectrum, further confirming this hypothesis (Alvaro-Benito et al. 2012).

Two GH18 family chitinases were converted into transglycosylases: *Bacillus circulans* chitinase A1 (*Bc* ChiA1) and *Trichoderma harzianum* chitinase 42 (*Th* Chit42) were mutated at the conserved aspartates that act as stabilizers of the general acid/base glutamate. Mutants D200A and D202A of *Bc* ChiA1, as well as D170N and D170A of *Th* Chit42, retained the ability to catalyze the transglycosylation reaction but lost the hydrolytic activity (Martinez et al. 2012).

Glycosyltransferases have an enormous potential for the synthesis of novel, nonnatural, relevant carbohydrate structures (Hancock et al. 2006). However, methodological difficulties in synthesizing the required activated substrates hamper the production of oligosaccharides by this procedure. Few studies have dealt the molecular engineering of glycosyltransferases, aiming to change the donor specificity of the enzymes. Mutation Y289L in bovine ß-(1,4)-galactosyltransferase (GT7) broadened the profile of donor substrates from a strict specificity for transglycosylation of UDP-galactose to include also UDP-N-acetyl-galactosamine and UDP-*N*-acetylglucosamine, without compromising galactosyltransferase activity (Ramakrishnan and Qasba 2002). Mutation of residue R228K in the same enzyme increased glucosyltransferase activity (15-fold) and decreased galactosyltransferase activity (30-fold) (Ramakrishnan et al. 2005). Similarly, the P234S mutation of a human α-(1,3)-galactosyltransferase (GT6) changed the donor specificity from UDP-galactose to UDP-N-acetyl-galactosamine (Ramakrishnan and Qasba 2002).

2.3.3 Glycosynthases

Withers and colleagues (1998) mutated the nucleophile of a retaining exo-acting glycosidase (a β-glucosidase) belonging to the GH7 family. The resulting mutant was inactive as hydrolase. Using as substrate an activated glycoside fluoride donor, such as α-glucosyl or α-galactosyl-fluoride, with anomeric conformation opposite to that of the substrate of the wild-type enzyme, the mutant enzyme was able to carry out transglycosylation. These new types of enzymes were called glycosynthases (GS). The same year, using the same approach, an endo-acting glycosynthase, able to catalyze the glycosylation of different glucoside acceptors with α-laminaribiosyl fluoride, was constructed from a GH16 β-endoglucanase (Malet and Planas 1998). This methodology has been applied to many other enzymes, belonging to different GHs families, with good results in improving the yield of oligosaccharide synthesis (Cobucci-Ponzano et al. 2011). The mutated enzymes include exo-acting glycosidases as glucosidases, mannosidases, galactosidases, fucosidases, and xylosidases and endo-acting glycosidases as endoglucanases, endoxylanases, laminarinases, and mannanases. In all these cases, substitution of the nucleophile by a residue with a shorter side chain (such as Ala, Gly, or Ser) made the resulting mutant unable to hydrolyze glycosidic bonds but still performed transglycosylation when incubated with activated glycosyl donors. Most GS derived from β-retaining GHs prefer α-glycosyl-fluorides donors and proceed through a single-displacement mechanism (α→β). Hyperthermophylic GS obtained from β-retaining GHs use aryl β-glycosides and require an excess of formate that acts as a nucleophile, forming an α-glycosyl formate intermediate in a double-displacement mechanism (β→α→β). In contrast, for GS derived from α-retaining GHs, β-glycosyl-azides work better as activated donors than β-glycosyl-fluorides (Cobucci-Ponzano et al. 2011). Some more recent examples are the GH13 cyclodextrin glucanotransferase E284G from *Bacillus* sp. (Li et al. 2014) and the GH35 ß-galactosidase E233G from *Bacillus circulans* (Henze et al. 2014) which used α-maltosyl-fluoride and α-D-galactopyranosyl-fluoride as donors, respectively. The latter mutant has been used to produce galacto-*N*-biose or lacto-*N*-biose with good yields (40–90 %) (Henze et al. 2014).

In some instances, consecutive rounds of directed mutagenesis have been applied to further improve the properties of a first generation of GS. One approach has been site-saturation mutagenesis of the nucleophilic residue to select the mutation (Ala, Ser, or Gly) that leads to higher transglycosylating efficiencies (Cobucci-Ponzano et al. 2011). Random mutagenesis has also been used to improve the transglycosylating activity of glycosynthases from families GH1 and GH52 up to 27- and 45-fold, respectively (Ben-David et al. 2007; Kim et al. 2004). Other studies have pursued more complex goals, aiming to change the substrate specificity of the enzyme through site-directed mutagenesis. This was the case of *Humicola insolens* Cel7B E197A mutant, a glycosynthase whose properties were altered by additional mutations at the +1 catalytic subsite. Mutations H209A, H209G, and A211T, combined with the original E197A mutation, were designed in order to widen the substrate specificity to include N-acetyl-glucosaminyl acceptors. However, the resulting mutants were unable to transfer α-lactosyl fluoride onto *N*′,*N*″-diacetylchitobiose, although cellobiose and *N*′-acetylchitobiose still remained as good acceptors (Blanchard et al. 2007a). In contrast, a change in the regioselectivity was observed using α-lactosyl fluoride as substrate donor and *O*-allyl N′-acetyl-2″-azido-ß-chitobioside as acceptor. While E197A and the triple mutant E197A/H209A/A211T preferentially catalyzed the formation of a ß-(1,3)-linked tetrasaccharide between the two disaccharides, the product specificity of the double mutants E197A/H209A and E197A/H209G was switched to a ß-(1,4) linkage (Blanchard et al. 2007b).

Honda and Kitaoka described in 2005 the first glycosynthase derived from an inverting glycosidase, the exo-oligoxilanase (GH8) from *Bacillus halodurans*. This glycosynthase was constructed by saturation random mutagenesis at the catalytic base D263. Nine of these mutants synthesized a xylotrisaccharide from xylose and α-xylobiose fluoride, with the D263C mutant showing the highest yield (Honda and Kitaoka 2006). An alternative method to convert the *Bacillus halodurans* exo-oligoxilanase into a glycosynthase was the mutagenesis of the Y198 residue, which is known to bind the water molecule acting as nucleophile in the hydrolytic reaction. This caused a drastic decrease in the hydrolytic activity and a small increase in the fluoride-releasing activity, which made the transglycosylation catalyzed by the Y198F mutant accumulate a larger amount of product than that achieved with the D263C mutant (Honda et al. 2008; Kitaoka et al. 2008). Another glycosynthase has been produced from the inverting 1,2-α-L-fucosidase (GH95) from *Bifidobacterium bifidum* by mutating the catalytic acid residue. The D766G mutant synthesized 2′-fucosyllactose from β-fucosyl-fluoride and lactose (Wada et al. 2008).

2.3.4 Enzyme Immobilization for Oligosaccharide Synthesis

Enzyme production is expensive for most industrial applications. Enzymes are labile and sensitive to denaturation by chemical agents and physical conditions, which hinders their use. Moreover, unlike conventional chemical catalysts, most enzymes operate in aqueous solutions making it difficult to separate them from the final product for convenience or to reuse the enzyme in a new reaction (Homaei et al. 2013). Immobilization of enzymes offers a possible solution to these problems and allows their implementation in continuous flow reactors. There is an ample variety of immobilization methods, generally classified in three main types: covalent or noncovalent surface binding, physical entrapment, and self-aggregation by cross-linking (Mateo et al. 2007; Sheldon and van Pelt 2013). There is no universal method, or support to immobilization, because it depends on the nature of the enzyme, the chemical characteristics of the substrate, and the operating conditions. Compared to the free form, immobilized enzymes are generally more stable (to temperature, pH, etc.) and easier to manipulate (Homaei et al. 2013). Sometimes, immobilization changes the kinetic properties of an enzyme altering its activity or substrate specificity (Rodrigues et al. 2013). The immobilization of oligosaccharide-producing enzymes is a promising strategy for large-scale production of these compounds, and there are some examples in bibliography with different supports and enzymes.

Because of the importance of GOS in the food market, immobilization of ß-galactosidases to improve the yield of GOS production has been assayed by different groups (Panesar et al. 2010). *Aspergillus oryzae* ß-galactosidase was immobilized using three different techniques: adsorption on celite, covalent coupling to chitosan with glutaraldehyde, and self-aggregation by cross-linking. The best activity was reported by covalent binding to chitosan, yielding a higher GOS yield (17.3 % of the total sugar) than the free enzyme (10 %) (Gaur et al. 2006). In other instances, total GOS production remained unchanged, but the possibility of reusing the enzyme made the process interesting. That was the case of *Bacillus circulans* ß-galactosidase immobilized in glyoxal-derivatized agarose, which yielded the same GOS production and profile than the free enzyme but allowed carrying out ten repeated batches (Urrutia et al. 2013), or covalent immobilization of *Thermotoga maritima* ß-galactosidase on the surface of epoxy-activated magnetic beads which allowed four cycles of reutilization without a significant degree of inactivation (Marín-Navarro et al. 2014). Immobilized β-galactosidases have also been used in continuous operation systems for GOS production (Klein et al. 2013).

FOS-, COS-, IMOS-, and XOS-producing enzymes have also been subjected to immobilization by different methods. *Aspergillus flavus* fructosyltransferase was entrapped on chitosan and alginate showing better performance with the latter support. The thermal stability of the enzyme increased, and immobilization allowed work with a continuous system, accumulating up to 63 % (w/w) of FOS after 7 days of reaction (Ganaie et al. 2014). Immobilization of chitosanase from *Bacillus pumilus* by multipoint covalent attachment with glutaraldehyde, using as support amylose-coated magnetic nanoparticles, allowed reusability of the enzyme, producing COS with a yield of 40 % of the used chitosan (Kuroiwa et al. 2008). Endo-dextranase from *Penicillium* sp. was covalently attached to an epoxy-activated disk and used to produce IMOS. The immobilized enzyme retained more than 77 % of its activity for 78 days including storage time and repeated uses every 10 days, with a total operational time of 90 h (Bertrand et al. 2014). Alginate fiber entrapment enabled repeated use during ten batches of the *Leuconostoc mesenteroides* dextransucrase for IMOS production (Tanriseven and Doğan 2002). Production of XOS by xylan hydrolysis, using immobilized xylanases, has been successfully assayed. *Talaromyces thermophilus* xylanase, entrapped in gelatin and cross-linked by glutaraldehyde, yielded the same product profile than the free enzyme but allowed up to 13 successive cycles of hydrolysis (Maalej-Achouri et al. 2009). Xylanase B from *Thermotoga maritima* fused to a histidine tag was fixed covalently to a nickel-chelate epoxy support. The double functionality of this resin allows a specific attachment of the enzyme to the support through the histidine-nickel interaction and also has the advantage of a stable, covalent binding mediated by the epoxy groups. Immobilization improved the thermostability of the xylanase and allowed its use in a continuous process (Li et al. 2007a).

Acknowledgments The authors have been funded by grant BIO2013-48779-C4-3-R, from Spain's "Secretaría de Estado de Investigación, Desarrollo e Innovación." D T-P held a FPU fellowship from "Ministerio de Economía y Competitividad."

References

Aachary AA, Prapulla SG (2011) Xylooligosaccharides (XOS) as an emerging prebiotic: microbial synthesis, utilization, structural characterization, bioactive properties, and applications. Compr Rev Food Sci Food Saf 10:2–16

Ahmed S, Riaz S, Jamil A (2009) Molecular cloning of fungal xylanases: an overview. Appl Microbiol Biotechnol 84:19–35

Aisaka K, Masuda T, Chikamune T (1996) Properties of maltose phosphorylase from *Propionibacterium freudenreichii*. J Ferment Bioeng 82:171–173

Akiyama K, Takase M, Horikoshi K, Okonogi S (2001) Production of galactooligosaccharides from lactose using a beta-glucosidase from *Thermus* sp. Z-1. Biosci Biotechnol Biochem 65:438–441

Albermann C, Piepersberg W, Wehmeier UF (2001) Synthesis of the milk oligosaccharide 2′-fucosyllactose using recombinant bacterial enzymes. Carbohydr Res 334:97–103

Altenbach D, Rudino-Pinera E, Olvera C, Boller T, Wiemken A, Ritsema T (2009) An acceptor-substrate binding site determining glycosyl transfer emerges from mutant analysis of a plant vacuolar invertase and a fructosyltransferase. Plant Mol Biol 69:47–56

Álvaro-Benito M, de Abreu M, Fernández-Arrojo L, Plou FJ, Jiménez-Barbero J, Ballesteros A, Fernández-Lobato M (2007) Characterization of a β-fructofuranosidase from *Schwanniomyces occidentalis* with transfructosylating activity yielding the prebiotic 6-kestose. J Biotechnol 132:75–81

Alvaro-Benito M, de Abreu M, Portillo F, Sanz-Aparicio J, Fernandez-Lobato M (2010) New insights into the fructosyltransferase activity of *Schwanniomyces occidentalis* β-fructofuranosidase, emerging from nonconventional codon usage and directed mutation. Appl Environ Microbiol 76:7491–7499

Alvaro-Benito M, Sainz-Polo MA, Gonzalez-Parez D, Gonzalez B, Plou FJ, Fernández-Lobato M, Sanz-Aparicio J (2012) Structural and kinetic insights reveal that the amino acid pair Gln-228/Asn-254 modulates the transfructosylating specificity of *Schwanniomyces occidentalis* β-fructofuranosidase, an enzyme that produces prebiotics. J Biol Chem 287:19674–19686

Apolinário AC, de Lima Damasceno BPG, de Macêdo Beltrão NE, Pessoa A, Converti A, da Silva JA (2014) Inulin-type fructans: a review on different aspects of biochemical and pharmaceutical technology. Carbohydr Polym 101:368–378

Arab-Jaziri F, Bissaro B, Dion M, Saurel O, Harrison D, Ferreira F, O'Donohue MJ (2013) Engineering transglycosidase activity into a GH51 α-l-arabinofuranosidase. New Biotechnol 30:536–544

Arnold K, Kiefer F, Kopp J, Battey JN, Podvinec M, Westbrook JD, Schwede T (2009) The protein model portal. J Struct Funct Genom 10:1–8

Ben-David A, Bravman T, Balazs YS, Czjzek M, Schomburg D, Shoham G, Shoham Y (2007) Glycosynthase activity of *Geobacillus stearothermophilus* GH52 beta-xylosidase: efficient synthesis of xylooligosaccharides from alpha-D-xylopyranosyl fluoride through a conjugated reaction. ChemBioChem 8:2145–2151

Bertrand E, Pierre G, Delattre C, Gardarin C, Bridiau N, Maugard T, Michaud P (2014) Dextranase immobilization on epoxy CIM® disk for the production of isomaltooligosaccharides from dextran. Carbohydr Polym 111:707–713

Blanchard S, Cottaz S, Coutinho PM, Patkar S, Vind J, Boerd H, Koivula A, Driguez H, Armand S (2007a) Mutation of fungal endoglucanases into glycosynthases and characterization of their acceptor substrate specificity. J Mol Catal B: Enzym 44:106–116

Blanchard S, Armand S, Couthino P, Patkar S, Vind J, Samain E, Cottaz S (2007b) Unexpected regioselectivity of *Humicola insolens* Cel7B glycosynthase mutants. Carbohydr Res 342:710–716

Bode L (2009) Human milk oligosaccharides: prebiotics and beyond. Nutr Rev 67:S183–S191

Bornscheuer U (2013) Protein engineering as a tool for the development of novel bioproduction systems. In: Zeng AP (ed) Fundamentals and application of new bioproduction systems, vol 137. Springer, Berlin/Heidelberg, pp 25–40

Bruins ME, Strubel M, van Lieshout JFT, Janssen AEM, Boom RM (2003) Oligosaccharide synthesis by the hyperthermostable β-glucosidase from *Pyrococcus furiosus*: kinetics and modelling. Enzym Microb Technol 33:3–11

Cardelle-Cobas A, Corzo N, Olano A, Pelaez C, Requena T, Avila M (2011) Galactooligosaccharides derived from lactose and lactulose: influence of structure on *Lactobacillus*, *Streptococcus* and *Bifidobacterium* growth. Int J Food Microbiol 149:81–87

Carvalheiro F, Esteves MP, Parajó JC, Pereira H, Gírio FM (2004) Production of oligosaccharides by autohydrolysis of Brewery's spent grain. Bioresour Technol 91:93–100

Chaen H, Nishimoto T, Nakada T, Fukuda S, Kurimoto M, Tsujisaka Y (2001) Enzymatic synthesis of novel oligosaccharides from l-sorbose, maltose, and sucrose using kojibiose phosphorylase. J Biosci Bioeng 92:173–176

Chapla D, Divecha J, Madamwar D, Shah A (2010) Utilization of agro-industrial waste for xylanase production by *Aspergillus foetidus* MTCC 4898 under solid state fermentation and its application in saccharification. Biochem Eng J 49:361–369

Chapla D, Pandit P, Shah A (2012) Production of xylooligosaccharides from corncob xylan by fungal xylanase and their utilization by probiotics. Bioresour Technol 115:215–221

Chen M-H, Bowman MJ, Dien BS, Rausch KD, Tumbleson ME, Singh V (2014) Autohydrolysis of *Miscanthus giganteus* for the production of xylooligosaccharides (XOS): kinetics, characterization and recovery. Bioresour Technol 155:359–365

Closa-Monasterolo R, Gispert-Llaurado M, Luque V, Ferre N, Rubio-Torrents C, Zaragoza-Jordana M, Escribano J (2013) Safety and efficacy of inulin and oligofructose supplementation in infant formula: results from a randomized clinical trial. Clin Nutr 32:918–927

Cobucci-Ponzano B, Strazzulli A, Rossi M, Moracci M (2011) Glycosynthases in biocatalysis. Adv Synth Catal 353:2284–2300

Côté GL (2009) Acceptor products of alternansucrase with gentiobiose. Production of novel oligosaccharides for food and feed and elimination of bitterness. Carbohydr Res 344:187–190

Courtois J (2009) Oligosaccharides from land plants and algae: production and applications in therapeutics and biotechnology. Curr Opin Microbiol 12:261–273

Crittenden RG, Playne MJ (1996) Production, properties and applications of food-grade oligosaccharides. Trends Food Sci Technol 7:353–361

da Silva AF, García-Fraga B, López-Seijas J, Sieiro C (2014) Characterization and optimization of heterologous expression in *Escherichia coli* of the chitinase encoded by the chiA gene of *Bacillus halodurans* C-125. Process Biochem 49:1622–1629

Davies G, Henrissat B (1995) Structures and mechanisms of glycosyl hydrolases. Structure 3:853–859

Davies GJ, Wilson KS, Henrissat B (1997) Nomenclature for sugar-binding subsites in glycosyl hydrolases. Biochem J 321:557–559

de Abreu MA, Alvaro-Benito M, Plou FJ, Fernandez-Lobato M, Alcalde M (2011) Screening beta-fructofuranosidases mutant libraries to enhance the transglycosylation rates of beta-(2→6) fructooligosaccharides. Comb Chem High Throughput Screening 14:730–738

Del Valle EMM (2004) Cyclodextrins and their uses: a review. Process Biochem 39:1033–1046

Dominguez A, Rodrigues L, Lima N, Teixeira J (2014) An overview of the recent developments on fructooligosaccharide production and applications. Food Bioproc Technol 7:324–337

Du Y, Shi P, Huang H, Zhang X, Luo H, Wang Y, Yao B (2013) Characterization of three novel thermophilic xylanases from *Humicola insolens* Y1 with application potentials in the brewing industry. Bioresour Technol 130:161–167

Duan KJ, Sheu DC, Lin CT (1995) Transglucosylation of a fungal alpha-glucosidase. The enzyme properties and correlation of isomaltooligosaccharide production. Ann N Y Acad Sci 31:750:325-328

Endo T, Koizumi S, Tabata K, Ozaki A (2000) Large-scale production of CMP-NeuAc and sialylated oligosaccharides through bacterial coupling. Appl Microbiol Biotechnol 53:257–261

Feng HY, Drone J, Hoffmann L, Tran V, Tellier C, Rabiller C, Dion M (2005) Converting a β-glycosidase into a β-transglycosidase by directed evolution. J Biol Chem 280:37088–37097

Fernández-Arrojo L, Marín D, Gómez De Segura A, Linde D, Alcalde M, Gutiérrez-Alonso P, Ballesteros A (2007) Transformation of maltose into prebiotic isomaltooligosaccharides by a novel α-glucosidase from *Xantophyllomyces dendrorhous*. Process Biochem 42:1530–1536

Fraser-Reid B, Wu Z, Udodong UE, Ottosson H (1990) Armed/disarmed effects in glycosyl donors: rationalization and sidetracking. J Organ Chem 55:6068–6070

Fujimoto Y, Hattori T, Uno S, Murata T, Usui T (2009) Enzymatic synthesis of gentiooligosaccharides by transglycosylation with β-glycosidases from *Penicillium multicolor*. Carbohydr Res 344:972–978

Ganaie MA, Gupta US, Kango N (2013) Screening of biocatalysts for transformation of sucrose to fructooligosaccharides. J Mol Catal B Enzym 97:12–17

Ganaie MA, Rawat HK, Wani OA, Gupta US, Kango N (2014) Immobilization of fructosyltransferase by chitosan and alginate for efficient production of fructooligosaccharides. Process Biochem 49:840–844

Gänzle MG (2012) Enzymatic synthesis of galacto-oligosaccharides and other lactose derivatives (hetero-oligosaccharides) from lactose. Int Dairy J 22:116–122

Gaur R, Pant H, Jain R, Khare SK (2006) Galacto-oligosaccharide synthesis by immobilized *Aspergillus oryzae* β-galactosidase. Food Chem 97:426–430

Goffin D, Delzenne N, Blecker C, Hanon E, Deroanne C, Paquot M (2011) Will isomalto-oligosaccharides, a well-established functional food in Asia, break through the European and American market? The status of knowledge on these prebiotics. Crit Rev Food Sci Nutr 51:394–409

Goo BG, Hwang YJ, Park JK (2014) *Bacillus thuringiensis*: a specific gamma-cyclodextrin producer strain. Carbohydr Res 386:12–17

Goulas T, Goulas A, Tzortzis G, Gibson GR (2009) Expression of four beta-galactosidases from *Bifidobacterium bifidum* NCIMB41171 and their contribution on the hydrolysis and synthesis of galactooligosaccharides. Appl Microbiol Biotechnol 84:899-907

Han WC, Byun SH, Kim MH, Sohn EH, Lim JD, Um BH, Jang KH (2009) Production of lactosucrose from sucrose and lactose by a levansucrase from *Zymomonas mobilis*. J Microbiol Biotechnol 19:1153–1160

Han NS, Kim TJ, Park YC, Kim J, Seo JH (2012) Biotechnological production of human milk oligosaccharides. Biotechnol Adv 30:1268–1278

Hancock SM, Vaughan MD, Withers SG (2006) Engineering of glycosidases and glycosyltransferases. Curr Opin Chem Biol 10:509–519

Hansson T, Kaper T, van Der Oost J, de Vos WM, Adlercreutz P (2001) Improved oligosaccharide synthesis by protein engineering of beta-glucosidase CelB from hyperthermophilic *Pyrococcus furiosus*. Biotechnol Bioeng 73:203–210

Hayashi S, Yoshiyama T, Fujii N, Shinohara S (2000) Production of a novel syrup containing neofructo-oligosaccharides by the cells of *Penicillium citrinum*. Biotechnol Lett 22:1465–1469

Henze M, You DJ, Kamerke C, Hoffmann N, Angkawidjaja C, Ernst S, Pietruszka J, Kanaya S, Elling L (2014) Rational design of a glycosynthase by the crystal structure of beta-galactosidase from *Bacillus circulans* (BgaC) and its use for the synthesis of N-acetyllactosamine type 1 glycan structures. J Biotechnol 191:78–85

Hernandez-Hernandez O, Marin-Manzano MC, Rubio LA, Moreno FJ, Sanz ML, Clemente A (2012) Monomer and linkage type of galacto-oligosaccharides affect their resistance to ileal digestion and prebiotic properties in rats. J Nutr 142:1232–1239

Hernot DC, Boileau TW, Bauer LL, Middelbos IS, Murphy MR, Swanson KS, Fahey GC (2009) In vitro fermentation profiles, gas production rates, and microbiota modulation as affected by certain fructans, galactooligosaccharides, and polydextrose. J Agric Food Chem 57:1354–1361

Homaei A, Sariri R, Vianello F, Stevanato R (2013) Enzyme immobilization: an update. J Chem Biol 6:185–205

Honda Y, Kitaoka M (2006) The first glycosynthase derived from an inverting glycoside hydrolase. J Biol Chem 281:1426–1431

Honda Y, Kitaoka M, Hayashi K (2004) Reaction mechanism of chitobiose phosphorylase from *Vibrio proteolyticus*: identification of family 36 glycosyltransferase in Vibrio. Biochem J 377:225–232

Honda Y, Fushinobu S, Hidaka M, Wakagi T, Shoun H, Taniguchi H, Kitaoka M (2008) Alternative strategy for converting an inverting glycoside hydrolase into a glycosynthase. Glycobiology 18:325–330

Ishii T, Tobita T (1993) Structural characterization of feruloyl oligosaccharides from spinach-leaf cell walls. Carbohydr Res 248:179–190

Ishikawa M, Koizumi S (2010) Microbial production of N-acetylneuraminic acid by genetically engineered *Escherichia coli*. Carbohydr Res 345:2605–2609

Ji ES, Park NH, Oh DK (2005) Galacto-oligosaccharide production by a thermostable recombinant β-galactosidase from *Thermotoga maritima*. World J Microbiol Biotechnol 21:759–764

Katapodis P, Christakopoulos P (2008) Enzymic production of feruloyl xylo-oligosaccharides from corn cobs by a family 10 xylanase from *Thermoascus aurantiacus*. LWT Food Sci Technol 41:1239–1243

Kim YS, Oh DK (2012) Lactulose production from lactose as a single substrate by a thermostable cellobiose 2-epimerase from *Caldicellulosiruptor saccharolyticus*. Bioresour Technol 104:668–672

Kim S-K, Rajapakse N (2005) Enzymatic production and biological activities of chitosan oligosaccharides (COS): a review. Carbohydr Polym 62:357–368

Kim YW, Lee SS, Warren RA, Withers SG (2004) Directed evolution of a glycosynthase from *Agrobacterium sp.* increases its catalytic activity dramatically and expands its substrate repertoire. J Biol Chem 279:42787–42793

Kim Y-S, Park C-S, Oh D-K (2006) Lactulose production from lactose and fructose by a thermostable β-galactosidase from *Sulfolobus solfataricus*. Enzym Microb Technol 39:903–908

Kitaoka M, Honda Y, Hidaka M, Fushinobu S (2008) Strategy for converting an inverting glycoside hydrolase into a glycosynthase. In: Park KH (ed) Carbohydrate-active enzymes. Woodhead Publishing, Cambridge, UK, pp 193–205

Kittl R, Withers SG (2010) New approaches to enzymatic glycoside synthesis through directed evolution. Carbohydr Res 345:1272–1279

Klein MP, Fallavena LP, Schöffer JN, Ayub MA, Rodrigues RC, Ninow JL, Hertz PF (2013) High stability of immobilized β-D-galactosidase for lactose hydrolysis and galactooligosaccharides synthesis. Carbohydr Polym 95:465-70

Koizumi S, Endo T, Tabata K, Nagano H, Ohnishi J, Ozaki A (2000) Large-scale production of GDP-fucose and Lewis X by bacterial coupling. J Ind Microbiol Biotechnol 25:213–217

Kuroiwa T, Noguchi Y, Nakajima M, Sato S, Mukataka S, Ichikawa S (2008) Production of chitosan oligosaccharides using chitosanase immobilized on amylose-coated magnetic nanoparticles. Process Biochem 43:62–69

Lafraya A, Sanz-Aparicio J, Polaina J, Marin-Navarro J (2011) Fructo-oligosaccharide synthesis by mutant versions of *Saccharomyces cerevisiae* invertase. Appl Environ Microbiol 77:6148–6157

Lairson LL, Henrissat B, Davies GJ, Withers SG (2008) Glycosyltransferases: structures, functions, and mechanisms. Annu Rev Biochem 77:521–555

Lasseur B, Schroeven L, Lammens W, Le Roy K, Spangenberg G, Manduzio H, Van den Ende W (2009) Transforming a fructan: fructan 6G-fructosyltransferase from perennial ryegrass into a sucrose: sucrose 1-fructosyltransferase. Plant Physiol 149:327–339

Lee MS, Cho SK, Eom HJ, Kim SY, Kim TJ, Han NS (2008) Optimized substrate concentrations for production of long-chain isomaltooligosaccharides using dextransucrase of *Leuconostoc mesenteroides* B-512F. J Microbiol Biotechnol 18:1141–1145

Lee YS, Zhou Y, Park DJ, Chang J, Choi YL (2013) beta-cyclodextrin production by the cyclodextrin glucanotransferase from *Paenibacillus illinoisensis* ZY-08:

cloning, purification, and properties. World J Microbiol Biotechnol 29:865–873

Leemhuis H, Kelly R, Dijkhuizen L (2010) Engineering of cyclodextrin glucanotransferases and the impact for biotechnological applications. Appl Microbiol Biotechnol 85:823–835

Li L, Zhu Y, Huang Z, Jiang Z, Chen W (2007a) Immobilization of the recombinant xylanase B (XynB) from the hyperthermophilic *Thermotoga maritima* on metal-chelate Eupergit C 250L. Enzym Microb Technol 41:278–285

Li Z, Wang M, Wang F, Gu Z, Du G, Wu J, Chen J (2007b) γ-Cyclodextrin: a review on enzymatic production and applications. Appl Microbiol Biotechnol 77:245–255

Li C, Ahn H-J, Kim J-H, Kim Y-W (2014) Transglycosylation of engineered cyclodextrin glucanotransferases as O-glycoligases. Carbohydr Polym 99:39–46

Linde D, Rodriguez-Colinas B, Estevez M, Poveda A, Plou FJ, Fernandez Lobato M (2012) Analysis of neo-fructooligosaccharides production mediated by the extracellular beta-fructofuranosidase from *Xanthophyllomyces dendrorhous*. Bioresour Technol 109:123–130

Liu C-L, Lan C-Y, Fu C-C, Juang R-S (2014) Production of hexaoligochitin from colloidal chitin using a chitinase from *Aeromonas schubertii*. Int J Biol Macromol 69:59–63

Lombard V, Golaconda Ramulu H, Drula E, Coutinho PM, Henrissat B (2014) The carbohydrate-active enzymes database (CAZy) in 2013. Nucleic Acids Res 42:D490–D495

Maalej-Achouri I, Guerfali M, Gargouri A, Belghith H (2009) Production of xylo-oligosaccharides from agro-industrial residues using immobilized *Talaromyces thermophilus* xylanase. J Mol Catal B Enzym 59:145–152

Macfarlane GT, Steed H, Macfarlane S (2008) Bacterial metabolism and health-related effects of galacto-oligosaccharides and other prebiotics. J Appl Microbiol 104:305–344

Mackenzie LF, Wang Q, Warren RAJ, Withers SG (1998) Glycosynthases: mutant glycosidases for oligosaccharide synthesis. J Am Chem Soc 120:5583–5584

Malet C, Planas A (1998) From beta-glucanase to beta-glucansynthase: glycosyl transfer to alpha-glycosyl fluorides catalyzed by a mutant endoglucanase lacking its catalytic nucleophile. FEBS Lett 440:208–212

Manisseri C, Gudipati M (2010) Bioactive xylo-oligosaccharides from wheat bran soluble polysaccharides. LWT Food Sci Technol 43:421–430

Marín-Navarro J, Talens-Perales D, Oude-Vrielink A, Cañada F, Polaina J (2014) Immobilization of thermostable β-galactosidase on epoxy support and its use for lactose hydrolysis and galactooligosaccharides biosynthesis. World J Microbiol Biotechnol 30:989–998

Mark BL, Vocadlo DJ, Knapp S, Triggs-Raine BL, Withers SG, James MNG (2001) Crystallographic evidence for substrate-assisted catalysis in a bacterial β-hexosaminidase. J Biol Chem 276:10330–10337

Martinez EA, Boer H, Koivula A, Samain E, Driguez H, Armand S, Cottaz S (2012) Engineering chitinases for the synthesis of chitin oligosaccharides: catalytic amino acid mutations convert the GH-18 family glycoside hydrolases into transglycosylases. J Mol Catal B Enzym 74:89–96

Martinez-Villaluenga C, Cardelle-Cobas A, Olano A, Corzo N, Villamiel M, Jimeno ML (2008) Enzymatic synthesis and identification of two trisaccharides produced from lactulose by transgalactosylation. J Agric Food Chem 56:557–563

Mateo C, Palomo JM, Fernandez-Lorente G, Guisan JM, Fernandez-Lafuente R (2007) Improvement of enzyme activity, stability and selectivity via immobilization techniques. Enzym Microb Technol 40:1451–1463

Millqvist-Fureby A, MacManus DA, Davies S, Vulfson EN (1998) Enzymatic transformations in supersaturated substrate solutions: II. Synthesis of disaccharides via transglycosylation. Biotechnol Bioeng 60:197–203

Monsan P, Ouarné F (2009) Oligosaccharides derived from sucrose. In: Charalampopoulos D, Rastall R (eds) Prebiotics and probiotics science and technology. Springer, New York, pp 293–336

Moon I-S, Cho G (1997) Production of maltooligosaccharides from starch and separation of maltopentaose by adsorption of them on activated carbon (I). Biotechnol Bioprocess Eng 2:19–22

Moreno FJ, Montilla A, Villamiel M, Corzo N, Olano A (2014) Analysis, structural characterization, and bioactivity of oligosaccharides derived from lactose. Electrophoresis 35:1519–1534

Mu W, Chen Q, Wang X, Zhang T, Jiang B (2013) Current studies on physiological functions and biological production of lactosucrose. Appl Microbiol Biotechnol 97:7073–7080

Mutanda T, Mokoena MP, Olaniran AO, Wilhelmi BS, Whiteley CG (2014) Microbial enzymatic production and applications of short-chain fructooligosaccharides and inulooligosaccharides: recent advances and current perspectives. J Ind Microbiol Biotechnol 41:893–906

Nakai H, Kitaoka M, Svensson B, Ohtsubo KI (2013) Recent development of phosphorylases possessing large potential for oligosaccharide synthesis. Curr Opin Chem Biol 17:301–309

Ning Y, Wang J, Chen J, Yang N, Jin Z, Xu X (2010) Production of neo-fructooligosaccharides using free-whole-cell biotransformation by *Xanthophyllomyces dendrorhous*. Bioresour Technol 101:7472-7478

Nishimoto M, Kitaoka M (2007) Practical preparation of lacto-N-biose I, a candidate for the bifidus factor in human milk. Biosci Biotechnol Biochem 71:2101–2104

Osanjo G, Dion M, Drone J, Solleux C, Tran V, Rabiller C, Tellier C (2007) Directed evolution of the alpha-L-fucosidase from *Thermotoga maritima* into an alpha-L-transfucosidase. Biochemistry 46:1022–1033

Otaka K (2006) Functional oligosaccharide and its new aspect as immune modulation. J Biol Macromol 6:3–9

Paixão SM, Teixeira PD, Silva TP, Teixeira AV, Alves L (2013) Screening of novel yeast inulinases and further application to bioprocesses. New Biotechnol 30:598–606

Panesar PS, Kumari S, Panesar R (2010) Potential applications of immobilized β-galactosidase in food processing industries. Enzyme Res 27:473137

Panesar PS, Kumari S (2011) Lactulose: production, purification and potential applications. Biotechnol Adv 29:940–948

Park A-R, Oh D-K (2010) Galacto-oligosaccharide production using microbial β-galactosidase: current state and perspectives. Appl Microbiol Biotechnol 85:1279–1286

Park S, Gildersleeve JC, Blixt O, Shin I (2013) Carbohydrate microarrays. Chem Soc Rev 42:4310–4326

Patel S, Goyal A (2011) Functional oligosaccharides: production, properties and applications. World J Microbiol Biotechnol 27:1119–1128

Placier G, Watzlawick H, Rabiller C, Mattes R (2009) Evolved beta-galactosidases from *Geobacillus stearothermophilus* with improved transgalactosylation yield for galacto-oligosaccharide production. Appl Environ Microbiol 75:6312–6321

Porras-Domínguez JR, Ávila-Fernández Á, Rodríguez-Alegría ME, Miranda-Molina A, Escalante A, González-Cervantes R, López Munguía A (2014) Levan-type FOS production using a *Bacillus licheniformis* endolevanase. Process Biochem 49:783–790

Prasad S, Dhiman RK, Duseja A, Chawla YK, Sharma A, Agarwal R (2007) Lactulose improves cognitive functions and health-related quality of life in patients with cirrhosis who have minimal hepatic encephalopathy. Hepatology 45:549–559

Ramakrishnan B, Qasba PK (2002) Structure-based design of beta 1,4-galactosyltransferase I (beta 4Gal-T1) with equally efficient N-acetylgalactosaminyltransferase activity: point mutation broadens beta 4Gal-T1 donor specificity. J Biol Chem 277:20833–20839

Ramakrishnan B, Boeggeman E, Qasba PK (2005) Mutation of arginine 228 to lysine enhances the glucosyltransferase activity of bovine beta-1,4-galactosyltransferase I. Biochemistry 44:3202–3210

Raman R, Raguram S, Venkataraman G, Paulson JC, Sasisekharan R (2005) Glycomics: an integrated systems approach to structure-function relationships of glycans. Nat Methods 2:817–824

Ritsema T, Hernandez L, Verhaar A, Altenbach D, Boller T, Wiemken A, Smeekens S (2006) Developing fructan-synthesizing capability in a plant invertase via mutations in the sucrose-binding box. Plant J 48:228–237

Rodrigues RC, Ortiz C, Berenguer-Murcia A, Torres R, Fernandez-Lafuente R (2013) Modifying enzyme activity and selectivity by immobilization. Chem Soc Rev 42:6290–6307

Rodriguez-Colinas B, de Abreu MA, Fernandez-Arrojo L, de Beer R, Poveda A, Jimenez-Barbero J, Haltrich D, Ballesteros-Olmo AO, Fernandez-Lobato M, Plou FJ (2011) Production of galacto-oligosaccharides by the β-Galactosidase from *Kluyveromyces lactis*: comparative analysis of permeabilized cells versus soluble enzyme. J Agric Chem 59:10477–10484

Rodriguez-Diaz J, Monedero V, Yebra MJ (2011) Utilization of natural fucosylated oligosaccharides by three novel alpha-L-fucosidases from a probiotic *Lactobacillus casei* strain. Appl Environ Microbiol 77:703–705

Roos YH (2009) Solid and liquid states of lactose. In: McSweeney P, Fox PF (eds) Advanced dairy chemistry. Springer, New York, pp 17–33

Rycroft CE, Jones MR, Gibson GR, Rastall RA (2001) Fermentation properties of gentio-oligosaccharides. Lett Appl Microbiol 32:156–161

Sabater-Molina M, Larque E, Torrella F, Zamora S (2009) Dietary fructooligosaccharides and potential benefits on health. J Physiol Biochem 65:315–328

Samala A, Srinivasan R, Yadav M (2015) Comparison of xylo-oligosaccharides production by autohydrolysis of fibers separated from ground corn flour and DDGS. Food Bioprod Process 94:354–364

Samanta AK, Jayapal N, Kolte AP, Senani S, Sridhar M, Suresh KP, Sampath KT (2012) Enzymatic production of xylooligosaccharides from alkali solubilized xylan of natural grass (*Sehima nervosum*). Bioresour Technol 112:199–205

Schroeven L, Lammens W, Van Laere A, Van den Ende W (2008) Transforming wheat vacuolar invertase into a high affinity sucrose: sucrose 1-fructosyltransferase. New Phytol 180:822–831

Schuster-Wolff-Bühring R, Fischer L, Hinrichs J (2010) Production and physiological action of the disaccharide lactulose. Int Dairy J 20:731–741

Sheldon RA, van Pelt S (2013) Enzyme immobilisation in biocatalysis: why, what and how. Chem Soc Rev 42:6223–6235

Singh M, Sharma R, Banerjee UC (2002) Biotechnological applications of cyclodextrins. Biotechnol Adv 20:341–359

Song J, Imanaka H, Imamura K, Minoda M, Katase T, Hoshi Y, Yamaguchi S, Nakanishi K (2011) Cloning and expression of a β-galactosidase gene of *Bacillus circulans*. Biosci Biotechnol Biochem 75:1194-1197

Szymanski CM, Wren BW (2005) Protein glycosylation in bacterial mucosal pathogens. Nat Rev Microbiol 3:225–237

Tanriseven A, Doğan Ş (2002) Production of isomalto-oligosaccharides using dextransucrase immobilized in alginate fibres. Process Biochem 37:1111–1115

Terada Y, Yanase M, Takata H, Takaha T, Okada S (1997) Cyclodextrins are not the major cyclic alpha-1,4-glucans produced by the initial action of cyclodextrin glucanotransferase on amylose. J Biol Chem 272:15729–15733

Torres DPM, Gonáalves M d P, Teixeira JA, Gonáalves M d P, Rodrigues L g R (2010) Galacto-oligosaccharides: production, properties, applications, and significance as prebiotics. Compr Rev Food Sci Food Saf 9:438–454

Urrutia P, Mateo C, Guisan JM, Wilson L, Illanes A (2013) Immobilization of *Bacillus circulans* β-galactosidase and its application in the synthesis of galacto-oligosaccharides under repeated-batch operation. Biochem Eng J 77:41–48

Van den Ende W, Lammens W, Van Laere A, Schroeven L, Le Roy K (2009) Donor and acceptor substrate selectivity among plant glycoside hydrolase family 32 enzymes. Febs J 276:5788–5798

Vigants A, Upite D, Scherbaka R, Lukjanenko J, Ionina R (2013) An influence of ethanol and temperature on products formation by different preparations of *Zymomonas mobilis* extracellular levansucrase. Folia Microbiol 58:75–80

Wada J, Honda Y, Nagae M, Kato R, Wakatsuki S, Katayama T, Taniguchi H, Kumagai H, Kitaoka M, Yamamoto K (2008) 1,2-alpha-l-Fucosynthase: a glycosynthase derived from an inverting alpha-glycosidase with an unusual reaction mechanism. FEBS Lett 582:3739–3743

Wang S-L, Liu C-P, Liang T-W (2012) Fermented and enzymatic production of chitin/chitosan oligosaccharides by extracellular chitinases from Bacillus cereus TKU027. Carbohydr Polym 90:1305–1313

Wang H, Yang R, Hua X, Zhao W, Zhang W (2013) Enzymatic production of lactulose and 1-lactulose: current state and perspectives. Appl Microbiol Biotechnol 97:6167-6180

Weijers CAGM, Franssen MCR, Visser GM (2008) Glycosyltransferase-catalyzed synthesis of bioactive oligosaccharides. Biotechnol Adv 26:436–456

Weishaupt M, Eller S, Seeberger PH (2010) Chapter Twenty-Two: solid phase synthesis of oligosaccharides. In: Minoru F (ed) Methods in enzymology, vol 478. Academic, San Diego, pp 463–484

Whisner CM, Martin BR, Schoterman MH, Nakatsu CH, McCabe LD, McCabe GP, Weaver CM (2013) Galacto-oligosaccharides increase calcium absorption and gut bifidobacteria in young girls: a double-blind cross-over trial. Br J Nutr 110:1292–1303

Wong TS, Roccatano D, Zacharias M, Schwaneberg U (2006) A statistical analysis of random mutagenesis methods used for directed protein evolution. J Mol Biol 355:858–871

Wu D, Chen S, Wang N, Chen J, Wu J (2012) Gamma-cyclodextrin production using cyclodextrin glycosyltransferase from *Bacillus clarkii* 7364. Appl Biochem Biotechnol 167:1954–1962

Wu Y, Yuan S, Chen S, Wu D, Chen J, Wu J (2013) Enhancing the production of galacto-oligosaccharides by mutagenesis of *Sulfolobus solfataricus* β-galactosidase. Food Chem 138:1588–1595

Yamamoto T, Nishimoto T, Chaen H, Fukuda S (2006) Improvement of the enzymatic properties of kojibiose phosphorylase from *Thermoanaerobacter brockii* by random mutagenesis and chimerization. J Appl Glycosci 53:123–129

Xu Q, Zheng X, Huang M, Wu M, Yan Y, Pan J, Yang Q, Duan CJ, Liu JL, Feng JX (2015) Purification and biochemical characterization of a novel fructofuranosidase from *Penicillium oxalicum* with transfructosylating activity producing neokestose. Process Biochem 50:1237–1246

Microencapsulation of Probiotics and Its Applications

3

Kumar Satya Prakash, Rupesh Chavan, and Vijendra Mishra

Abstract

Probiotics are the live microorganisms which when administered in adequate amounts confer a health benefit on the host. Research has indicated that probiotics provide various health benefits to consumers such as reduction in different forms of diarrhoea, antimutagenic effect, alleviation of lactose intolerance and immunomodulation. But these effects are considered to be strain dependent and also depend on a number of other technological factors. For any organism to be considered as probiotic, it must survive the upper gastrointestinal (GI) passage tract and must remain viable at the site of its action, and it must be able to function in the gut environment. Various techniques have been applied to enhance viability of probiotics, and encapsulation is a method whereby the organism is entrapped in a matrix so that it can withstand passage of GI tract. The chapter provide various aspects of microencapsulation in probiotics.

Keywords

Probiotics • Health benefits • Microencapsulation • Drying

3.1 Introduction

Hippocrates, a Greek philosopher (and father of medicine), introduced the concept that food could serve as medicine. During the last decade, the demand for such foods has increased remarkably because of changes in the lifestyle, eating habits, urbanisation and globalisation. This has led to the shift in eating habits from ordinary food to the foods which besides providing nutrition also promote health benefits, for which the names *func-*

K.S. Prakash • R. Chavan • V. Mishra (✉)
National Institute of Food Technology Entrepreneurship and Management, Plot 97 Sector 56, HSIIDC, Kundli, Sonipat, Haryana, India
e-mail: vijendramishra.niftem@gmail.com

P. Shukla (ed.), *Frontier Discoveries and Innovations in Interdisciplinary Microbiology*, DOI 10.1007/978-81-322-2610-9_3

tional foods, *probiotics*, *synbiotics*, *etc*. have been given (Sharma and Devi 2014).

The word *probiotic* refers to the single or diverse cultures of the live beneficial microorganisms (usually *bifidobacteria*, *lactobacilli* and *streptococci*) which, when consumed, produce a favourable effect on the host by convalescing the properties of the indigenous microflora (Holzapfel et al. 1998). As per FAO/WHO (2002) guidelines, probiotics are the "live microorganisms which when administered in adequate amounts confer a health benefit on the host". The word "probiotic" is derived from Greek words "pro" and "bios" translated as "for life" and was initially used as an antonym of the word "antibiotic" (Hamilton-Miller et al. 2003). It is believed that the word was first used in 1954 by Vergio (Holzapfel and Schillinger 2002). However, there exists sufficient literature that suggests that fermented milk was consumed in India long before for promoting the health benefits. It was at the beginning of this century a Russian Nobel laureate Elie Metchnikoff, at the Pasteur Institute, linked health and longevity with the consumption of fermented milk products. He postulated that the bacillus present could positively modify the bacterial community structure of the colon, thus contributing to human health status (*monograph reprinted in* Metchnikoff 2004; Vasiljevic and Shah 2008).

3.2 Desired Characteristics of Probiotic

The probiotics ought to have superior technological properties accordingly; it can be contrived and unified into food products without losing its viability and functionality and without creating unpleasant flavours or textures, it also must survive the upper gastrointestinal (GI) passage tract and must remain viable at the site of its action, and it must be able to function in the gut environment. In order to check the characteristics, the following tests are carried out to check its effectiveness:

1. Acid and bile salt tolerance
2. Antimicrobial action
3. Adherence properties

3.3 Health Benefit of Probiotics

Probiotics are a group of good bacteria which gives strength to our intestinal function. King et. al. (2003) reported that probiotic might reduce both the duration of illness and the frequency of stools because probiotic prohibits colonisation by the invader and controls the intestinal pH through the release of acetic acid and lactic acid. These bacteria could effectively prevent the constipation and diarrhoea caused by pathogenic bacteria. Saxelin et al. (2003) concluded that the probiotic foods contain living probiotic microorganism in ample concentration, so that the proposed effect is achieved after their ingestion. Parvez et al. (2006) reported that probiotics can be used as food supplement which provides the protection against the gastrointestinal (GI) disorders, viz. infections and bowel syndromes, by favouring and colonising the healthy gut microflora. Some health benefits of probiotics are summarised in Table 3.1.

Kailasapathy and Chin (2000) and Sanders et al. (2007) concluded that probiotics possess therapeutic role as they modulate the immunity, prevent lactose intolerance, lower the cholesterol level and prevent some cancers. Weichselbaum (2009), Patel et al. (2010) and Tuomola et al. (1999) also revealed that probiotics have the possible beneficial effects on our health. He classified the potential beneficial effects of probiotics into three-way action, which are listed:

I. The first effect is due to the immunity modulation within the host, which possess importance as it is responsible for the prevention of infectious disease of GI tract. Probiotics modulate the immune system as a result of products like microbial metabolites, components of the cell wall or deoxyribonucleic acids (DNA), which are recognisable to the host cells due to the presence of particular receptors on the epithelial cells of the gut. The possible interaction between these metabolites and host immune cells via adhesion leads to the immunity modulation.
II. The second mechanism suggests that there exists direct interaction between probiotics and other microorganisms (which may be patho-

Table 3.1 Some of the established effects of probiotics

S. no.	Target health benefit	Postulated mechanism
01	Anti-colon cancer	Deactivation of carcinogenic compounds, inhibition of carcinogen-producing enzymes of colonic microbes, immune modulation, mutagen binding
02	Aid in lactose digestion, resistance to enteric pathogens	Bacterial lactase, secretory immune effect, alteration of intestinal conditions to be less favourable for pathogenicity (pH, SCFAs, bacteriocins), alteration of toxin-binding sites, influence on gut flora populations
03	Allergy	Prevention of antigen translocation into bloodstream
04	Blood lipids, heart disease	Assimilation of cholesterol within bacterial cell, increased excretion of bile salts due to deconjugation by bile salt hydrolase
05	Small bowel bacterial overgrowth	Influence on activity of overgrowth flora, decreasing toxic metabolite production, modification of intestinal environment
07	Crohn's disease	Reduced bowel movements
08	*Helicobacter pylori* infections	Lactic acid production

Source: Sanders and Huisin't Veld (1999), Sharma and Devi (2014)

genic). Here the restoration of the good microorganisms in the gut results in the treatment of the infections like inflammation. This idea is supported by the fact that probiotics have the capacity to compete with the pathogens in terms of their adhesion to the intestines.

III. The third mechanism of action has been summarised as the ability of the probiotics to affect the secondary metabolites (toxins) of the pathogenic microbes and also the host products like bile salts and some food ingredients.

3.4 Microencapsulation

Microencapsulation can be defined as a technology in which the liquids, volatiles (gases) or solid materials (including flavours, enzymes, cells, medicine or other active ingredients) are packed within a continuous polymer film, so that the contents are released at predictable and controlled manner under specific conditions (Desai and Park 2005). Microencapsulation dates back to the early 1930s, and this technology has boosted during the past decade due to the advancement in technologies and modifications in the entrapping materials. Champagne and Fustier (2007) found that there are various points to be taken into consideration while microencapsulating probiotic cells such as strain of probiotics, their beneficial effects on host when reached in gastrointestinal (GI) tract, quantity required to give beneficial effects and viability of probiotic cells during processing, storage and sensory properties of product.

3.4.1 Beneficial Effects of Microencapsulation

Microencapsulation has resulted in various advantages towards the probiotics like prevention against the bacteriophages, contamination, etc. Steenson et al. (1987) found that when the probiotics were encapsulated into the matrix of alginate beads, the bacteriophage attack was avoided because of the small pore size. Kearney et al. (1990) demonstrated that the survivability of the *L. plantarum* in alginate beads was enhanced by 30 % after lyophilisation. Similar results were reported by Kim and Yoon (1995); they found that the survivability of the probiotic cultures was enhanced up to 40 % when encapsulated in the calcium-alginate beads after of the milk.

Several reports have shown that the stability of the probiotic cultures can be increased by encapsulation, for example, the probiotic bacteria when encapsulated in the gel matrix of calcium alginate showed much stability during the storage (Sheu and Marshall 1993; Kebary et al. 1998). Woo et al. (1999) reported that the survival of the *B. longum* when entrapped into the matrix of 3 % alginate and 0.15 % xanthan gum enhanced. Similarly, the survivability of the bacterium *L. casei* when entrapped in the chitosan-coated alginate beads confirmed higher survivability upon storage (Koo et al. 2001). Microencapsulation also prevents the probiotics from the harsh environment of gastric solution. Several studies have shown that the survival of the probiotic strains like *L. casei*, *B. longum*, *L. gasseri*, *L. acidophilus*, *B. bifidum*, *B. infantis*, *B. breve* and *B. pseudolongum* inside the gastric environment enhanced when entrapped within the wall materials like chitosan and alginates in comparison to the non-entrapped cells (Rao et al. 1989; Urbanska et al. 2007; Chávarri et al. 2010).

Soma et al. (2009) reported an increase in the number of *Lactobacillus acidophilus* ATCC 43121 by just about 3 logs, after they were entrapped within the beads of xanthan–chitosan–xanthan network. Other researchers reported similar results when they encapsulated the strains of *L. rhamnosus* and *B. longum* with a blend of gellan and xanthan gum (Jiménez-Pranteda et al. 2009; Ding and Shah 2007). Kharter et al. (2010) and Zou et al. (2011) carried out research on the some strains of the LAB and bifidobacteria after encapsulating them in chitosan–alginate beads and alginate-maize resistant starch; they concluded that protection against gastric juices increased to a greater extent.

3.4.2 Methods of Microencapsulation

In microencapsulation a thin layer coating is done. Different techniques have been developed so far for the successful entrapment of the probiotic cells which have been classified into three categories, viz. (1) physico-mechanical, (2) physico-chemical and (3) chemical methods.

The different types of encapsulation techniques and the wall materials have been listed in Figs. 3.1 and 3.2, respectively. In physical

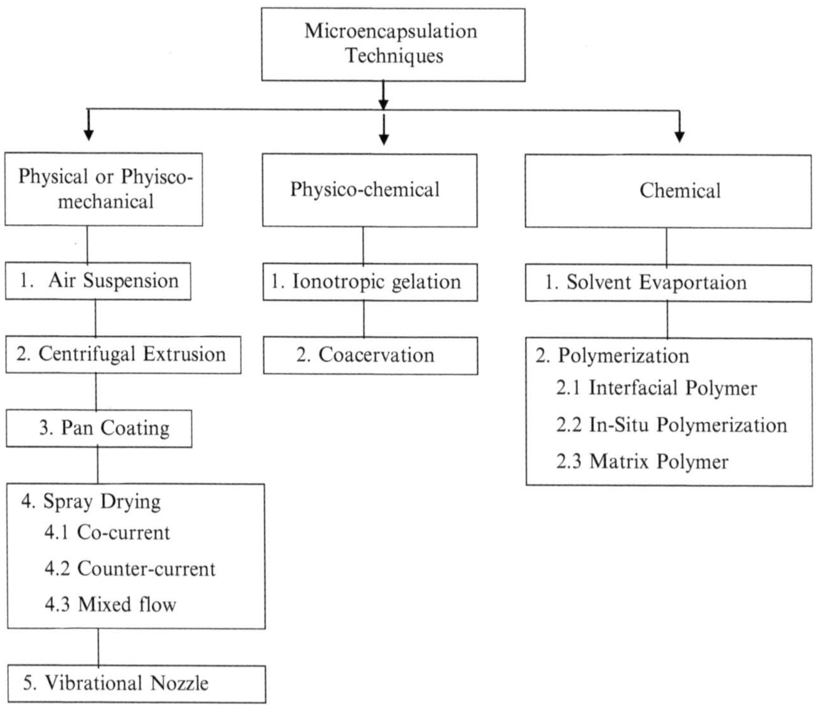

Fig. 3.1 General overview of encapsulation techniques

methods there is conversion of liquid phase into solid immobile phase in the form of powder, in which the probiotic is entrapped. The probiotics encapsulated in this method are released when the powder gains moisture, while in the case of chemical methods, there is the formation of the hydrocolloidal gel phase via cross-linking or chemical reactions. Here the probiotics are released only when there is change in the pH or the ionic conditions of the wall material change.

3.4.2.1 Spray-Drying

It (Fig. 3.3) is the most widely used stabilisation procedure in food/pharma/chemical industries due to high throughput and low costs. During spray-drying, a matrix material is first dissolved in the continuous phase which surrounds the probiotic cells inside the spray droplet. This solution is rapidly atomised in heated air and shrinks surrounding the cells in an envelope. Petrovic et al. (2007) observed that in spray-drying technique, the centrifuged cell mass of probiotic bacteria is dispersed in suitable encapsulating material like any food polymeric solution, gums (hydrocolloids), modified starches, non-gelling proteins and dextrin to form an emulsion or dispersion. These encapsulating materials give the protective coating to the central probiotic bacteria and

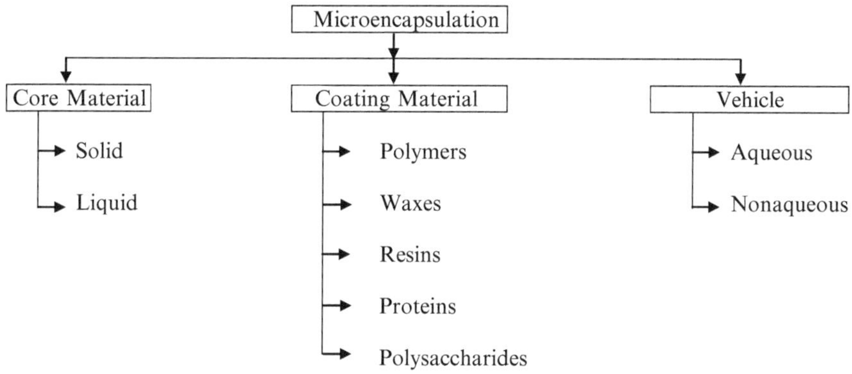

Fig. 3.2 Type of core materials, coating materials and vehicles used in microencapsulation

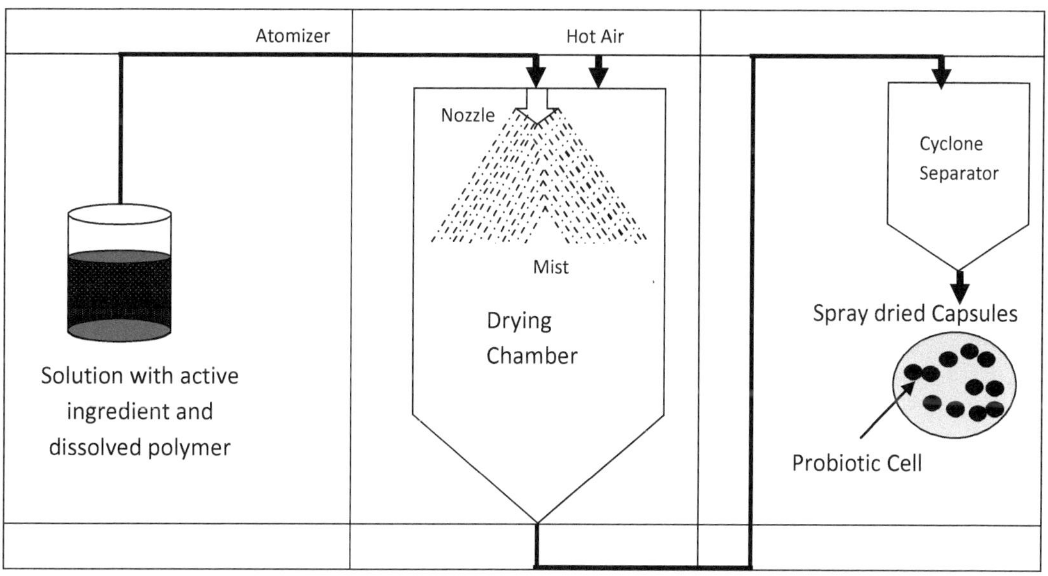

Fig. 3.3 Outline of spray-drying for probiotics

provide oxygen barrier. Fu and Etzel (1995) studied that encapsulation by spray-drying is relatively cheap and can be operated on large scale and on continuous mode, but the disadvantage of spray-drying is that there is huge loss in viability of probiotic bacteria because of dehydration and high inlet temperature, but by adjusting the proper inlet and outlet temperatures, the viable count of encapsulated cells can be obtained (Kailasapathy 2002). Crittenden et al. (2006) used protein–carbohydrate–oil film for the successful encapsulation of the *B. infantis* by spray-drying at an inlet and outlet temperatures of 160 and 65 °C, respectively. Some of the researchers were successful even encapsulating the probiotics at higher temperatures like Zhao et al. (2008) found that strains of *L. acidophilus* when encapsulated with gum acacia and cyclodextrin at inlet and outlet temperatures of 170°C and 90 °C, respectively, the post spray drying survival of the encapsulated *L. acidophilus* was 1.50×10^{9}c.f.u./ml and the survival after 8 weeks of storage at 4°C, was more than 10^{7}c.f.u./ml. Ying et al. (2011) found that encapsulated *L. rhamnosus* by spray-drying in the matrix of resistant starch and whey protein had a superior storage stability in contrast to freeze-drying technique.

Jantzen et al. (2013) observed that the spray-drying of probiotic *L. reuteri* with whey at outlet temperature (55±2 and 65±2 °C) and viability of microencapsulated probiotics has decreased by 2 log cycles after drying. Pereira et al. (2014) found that the spray-dried probiotic cashew apple juice showed levels of *L. casei* NRRL B-442 viable cells higher than the recommended minimum level for probiotic bacteria after the drying process; it is indicating that the spray-drying can be used for producing probiotic cashew apple powder.

3.4.2.2 Freeze-Drying

It is a drying process in which the water is removed from the food system/matrix in the form of vapours via sublimation process. In freeze-drying the food material is subjected to freezing under vacuum. The reduced temperature freezes the free water and low pressure results in the conversion of this ice into vapours directly. The process has been classified into three stages which include:

(a) Freezing
(b) Primary drying
(c) Secondary drying

In case of microencapsulation by freeze-drying, both core and the wall (i.e. probiotics and entrapping material) along with the water present in the matrix are frozen to temperature as low as −20–−90 °C and then dried by direct sublimation process under reduced pressures, i.e. below 4.7 mmHg. As freeze-drying occurs at both reduced temperature and pressure, a porous and unshrinking edifice culture of the probiotics is obtained. Although freeze-drying, being the nonthermal process, has emerged as a most standard technique for the drying process in the microbiological industries, there occurs some loss in cell viability, and also the energy consumption is very high as compared to the spray-drying technique (30–50 times more expensive). In order to prevent the loss during the freezing and to improve the survivability of the probiotics, the demand for the cryoprotectants (like proteins, maltodextrins, etc.) has increased (Gharsallaoui et al. 2007; Morgan et al. 2006). Carvalho et al. (2004) reported that the probiotic cultures of the strains of *L. bulgaricus* survived for over 10 months at −20 °C when sorbitol and fructose were used as cryoprotectants during freeze-drying. The various cryoprotectants used for the freeze-drying of the probiotics are listed in Table 3.2.

Table 3.2 List of cryoprotectants used in freeze-drying of probiotics (Krasaekoopt 2012)

Microorganism	Cryoprotectant	References
L. acidophilus, B. bacterium	Skim milk	Goderska and Czarnecki (2008)
L. rhamnosus	Glycerine	Savini et al. (2010)
L. casei	High-resistance maize starch and inulin	Babu et al. (2011)

3.4.2.3 Spray-Freeze-Drying

It is a drying technique which involves the spraying of a liquid solution into a freezing medium like liquid nitrogen, resulting in the freezing of water within it. After that the frozen material is subjected to conventional vacuum freeze-drying for the removal of water via sublimation process (Kailasapathy 2009; Semyonov et al. 2010). This process has advantages that the size of the product is uniform and controlled with larger surface area as compared to the spray-dried powder. Dolly et al. (2011) encapsulated *L. plantarum* by spray-freeze-drying using whey protein as an encapsulating agent; they found that the final product has 20 % higher viability as compared to the spray-dried product. They also demonstrated that it had more tolerance to acidic conditions.

3.4.2.4 Extrusion Technique

Extrusion is a more laborious, simple, economic but less hazardous method of encapsulation and consists of projecting a solution consisting of the encapsulation matrix and the core materials through a spout at a reasonably high pressure. The resultant droplets free-fall from the nozzle into a hardening solution (Krasaekoopt et al. 2003; Lakkis 2007). Capsule size is influenced by the nozzle size (De Vos et al. 2010). If droplet formation is performed in a controlled environment, this technique is referred to as prilling. To achieve this control, the jet nozzle is subjected to pulsation (Burgain et al. 2011). Some of the main technologies to ensure fluid dispersion into droplets and subsequent capsules are coaxial airflow, use of an electrostatic field, jet-cutting and spinning disc atomisation (Whelehan and Marison 2011; Cook et al. 2012). Due to the gentle operation of the extrusion process and the fact that deleterious solvents are not required, high cell viability can be maintained (Burgain et al. 2011; Krasaekoopt et al. 2003). Also the viscosity of the fluid used does not limit production of capsules (Lakkis 2007). Extrusion can be performed under both the aerobic and anaerobic conditions, an advantage when using the predominantly oxygen-sensitive LABs and bifidobacteria.

3.4.2.5 Emulsion Method

It is a type of chemical technique, employed for the encapsulation of probiotic cells on the basis of interaction between the continuous phase, i.e. vegetable oil (corn oil, sunflower oil, etc.) or mineral oil, and discontinuous phase, i.e. cell-polymer suspension. Following homogenisation to generate a water-in-oil emulsion, the water-soluble polymer is cross-linked to form small-gel particles within the continuous oil phase. Microencapsulation can be down in three steps: in the first step we disperse droplets with the emulsification, in the second case we add calcium chloride ($CaCl_2$), and in final step we got encapsulated probiotic. To make the emulsification process more effective, some emulsifiers may be added like Tween 80 (Krasaekoopt et al. 2003). Figure 3.4 depicts diagrammatically the process of emulsification. Oliveira et al. (2007) demonstrated that *L. acidophilus* and *B. lactis* when entrapped within the matrix of pectin–casein complex by emulsion technique produced a higher viability at 7 °C storage temperature; however, protection against the low pH conditions was not up to the mark.

3.5 Importance of Microencapsulated Probiotics in the Ready-to-Eat Foods

The incorporation of the probiotics into the ready-to-eat foods is not a new concept, but the concept has advanced to a greater extent in recent years due to the emergence of new techniques and the identification of new probiotic strains. The various foods that can be used as a carrier for probiotics include frozen dairy desserts, yogurt, cheese, certain fruit juices, etc. And as per the WHO guidelines, the minimum level of the probiotic microorganisms at the time of consumption shall be above the minimum level, i.e. 10^7 cfu/g, and also the probiotic should not change the sensory quality of food.

Fig. 3.4 Diagrammatic representation of emulsion method

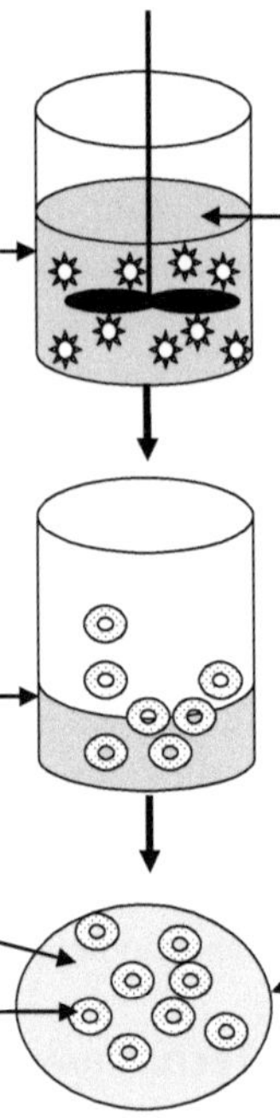

Table 3.3 List of some strains subjected to encapsulation in yogurt preparation (Krasaekoopt 2012)

Probiotic strain	Microencapsulation technology	Materials	References
L. acidophilus	Emulsification	Alginate–starch	Sultana et al. (2000)
B. longum	Emulsification	k-Carrageenan	Adhikari et al. (2003)
L. acidophilus and B. infantis	Emulsification	Alginate–starch	Godward and Kailasapathy (2003)
B. breve	Emulsification	Milk fat and whey protein	Picot and Lacroix (2003)
L. acidophilus	Extrusion	Ca-alginate and chitosan	Iyer and Kailasapathy (2005)
L. acidophilus and L. casei	Emulsification	Alginate	Capela et al. (2006)
L. acidophilus	Emulsification	Alginate–starch	Kailasapathy (2006)
L. acidophilus	Extrusion	Alginate–chitosan	Urbanska et al. (2007)
L. casei	Extrusion	Alginate–pectin	Sandoval-Castilla et al. (2010)

3.5.1 Yogurt

Yogurt is made from whole or partly defatted milk with small quantity of skim milk powder. The two microorganisms *L. bulgaricus* and *Streptococcus thermophilus* that grow together symbiotically are responsible for conversion of lactose to lactic acid. The therapeutic value of the yogurt can be enhanced by the incorporation of the probiotic cultures into it (Weichselbaum 2009). As yogurt has lactic acid that reduces its pH to around 4.2–4.6, which makes it a poor carrier for probiotics, there has been some study in which encapsulated cultures of probiotics were incorporated into the yogurt (Table 3.3) and their survival was much better (Kailasapathy 2009). But the problem arises in the low acetic acid production which is responsible for the characteristic flavour in yogurt. Adhikari et al. (2003) produced yogurt that had encapsulated *bifidobacteria*, but the problem in the stirred yogurt arose attributed to its poor sensory quality because of the development of grainy/sandy texture that reduced its consumer acceptance. Sultana et al. (2000) encapsulated the probiotic cells with prebiotic resistant starch and cryoprotectants (glycerol) to improve their viability. Krasaekoopt et al. (2011) demonstrated that encapsulated *L. acidophilus and L. casei* in the matrix of galacto-oligosaccharides produced the best protection against the simulated gastric (pH 1.55) and intestinal juice (0.6 % bile salt). They also concluded that the viability was higher as much as 0.8 logs in entrapped probiotics. Many other studies have confirmed the use of co-encapsulation (i.e. use of two different probiotic cultures) with a probiotic like raftilose, inulin, etc. (Iyer et al. 2005).

3.5.2 Cheese

Cheddar cheese has emerged as an excellent carrier for the probiotics because of its high pH, i.e. 5.5, high fat content (33 %) and excellent buffering activity. These properties protect the probiotics from the harsh environment in the stomach and intestines. Some studies have shown that the survivability of the *bifidobacterium* when used as probiotic in cheese extended up to 24 weeks; however, the cell number reduced significantly, which was attributed to the reduced temperature (around 7 °C) and decreased lactose content. Although the sensory properties remained unaffected (Dinakar and Mistry 1994; Stanton et al. 1998), Ozer et al. in 2009 demonstrated that the sensory quality of the cheese (white brined and kasar) remained unchanged when the entrapped probiotics *L. acidophilus and B. bifidum* were added into the cheese.

3.5.3 Fruit Juice

The consumption of fruit juices is increasing day by day and has already covered a large population already round the globe. So, the idea of using fruit juices as a carrier for encapsulated probiotics has flourished in recent years and has gained attention both for researchers and industries. Tourila and Cardello (2002) suggested that fruit juices could be used as a carrier for the probiotics. Yoon et al. (2004) developed a probiotic drink of tomato juice in which the probiotic cells *L. acidophilus* grow within the fruit juice. They reported this fermented juice could be used by the consumers having allergy to milk and milk products. Although probiotication reduced the sensory properties of the juice, Suomalainen et al. (2006) reported that the survivability of the *L. rhamnosus* was improved to a greater extent in the GI tract after mixing whey in the orange juice. Krasaekoopt et al. (2010) assessed the quality parameters of the fruit (orange and grape) juices containing the encapsulated probiotics within the matrix of the alginate beads with chitosan coating. They reported that the consumer acceptance was about 85 %. Krasaekoopt and Tandhanskul (2008) and Sohail et al. (2011) used alginate and calcium chloride as wall materials for the encapsulation of the probiotic cultures *L. casei*, L. *rhamnosus and L. acidophilus*; they concluded that survivability of the probiotic cells in the commercial fruit juice (orange) at low temperatures (4 °C) was enhanced by about 4 logs. Additionally, there were no changes in the pH of the juice upon storage for around 35 days.

3.5.4 Miscellaneous Foods

There exist many other foods that can be used as a carrier medium for the encapsulated probiotics like chocolates, bakery items and meat products. Muthukumarasamy and Holley (2007) successfully developed the fermented meat sausages containing the encapsulated probiotic bacteria cells. However, it was found that the probiotics were not able to inhibit the activity of the *E. coli*. Maillard and Landuyt (2008) encapsulated the probiotic cells by spray coating to be used in the chocolate. They concluded that the survivability in the chocolate was much higher (approximately three times) as compared to the dairy products. Lahtinen et al. (2007) demonstrated that the fat content in the chocolate acts as a shield for the probiotics towards the harsh conditions.

3.6 Conclusion

Microencapsulation has emerged as a wonderful technique to shield the sensitive food components and increase the survivability of the probiotic cells until they reach the target organ. The encapsulated materials are very easy to handle and their efficiency depends upon the encapsulating technique and wall materials employed. Although various encapsulating techniques, wall materials, and probiotic strains have been identified, most of them have limited usage because of the scaleup problems.

This chapter gives a brief outline of the probiotic cultures commonly employed along with the brief emphasis on the different techniques and foods employed for encapsulation of the probiotics. A lot more needs to be done to develop techniques which are economically feasible, and also the final size of the product is minimised as it is inversely related to the stability of the product.

References

Adhikari K, Mustapha A, Grun IU (2003) Survival and metabolic activity of microencapsulated *Bifidobacterium longum* in stirred yoghurt. J food Sci 68(1):275–280

Babu G, Rath S, Nithyalakshmi V (2011) Probiotic viability of freeze dried synbiotic microcapsules in skim milk powder at ambient storage condition. Int J Food Saf 13:62–68

Burgain J, Gaiani C, Linder M, And Scher J (2011) Encapsulation of probiotic living cells, from laboratory scale to industrial applications. J Food Eng 104:467–483

Capela P, Hay TKC, Shah NP (2006) Effect of cryoprotectants, prebiotics and microencapsulation on survival of probiotic organisms in yoghurt and freeze dried yoghurt. Food Res Int 39:203–211

Carvalho AS, Silva J, Ho P, Teixeira P, Malcata FX, Gibbs P (2004) Effects of various sugars added to growth and drying media upon thermotolerance and survival throughout storage of freeze-dried *Lactobacillus delbrueckii* ssp. *bulgaricus*. Biotechnol Prog 20(1): 248–254

Champagne CP, Fustier P (2007) Microencapsulation for the improved delivery of bioactive compounds into foods. Curr Opin Biotechnol 18:184–190

Chávarri M, Marañón I, Ares R, Ibáñez FC, Marzo F, And Villarán MDC (2010) Microencapsulation of a probiotic and prebiotic in alginate–chitosan capsules improves survival in simulated gastro-intestinal conditions. Int J Food Microbiol 142(1–2):185–189

Chen MJ, Chen KN (2007) Applications of probiotic encapsulation in dairy products. In: Lakkis JM (ed) Encapsulation and controlled release technologies in food systems. Wiley-Blackwell, New York, pp 83–107

Cook MT, Tzortzis G, Charalampopoulos D, Khutoryanskiy VV (2012) Microencapsulation of probiotics for gastrointestinal delivery. J Control Release 162(1):56–67

Crittenden R, Weerakkody R, Sanguansri L, Augustin MA (2006) Synbiotic microcapsules that enhance microbial viability during nonrefrigerated storage and gastrointestinal transit. Appl Environ Microbiol 72(3):2280–2282

De Vos P, Faas MM, Spasojevic M, Sikkema J (2010) Encapsulation for preservation of functionality and targeted delivery of bioactive food components. Int Dairy J 20(4):292–302

Desai KGH, Park HJ (2005) Recent developments in microencapsulation of food ingredients. Drying Technol 23(7):1361–1394

Dinakar P, Mistry VV (1994) Growth and viability of *Bifidobacterium bifidum* in cheddar cheese. J Dairy Sci 77(10):2854–2864

Ding WK, Shah NP (2007) Acid bile and heat tolerance of free and microencapsulated probiotic bacteria. J Food Sci 72(9):446–450

Dolly P, Anishaparvin A, Joseph GS, Anandharamakrishnan C (2011) Microencapsulation of Lactobacillus plantarum (mtcc 5422) by spray-freeze-drying method and evaluation of survival in simulated gastrointestinal condition. Informa Healthcare 568–574. http://www.informahealthcare.com/mnc

FAO/WHO (2002) Guidelines for the evaluation of probiotics in food. Report of a joint FAO/WHO working group on drafting guidelines for the evaluation of probiotics in foods, London

Fu WY, Etzel MR (1995) Spray-drying of *Lactococcus lactis* Ssp. *lactis* C2 and cellular injury. J Food Sci 60:195–200

Gharsallaoui A, Roudaut G, Chambin O, Voilley A, Saurel R (2007) Applications of spray-drying in microencapsulation of food ingredients: an overview. Food Res Int 40:1107–1121

Goderska K, Czarnecki Z (2008) Influence of micro encapsulation and spray drying on the viability of *Lactobacillus* and *Bifidobacterium strains*. Pol J Microbiol 57:135–140

Godward G, Kailasapathy K (2003) Viability and survival of free, encapsulated and co-encapsulated probiotic bacteria in yoghurt. Milchwissenschaft 58(7–8):396–399

Hamilton-Miller JMT, Gibson GR, Bruck W (2003) Some insight into the derivation and early uses of the word probiotic. Br J Nutr 90:845

Holzapfel WH, Schillinger U (2002) Introduction to pre- and probiotics. Food Res Int 35:109–116

Holzapfel WH, Habere P, Snel J, Schillinger U, Huis in't Veld JHJ (1998) Overview of gut flora and probiotics. Int J Food Microbiol 41(2):85–101

Iyer C, Kailasapathy K (2005) Effect of co-encapsulation of probiotics with prebiotics on increasing the viability of encapsulated bacteria under in vitro acidic and bile salt conditions and in yoghurt. J Food Sci 70:M18–M23

Jantzen M, Gödel A, Beermann C (2013) Direct spray drying and microencapsulation of probiotic Lactobacillus reuteri from slurry fermentation with whey. J Appl Microbiol 115:1029–1036

Jiménez-pranteda ML, Poncelet D, Ramos-cormenzana A, Monteoliva-sanchez M (2009) Viability of probiotics exopolymers beads exposed to specific in vitro conditions. In: XVIIth international conference on bioencapsulation, Groningen, 24–26 Sept 2009

Kailasapathy K (2002) Microencapsulation of probiotic bacteria: technology and potential applications. Curr Issues Intest Microbiol 3:39–48

Kailasapathy K (2006) Survival of free and encapsulated probiotic bacteria and their effect on the sensory properties of yoghurt. LWT 39:1221–1227

Kailasapathy K (2009) Encapsulation technologies for functional foods and nutraceutical product development. CAB Rev Perspect Agric Vet Sci Nutr Nat Res 4(033):1–19

Kailasapathy K, Chin J (2000) Survival and therapeutic potential of probiotics organisms with reference to Lactobacillus acidophilus and Bifidobacterium spp. Immunol Cell Biol 78(1):80–88

Kearney L, Upton M, M.C Laughlin A (1990) Enhancing the viability of Lactobacillus plantarum inoculum by immobilizing the cells in calcium-alginate beads incorporating cryoprotectants. Appl Environ Microbiol 56:3112–3116

Kebary KMK, Hussein SA, Badawi RM (1998) Improving viability of bifidobacteria and their effect on frozen milk. Egypt J Dairy Sci 26:319–337

Kharter KAA, Ali MA, Ahmad EAM (2010) Effect of encapsulation on some probiotic criteria. Am J Sci 6:810–819

Kim KI, Yoon YH (1995) A study on the preparation of direct vat lactic acid bacterial starter. Korean J Dairy Sci 17(2):129–134

King CK, Glass R, Bresee JS, Duggan C (2003) Managing acute gastroenteritis among children: oral rehydration, maintenance, and nutritional therapy. MMWR Recomm Rep 52(RR-16):1–16

Koo S, Cho Y, Huh C, Beak Y, Park J (2001) Improvement of the stability of Lactobacillus casei YIT 9081 by microencapsulation using alginate and chitosan. J Microbiol Biotechnol 11(3):376–383

Krasaekoopt W (2012) Properties and applications of different probiotic delivery systems, research book chapter. WP Woodhead Publishing

Krasaekoopt W, Tandhanskul A (2008) Sensory and acceptance assessment of yogurt containing probiotic beads in Thailand. KU J (Natural Science) 42:99–106

Krasaekoopt W, Bhandari B, Deeth H (2003) Evaluation of encapsulation techniques of probiotics for yoghurt. Int Dairy J 13:3–13

Lahtinen SJ, Ouwehand AC, Salminen SJ, Forssell P, Myllärinen P (2007) Effect of starch- and lipid-based encapsulation on the culturability of two *Bifidobacterium longum* strains. Lett Appl Microbiol 44(5):500–505

Lakkis JM (2007) Encapsulation and controlled release technologies in food systems, 1st edn. Blackwell Publishing, Ames

Maillard M, Landuyt A (2008) Chocolate: an ideal carrier for probiotics. Agro Food Ind Hi-Tec 19(3):13–15

Metchnikoff LE (2004) The prolongation of life: optimistic studies (reprinted edition 1907). Springer, New York

Morgan C, Herman N, White P, Vesey G (2006) Preservation of microorganisms by drying: a review. J Microbiol Methods 66(2):183–193

Muthukumarasamy P, Holley RA (2007) Survival of Escherichia coli O157:H7 in dry fermented sausages containing microencapsulated probiotic lactic acid bacteria. Food Microbiol 24:82–88

Oliveira AC, Moretti TS, Boschini C, Baliero JC, Freitas O, Favaro-Trindade CS (2007) Stability of microencapsulated *B. lactis* (BI 01) and *L. acidophilus* (LAC 4) by complex coacervation followed by spray drying. J Microencapsul 24(7):673–681

Ozer B, Kirmaci HA, Senel E, Atamer M, Hayaloglu A (2009) Improving the viability of *Bifidobacterium bifidum* BB-12 and *Lactobacillus acidophilus* LA-5 in white-brined cheese by microencapsulation. Int Dairy J 19:22–29

Parvez S, Malik KA, Kang SA, Kim HY (2006) Probiotics and their fermented food products are beneficial for health. J Appl Microbiol 100(6):1171–1185

Patel AK, Singhania RR, Pandey A, Chincholkar SB (2010) Probiotic bile salt hydrolase: current developments and perspectives. Appl Biochem Biotechnol 162(1):166–180

Pereira ALF, Almeida FDL, Lima MA, Costa JMC, Rodrigues, S (2013) Spray-drying of probiotic cashew apple juice. Food Bioprocess Technol 1935–5130

Petrovic T, Nedovic V, Dimitric-Brankovic S (2007) Protection of probiotic microorganisms by microencapsulation. CI CEQ 13(3):169–174

Picot A, Lacroix C (2003) Effect of micronization on viability and thermotolerance of probiotic freeze-dried cultures. Int Dairy J 13:455–462

Rao AV, Shiwnarain N, Maharaj I (1989) Survival of microencapsulated Bifidobacterium pseudolongum in simulated gastric and intestinal juices. Can Inst Food Sci Technol J 22:345–349

Sanders ME, Huisin't Veld J (1999) Bringing a probiotic-containing functional food to the market: microbiological, product, regulatory and labelling issues. Antonie Van Leeuwenhoek 76:293–315

Sanders ME, Gibson G, Gill HS, Guarner F (2007) Probiotics: their potential to impact human health. CAST issue paper no. 36, Oct 2007. http://www.cast-science.org/publications.asp

Sandoval-Castilla O, Lobato-Valleros C, Garcia-Galindo HS, Alvarez-Ramirez J, Vernom-Carter EJ (2010) Textural properties of alginate-pectin beads and survivability of entrapped Lb. casei in simulated gastrointestinal conditions and in yoghurt. Food Res Int 43:111–117

Savini M, Cecchini C, Verdenelli MC, Silvi S, Orpianesi C, Cresci A (2010) Pilotscale production and viability analysis of freeze-dried probiotic bacteria using different protective agents. Nutrients 2:330–339

Saxelin M, Korpela R, Mayra-Makinen A (2003) Introduction: classifying functional dairy products. In: Mattila-Sandholm T, Saarela M (eds) Functional dairy products. Woodhead Publishing Limited, Cambridge, pp 1–16

Semyonov D, Ramon O, Kaplun Z, Levin-Brener L, Gurevich N, And Shimoni E (2010) Microencapsulation of Lactobacillus paracasei by spray freeze drying. Food Res Int 43:193–202

Sharma M, Devi M (2014) Probiotics: a comprehensive approach toward health foods. Crit Rev Food Sci Nutr 54(4):537–552

Sheu TY, Marshall RT (1993) Microentrapment of lactobacilli in calcium alginate gels. J Food Sci 58:557–561

Sohail A, Turner MS, Coombes A, Bostrom T, Bhandari B (2011) Survivability of probiotics encapsulated in alginate gel microbeads using a novel impinging aerosols method. Int J Food Microbiol 145:162–168

Soma PK, Williams PD, Lo YM (2009) Advancements in non-starch polysaccharide research for frozen foods and microencapsulation of probiotics. Front Chem Eng Chin 3(4):413–426

Stanton C, Gardiner G, Lynch PB, Collins JK, Fitzgerald G, Ross RP (1998) Probiotic cheese. Int Dairy J 8:491–496

Steenson LR, Klaenhammer TR, Swaisgood HE (1987) Calcium alginate-immobilized cultures of lactic streptococci are protected from attack by lytic bacteriophage. J Dairy Sci 70:1121–1127

Sultana K, Godward G, Reynolds N, Arumugaswamy R, Peiris P (2000) Encapsulation of probiotic bacteria with alginate-starch and evaluation of survival in simulated gastrointestinal conditions and in yoghurt. Int J Food Microbiol 62:47–55

Suomalainen T, Lagstrom H, Matto J, Saarela M, Arvilommi H, Laitenen I, Ouwehand AC, Salmien S (2006) Influence of whey-based fruit juice containing Lactobacillus rhamnosus on intestinal well-being and humoral immune response in healthy adults. Food Sci Technol 39(7):788–795

Tuomola EM, Ouwehand AC, Salminen SJ (1999) The effect of probiotic bacteria on the adhesion of pathogens to human intestinal mucus. FEMS Immunol Med Microbiol 26(2):137–142

Tuorila H, Cardello AV (2002) Consumer responses to an off-flavor in juice in the presence of specific health claims. Food Qual Prefer 13:561–569

Urbanska AM, Bhathena J, Prakash S (2007) Live encapsulated Lactobacillus acidophilus cells in yogurt for therapeutic oral delivery: preparation and in vitro analysis of alginate–chitosan microcapsules. Can J Physiol Pharmacol 85(9):884–893

Vasiljevic T, Shah NP (2008) Probiotics-from Metchnikoff to bioactives. Int Dairy J 18:714–728

Weichselbaum E (2009) Probiotics and health: a review of the evidence. Nutr Bull 34(4):340–373

Whelehan M, Marison IW (2011) Microencapsulation using vibrating technology. J Microencapsul 28(8):669–688

Woo C, Lee K, Heo T (1999) Improvement of Bifidobacterium longum stability using cell-entrapment technique. J Microbiol Biotechnol 9(2):132–139

Ying D, Sun J, Sanguansri L, Weerakkody R, Augustin M (2011) Enhanced survival of spray dried microencapsulated *Lactobacillus rhamnosus* GG in the presence of glucose. J Food Eng 109(3):597–602

Yoon KY, Woodams EE, And Hang YD (2004) Probiotication of tomato juice by lactic acid bacteria. J Microbiol 42(4):315–318

Zhao R, Sun J, Torley P, Wang D, Niu S (2008) Measurement of particle diameter of *Lactobacillus acidophilus* microcapsule by spray drying and analysis on its microstructure. World J Microbiol Biotech 24(8):1349–1354

Zou Q, Zhao J, Liu X, Tian F, Zhang H, Zhang H, Chen W (2011) Microencapsulation of *Bifidobacterium bifidum* F-35 in reinforced alginate microspheres prepared by emulsification/internal gelation. Int Food Sci Technol 46(8):1672–1678

Cyanobacterial Phycobilins: Production, Purification, and Regulation

4

Hillol Chakdar and Sunil Pabbi

Abstract

The cyanobacteria are a diverse group of prokaryotic organisms that carry out oxygenic photosynthesis and are thought to be responsible for the oxygenation of our atmosphere. Like red algae and cryptomonads, cyanobacteria also contain phycobiliproteins (PBPs) which serve as major accessory pigments during photosynthesis. PBPs are large water-soluble supramolecular protein aggregates involved in light harvesting and can be divided broadly into three classes, viz., phycoerythrin (PE), phycocyanin (PC), and allophycocyanin (APC) based on their spectral properties. PBPs have been extracted and purified from *Spirulina* spp., *Synechococcus* sp., *Oscillatoria* sp., etc., and produced commercially from *Spirulina platensis, Anabaena* sp., and *Galdieria sulphuraria.* Since cyanobacteria exhibit wide variations in nutrient availability, light intensity, light quality (wavelength), temperature, water activity, etc., these variations also result in altered metabolic activity of these organisms as a result of differential expression of different genes. The expression of phycobiliprotein coding genes is also accordingly modulated to adapt to a particular condition. Many workers have reported changes in phycobilisome structure and expression of *cpc* genes in response to light quality, light quantity, and nutrients like nitrogen, sulfur, etc. The composition and function of phycobiliproteins in cyanobacteria have also been reported to change under stress conditions. In the present paper, we have reviewed the production, purification, and regulation of cyanobacterial phycobilins including their importance in the commercial sector, as they have several applications as natural dyes in food and cosmetic industry, immunological assays, health-promoting properties, and broad range of other pharmaceutical applications.

H. Chakdar
ICAR-National Bureau of Agriculturally Important Microorganisms (NBAIM), Kushmaur, Mau 275101, Uttar Pradesh, India

S. Pabbi (✉)
Centre for Conservation and Utilisation of Blue Green Algae (CCUBGA), ICAR-Indian Agricultural Research Institute (IARI), New Delhi 110012, India
e-mail: sunil.pabbi@gmail.com

P. Shukla (ed.), *Frontier Discoveries and Innovations in Interdisciplinary Microbiology*, DOI 10.1007/978-81-322-2610-9_4

Keywords
Cyanobacteria • Phycobilins • Complementary chromatic adaptation, gene regulation • Phycobiliprotein purification

The cyanobacteria constitute a kingdom of the domain bacteria whose members manifest a vast diversity of habitats and metabolisms. Species differ with respect to possession of panoply of specializations, a sampling of which includes diazotrophy, filamentous growth form, production of buoyancy organelles, and differentiation of specialized cell types. However, the group can be defined by the unifying property of chlorophyll *a*-based oxygenic photosynthesis which is considered as the source of the earth's oxidizing atmosphere and extends to plants, whose chloroplasts trace their ancestry to this kingdom (Olsen et al. 1994; Woese 1987). Unlike the higher plants, cyanobacteria possess phycobiliproteins as major light-harvesting pigments which are also found in red algae and glaucophytes. Due to the presence of blue-colored biliprotein phycocyanin, many of the cyanobacteria appear bluish-green in color and they have been named as blue-green algae despite their prokaryotic nature. Phycobiliproteins play important role in cyanobacterial metabolism as well as help them to adapt in a diverse environmental condition especially conditions with variable light quality and quantity. These proteins due to their multifunctional role in cyanobacterial survival and versatile chemical properties have been very interesting subject of research particularly for the algologists and biochemists. Because of their attractive color and stable fluorescence property, these proteins have been exploited in food as well as pharmaceutical industries, and a number multinational companies all over the world are engaged in commercial production of phycobiliproteins and related products.

Some of the major aspects of phycobiliproteins have been reviewed earlier by Bogorad (1975), Zilinskas and Greenwald (1986), Grossman et al. (1988), and Glazer (1994); however, these reviews did not discuss many of the important and interesting aspects of phycobiliproteins like their biosynthesis, modulation of their production by environmental conditions, their purification, and most importantly their application.

4.1 Phycobilisomes: Structure, Components, and Biosynthesis

Phycobilisomes (PBsomes) are protein assemblies that function in light harvesting and energy transfer, usually to PS II. When the energy absorbed by the PBsomes reaches the reaction centers of PS II, excitation energy is transduced to chemical energy. These biliproteins also known as phycobiliproteins absorb radiations in regions of the visible spectrum where *Chl*a has low absorptivity. Depending on the bilin energy, phycobiliproteins are categorized into three types: phycoerythrins with high energy (λ_{max} ~565 nm), phycocyanins with intermediate energy (λ_{max} ~620 nm), and allophycocyanins with low energy (λ_{max} ~650 nm). The energy flows from high-energy to low-energy pigments and phycobiliproteins are organized accordingly. Six rods in hemidiscoidal PBsomes have three cylinders in their core which is situated near the thylakoid membrane and PS II, where chlorophyll *a* is located. The rods have the phycoerythrin farthest from the core. The core has the allophycocyanins, of which there are two functional types: allophycocyanin with absorbance maxima of 650 nm and two other lower-energy allophycocyanins, the L_{CM} polypeptide and the α^B polypeptide, involved in energy transfer to chlorophyll (Gantt 1975; MacColl and Guard-Friar 1987; Bryant et al. 1979; Sidler 1994; Bald et al. 1996). The rods are built with stacks of disks, and the disk adjacent to the core is invariably phycocyanin.

Phycobiliproteins are commonly found to consist of hetero-monomers which are composed of two different subunits, α and β (Gantt 1980; Glazer 1986; Zilinskas and Greenwald 1986; Rowan 1989). These subunits are present in equimolar ratio (α:β::1:1) and differ in molecular mass, amino acid composition, and chromophore content. Fundamental configuration of the phycobiliprotein assembly is a stable trimer $(\alpha\beta)_3$ forming a toroidal-shaped aggregate (Sidler 1994; MacColl 1998; Tandeau de Marsac 2003; Adir 2005). The aggregate is 3–3.5 nm thick with a radius of about 5.5 nm and a central hole of 3 nm in diameter. Some biliproteins exist in hexamers, such as PE by face-to-face aggregation of two trimeric disks in the central cavity (Sidler 1994; MacColl 1998; Tandeau de Marsac 2003; Adir 2005; Glazer 1989; Ducret et al. 1996, 1998). Phycobiliproteins are brilliantly colored due to covalently bound open-chain tetrapyrrole chromophores named phycobilins. They are either blue-colored phycocyanobilin (PCB), red-colored phycoerythrobilin (PEB), the yellow-colored phycourobilin (PUB), or the purple-colored phycoviolobilin (PVB), also named cryptoviolin. These chromophores are attached to the polypeptide chain at conserved sites either by one or two cysteinyl thioester linkage (Glazer 1985).

4.1.1 Phycoerythrin (PE)

PE is the most abundant PBP in many red algae and in some cyanobacteria. These are characterized by strong absorption in the green region (480–570 nm) and by intensive fluorescence emissions at about 575–580 nm. Based on their absorption spectra, these red-colored proteins fall into three distinct species: (1) B-phycoerythrin (B-PE) (λ_{max} ~540–560 nm, shoulder ~495 nm), (2) R-phycoerythrin (R-PE) (λ_{max} ~565, 545 and 495 nm), and (3) C-phycoerythrin (C-PE) (λ_{max} ~563, 543 and ~492 nm). The prefixes B-, R-, and C- were used historically for indicating the type of organisms from which the pigment proteins were originally extracted, for example, Bangiales, Floridian Rhodophyceae, and Cyanophyceae. C-PEs are most abundant phycoerythrins in cyanobacteria. They are the products by some cyanobacteria that acclimate to their environment variations in light quality and intensity, especially green-to-red light ratios, through a process traditionally called complementary chromatic adaptation (Ohki et al. 1985; Toledo et al. 1999; Grossman et al. 2001; Palenik 2001; Misckiewicz et al. 2002; Grossman 2003; Everroad et al. 2006). These cyanobacteria make more of the light-harvesting protein phycocyanin in red light and more of the protein phycoerythrin in green light (Rowan 1989; Sidler 1994; MacColl 1998; Toledo et al. 1999; Grossman et al. 2001; Palenik 2001; Misckiewicz et al. 2002; Grossman 2003; Everroad et al. 2006). There are two subtypes of C-PE, viz., C-PE-I or PE (I) and C-PE-II or PE (II). The PEs from freshwater and soil cyanobacteria typically contain only PEBs and exhibit absorption spectra with maxima at about 565 nm and fluorescence emission spectra with peaks at about 575 nm. These are the spectral properties characteristic of C-PE-I (or PE (I)), also known as C-PE. C-PE-I usually carries five PEBs per monomer (αβ) (Rowan 1989; Sidler 1994; MacColl 1998; Ong and Glazer 1991). The five PEBs are covalently attached to the cysteine residues, in the way very similar to B-PE, at α -84, α -140/143, β -84/82, β -50/β -61, and β -155/159, respectively. The α and β subunits vary with their origins in molecular mass from 15 to 22 kDa. C-PE-I is a disk-shaped hexamer, which is constructed by two trimeric $(\alpha\beta)_3$ through face-to-face aggregation, commonly exists in form $(\alpha\beta)_6L_R$ where a rod-linker polypeptide (L_R) is bound to the central hole of the two stacking trimers. Unlike the γ subunit from B-PE and R-PE, however, the L_R of C-PE-I generally carries no phycobilins (Sidler 1994; MacColl 1998; Ong and Glazer 1991).

4.1.2 Phycocyanin

Phycocyanins are one of the most widespread phycobiliproteins observed in almost all phycobiliprotein-containing organisms, including cyanobacteria, red algae, glaucophytes, and some cryptophytes. They are most abundantly found in most cyanobacterial species that grow in natural environment. The phycocyanin from the PBsome-containing algae is subdivided into three types: (1) C-phycocyanin (C-PC, λ_{max} ~615–620 nm) exclusively existing in cyanobacteria, (2) phycoerythrocyanin (PEC, λ_{max} ~575 nm) inducible only in some cyanobacteria, and (3) R-phycocyanin (R-PC, λ_{max} ~615 nm) mainly in red algae (Gantt 1980; Glazer 1986; Zilinskas and Greenwald 1986; Rowan 1989; Reuter and Muller 1993; Sidler 1994; MacColl 1998; Tandeau de Marsac 2003; Adir 2005). These blue- or blue-purple-colored phycobiliproteins have strong light absorption ability mainly in range from 580 to 630 nm and emit intensive red fluorescence at 635–645 nm. Phycocyanins exist commonly as trimer $(\alpha\beta)_3$ from aggregation of three $(\alpha\beta)$ monomers. They have a thickness of 3 nm, a diameter of 11 nm, and a central hole in diameter of 3.5 nm. In PBsomes, phycocyanins commonly occur in disk-shaped hexamers, and a rod-linker polypeptide (L_R) or a rod-core-linker polypeptide is attached to the center cavity of hexamers to form $(\alpha\beta)_6L_R$ or $(\alpha\beta)_6L_{RC}$ complexes (Glazer 1989; Reuter and Nickel-Reuter 1993; Sidler 1994; Pizarro and Sauer 2001). The interaction of phycocyanins with the linker polypeptides may typically cause a large red shift of about 17 nm in their absorption and fluorescence emission maxima (Sidler 1994; Pizarro and Sauer 2001). C-phycocyanin (C-PC) is a blue-colored, deeply red-fluorescent phycobiliprotein and a predominant form among phycocyanins which is also one of the most studied PBPs. The C-PC subunits have globin-like folding, and according to the myoglobin nomenclature of helices, the eight helices of a C-PC subunit are denoted as X, Y, A, B, E, F, G, and H (Schirmer et al. 1985). During evolution, the oxygen-binding proteins of higher organisms (oxygen consumption in higher organisms) and PCs of oxygen-producing prokaryotes might have diverged from a common protein (Pastore and Lesk 1990). In a C-PC protein, three phycocyanobilin chromophores are carried by each $(\alpha\beta)$ monomer. The three PCBs have been determined to bind covalently to certain conserved cysteine residues at Cys α-84, β-84, and β-155. Thus one C-PC in trimer carries three α-PCBs and six β-PCBs, and in hexamer it contains six α-PCBs and 12 β-PCBs. Investigation of energy transfer between chromophores in C-PC trimers has demonstrated that α-84 PCB and β-155 PCB function as the two sensitizing chromophores located at the periphery of C-PC trimers, whereas β-84 PCB works as the fluorescing pigment located near the central cavity of C-PC trimers and hexamers. In other words, the α-84 PCB and β-155 PCB in C-PCs function as the excitation energy transfer donors and the β-84 PCB as the terminal acceptors (Sidler 1994; Debreczeny et al. 1995a, b; MacColl 1998; Eisele et al. 2000; Kikuchi et al. 2000).

Amino acid sequences of the apoproteins and/or chromophore composition have varied with adaptation process to light and nutrient conditions. One, or in some cases two, of the peripheral sensitizing chromophores (α-84 PCB and/or β-155 PCB) have been replaced by PVB, PEB, or PUB chromophores in each of the variants in order to adapt to conditions more enriched in blue and green wavelengths of light. The fluorescing β-84 PCB was, however, conserved in all variants.

4.1.3 Allophycocyanin

Allophycocyanins (APCs) are a type of core-constructing phycobiliproteins. These are a less-contained phycobiliprotein species with respect to PCs and PEs, but they exist in all PBsome-containing organisms, cyanobacteria, glaucophytes, and red algae, which grow in natural environment. Compared with red algae, cyanobacteria may generally contain higher amounts of APCs with respect to total phycobiliprotein content of the organisms. A PBsome core is commonly composed of three kinds of allophycocyanins that are known as allophycocyanin (APC), allophycocyanin-B (APC-B), and allophycocyanin core-membrane linker (L_{CM}) complex (APC-L_{CM}) (Gantt 1980; Glazer 1986, 1989; Zilinskas and Greenwald 1986; Rowan 1989; Sidler 1994; MacColl 1998). Among these three, APC constitutes the largest number of core biliprotein components, whereas APC-B and APC-L_{CM} have equal copies in the core. In the PBsome core construction, APC participates in the assembly of all the core cylinders, but APC-B and APC-L_{CM} take part merely in the two basal cylinders. Among the three types of allophycocyanins, the APC is the best described one. It exists in trimer of hetero-monomers (αβ) in PBsomes and in solution when extracted out of organisms. The APC trimers $(\alpha\beta)_3$, like PCs, show a threefold symmetry of the disk-shaped structure in crystal (Sidler 1994; Brejc et al. 1995; MacColl 1998; Liu et al. 1999). The APC $(\alpha\beta)_3$ in the disk-shaped conformation shows about 11 nm in diameter and 3 nm in thickness, and it has a central cavity of about 3.5 nm in diameter (Brejc et al. 1995; Liu et al. 1999). The α subunit consists of 160 amino acid residues, and the β subunit is composed of 161 amino acids (Troxler et al. 1980; Brejc et al. 1995; Liu et al. 1999); they exhibit apparent molecular masses in range from 17 to 20 kDa (Zilinskas 1982; Zilinskas and Greenwald 1986; Sidler 1994; MacColl 1998). One (αβ) monomer of the trimer contains two PCBs where one is attached at α-84 and the other at β-84. In the APC trimer, the β-84 PCB is situated near the central cavity, whereas the α-84 PCB is near the peripheral. The PCBs in the monomers were found to have an absorption maximum at 614 nm, but in the trimers they exhibited a sharp maximum at 650 nm and a prominent shoulder from 610 to 620 nm. It has been believed that one of the PCBs in the monomer, more probably β-84 PCB, is changed by interaction with the apoproteins in the trimer construction and that the PCB then creates the 650 nm absorption maximum owing to its unique conformation or a different environment (Sidler 1994; MacColl 1998, 2004). The other PCB, more possibly α-84 PCB that retains its original conformation similar to what it has in the monomer, gives the absorption shoulder from 610 to 620 nm. In this case, energy transfer in the trimer may be by Förster resonance from the α-84 PCB donor to the β-84 PCB acceptor (MacColl 2004; Loos et al. 2004).

4.1.4 Biosynthesis of Phycobiliproteins

Phycobiliproteins are a class of brilliantly colored proteins containing open-chain tetrapyrrole chromophoric groups apart from the apoprotein part. The biosynthesis of phycobiliproteins, viz., biosynthesis of tetrapyrrole phycobilins and their attachment to the apoproteins component are discussed here separately.

4.1.4.1 Biosynthesis of Phycobilins

Cyanobacteria can synthesize a vast array of tetrapyrroles like hemes, chlorophylls, phycobilins, siroheme, etc. As in case of most other organisms, δ-aminolevulinic acid (ALA) is the precursor of all tetrapyrroles in cyanobacteria. However, unlike animals and fungi, ALA in cyanobacteria (including all plants, algae, and most of the bacteria) is synthesized from glutamate (Glu). Glutamate is reduced to glutamate-1-semialdehyde (GSA) which supplies the C and N atoms of the tetrapyrrole nucleus. Asymmetric condensation of two molecules of ALA with the help of the enzyme porphobilinogen (PBG) synthase results in the formation of PBG. Uroporphyrinogen III (UPG III), the last common precursor of all end-product tetrapyrroles, is synthesized from PBG by two enzymes, viz., hydroxymethylbilane (HMB) synthase which converts PBG to HMB and uroporphyrinogen III synthase which catalyzes the formation of UPG III from HMB. UPG III can thus lead to the formation of protoporphyrin (precursor of chlorophylls, hemes, and phycobilins) or precorrins. Uroporphyrinogen decarboxylase enzyme decarboxylates all four acetate residues of UPG III to yield coproporphyrinogen (CPG). Another enzyme CPG oxidase oxidatively decarboxylates the propionate groups to produce protoporphyrinogen IX. This enzyme is specific for the III isomer of CPG over the nonphysiological I isomer. Protoporphyrinogen oxidase catalyzes the removal of six electrons from the tetrapyrrole macrocycle to form protoporphyrin IX which is the last biosynthetic step that is common to hemes, bilins, and chlorophylls (Fig. 4.1). Ferrochelatase catalyses the last step of protoheme formation: the insertion of Fe^{+2} into protoporphyrin IX. The increase in ferrochelatase activity is positively correlated with the accumulation which suggests that the level of this enzyme may be important rate-controlling factor in phycobilin synthesis (Brown et al. 1984). The opening of macrocycle by heme oxygenase enzyme results in conversion of protoheme to biliverdin IXα.

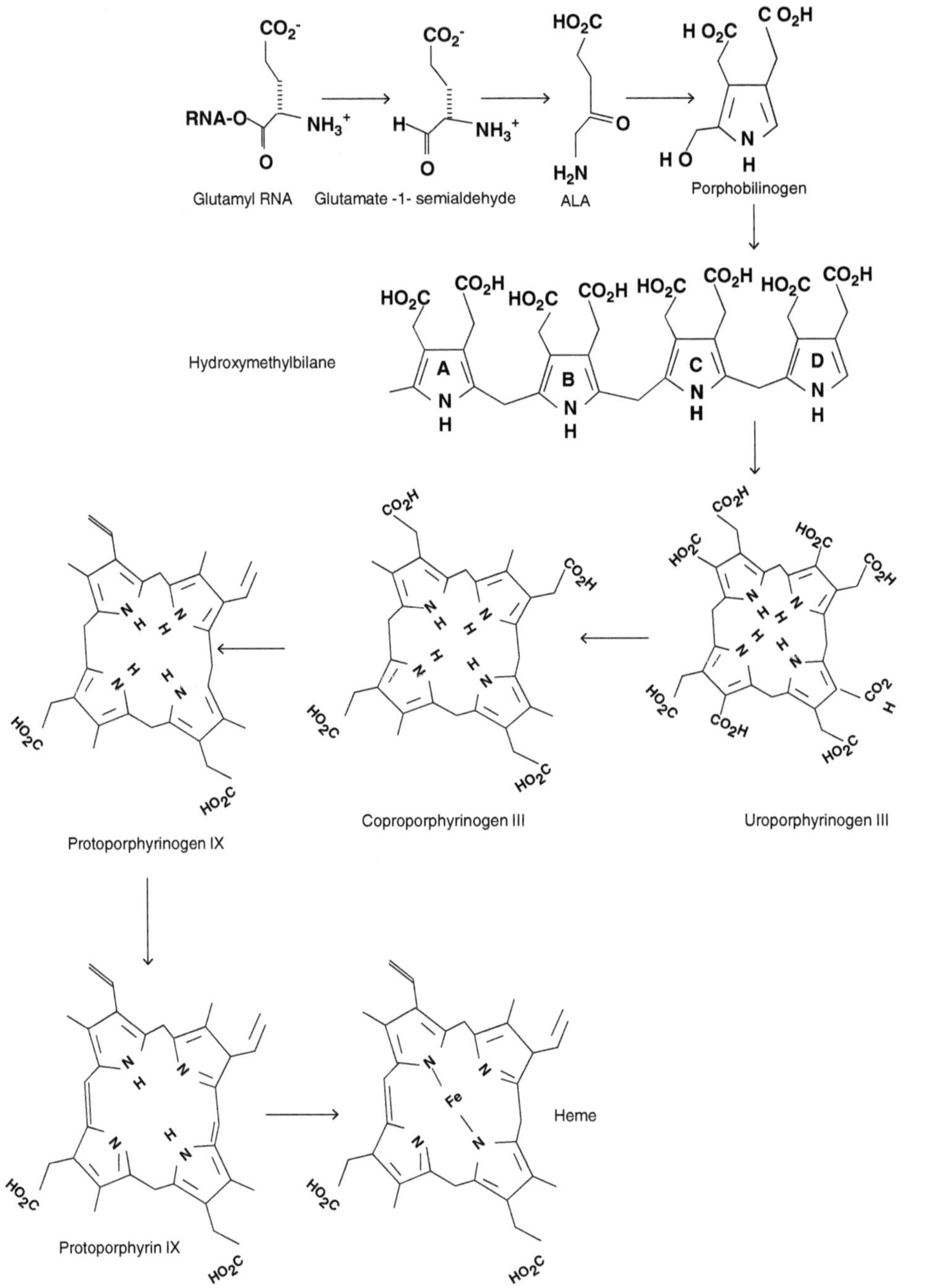

Fig. 4.1 Biosynthesis of heme from glutamate

Phycocyanobilin and phycoerythrobilin are isomers containing four more protons than their precursor, biliverdin. NADPH acts as electron donor for the reduction of biliverdin in the ferredoxin-linked reductase system (Beale and Cornejo 1991a). The reduction proceeds in two two-electron steps, with utilization of one NADPH molecule at each step (Beale and Weinstein 1990). Phycoerythrobilin is the first fully reduced bilin produced from biliverdin IXα which is subsequently isomerized to phycocyanobilin (Beale and Cornejo 1991b). Phytochromobilin and 15, 16-dihydrobiliverdin are thought to be intermediates of these two two-electron reduction steps. It is proposed that phycobilin biosynthesis might proceed through phytochromobilin (Beale and Cornejo 1984). However, Beale and Cornejo (1991a, b) concluded that 15, 16-dihydrobiliverdin, rather than phytochromobilin, is the partially reduced intermediate in phycobilin synthesis (Fig. 4.2). Once the pyrrole ring is reduced, the vinyl group

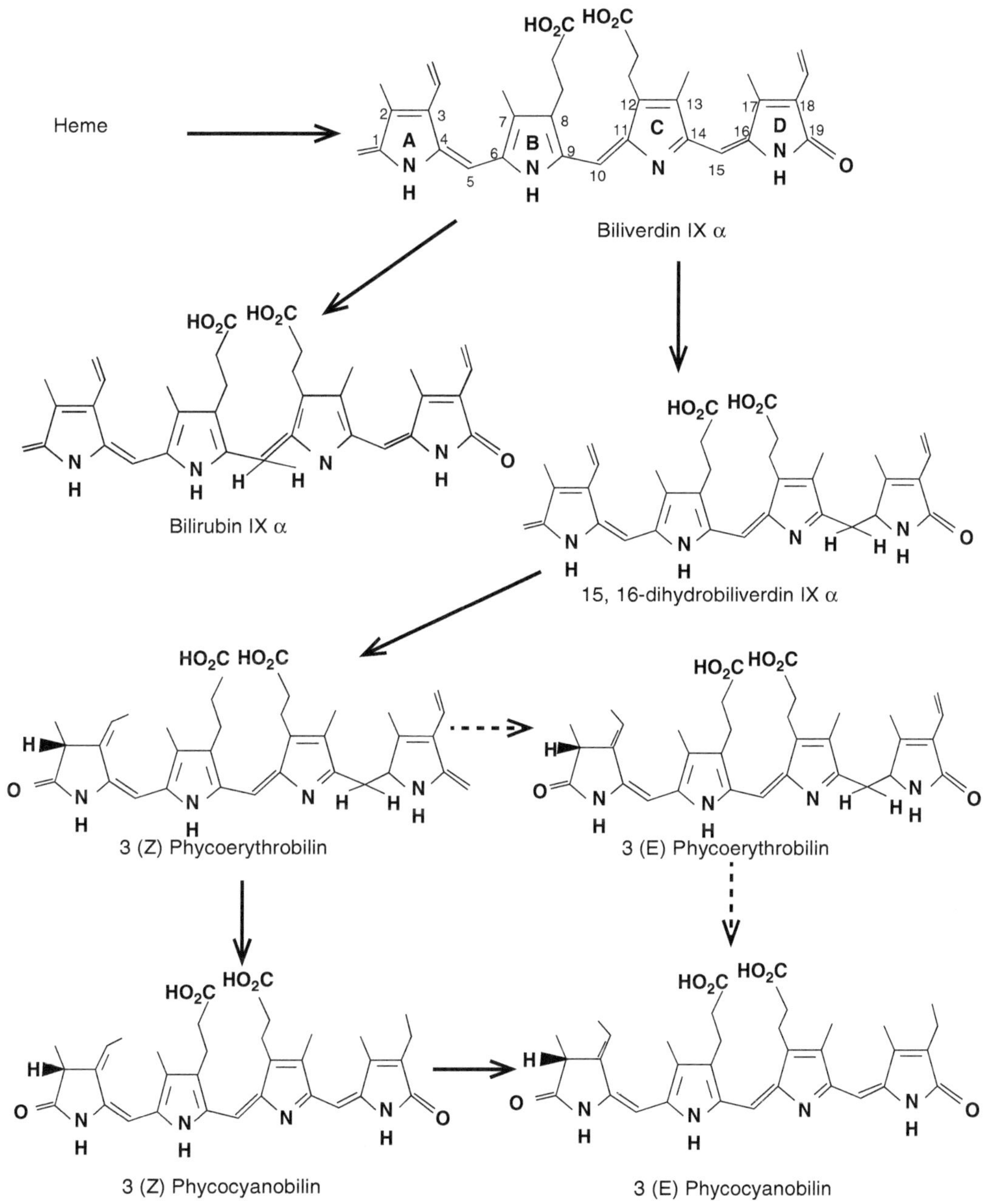

Fig. 4.2 Biosynthesis of phycobilins (Biliverdin IXα onwards) (Adopted from Schluchter and Glazer (1997). Originally referred by Schluchter and Glazer from Beale and Cornejo (1991a, b))

isomerizes to an ethylidene group to form 3(Z)-phycoerythrobilin. 3(Z)-phycocyanobilin dimethyl ester is the spontaneous isomerization product of synthetic 3-vinylphycocyanobilin dimethyl ester (Gossauer et al. 1989). The early phycobilin products are also formed with the ethylidene group in Z configuration. The 2-ethylidine isomer is reported to be the generally preferred initial product of vinyl isomerization although the E configuration is reported to be thermodynamically favored (Rüdiger et al. 1980; Weller and Gossauer 1980).

4.1.4.2 Ligation of Apoproteins to Phycobilins

Currently, there are four characteristic modes of chromophore attachment:

1. Spontaneous attachment: Many of previous studies reported that most apoproteins could bind phycobilins (phycocyanobilin, phycoerythrobilin) spontaneously in vitro; however, this process has quite low fidelity leading to mixed products (Arciero et al. 1988; Fairchild and Glazer 1994a; Schluchter and Glazer 1999). Spontaneous chromophore binding in vivo is unlikely as chromophores are reactive and present in very low concentration, and biliprotein synthesis is part of major metabolism in cyanobacteria.
2. Autocatalytic attachment: Autocatalytic attachment is a kind of spontaneous attachment leading to chromoproteins that are structurally and functionally indistinguishable from the respective native forms isolated from the parent organism (except for the lack of methylation of some β-subunits). Currently, correct chromophore binding, which is a true autocatalytic lyase activity, is mostly observed only among phytochromes (Wu and Lagarias 2000). Autocatalytic binding is also reported for ApcE, which could be reconstituted with phycocyanobilin (PCB) to give native-like $_{LCM}$ (Zhao et al. 2005).
3. E/F-type lyases: In *Synechococcus* PCC 7002, a heterodimeric lyase (CpcE/F) has been identified which catalyses both the forward (binding) and the reverse (releasing) reaction (Zhou et al. 1992; Fairchild et al. 1992). It also catalyses the addition of PEB to apoα-C-PC (CpcA), but with lower affinity as compared with PCB (Fairchild and Glazer 1994b). *pec* operons of *Nostoc* PCC7120 and *Mastigocladus laminosus* have been reported to have two homologous genes, *pecE/F* (Kufer et al. 1991), coding for a variant of this lyase (Zhao et al. 2002). Phycourobilin (PUB) is another example of such a chromophore where D5-to-D2 double-bond isomerization generates bound PUB from PEB catalyzed by such isomerizing lyase that uses PEB as substrate. Evidence for such a lyase has been found in marine cyanobacteria like *Synechococcus* sp. WH81024. Cyanobacterial S/U-type lyases: High substrate specificity and small number of E/F-type lyases prompted the search for some other lyases in cyanobacteria. Few new types of lyases in *Synechococcus* sp. PCC7002 were reported by Shen et al. (2004). One of them, CpcS, is coded by a homologue of a gene, *cpeS*, which is reported in *Fremyella diplosiphon*, on an operon with *cpeR* associated with gene regulation (Cobley et al. 2002). The S/U type of lyases comprises a new family of proteins unrelated to the E/F-type lyases and exhibit large variations. The main feature of the S/U lyases is high specificity for the binding site along with a very low specificity for the chromophore and the receptor apoprotein. CpcS from *Nostoc* sp. PCC 7120 is a nearly universal lyase for PCB attachment at Cys-84 of apo-phycobiliproteins (Zhao et al. 2007).

4.2 Organization and Regulation of Phycobiliprotein-Encoding Genes

Extensive structural and functional analyses of PBsome via molecular genetics have been performed for *Synechococcus* sp. strain 7002 (Bryant 1991) and for other cyanobacteria (Houmard et al. 1986; Grossman et al. 1995; Liu et al. 2005). Transcriptional units in different cyanobacteria vary in numbers as well as in terms of distribution of genes. The organization of complete set of genes encoding structural components of PBsome for *Mastigocladus laminosus* is shown in Fig. 4.3.

The *cpc* operons are composed of genes encoding PBsome rod components and polypeptides involved in chromophore attachment to the PC^{α} subunit (*cpcE* and *cpcF* genes; Bryant 1988, 1991; Zhou et al. 1992; Swanson et al. 1992; Fairchild et al. 1992). The *cpcG* gene(s), encoding the L_{RC}, is located on *pec-cpc* super operon in *Anabaena* sp. strain PCC 7120 and *Mastigocladus laminosus*, whereas the same gene is reported to be on other transcriptional unit in *Synechococcus* sp. strain PCC 7002. Liu et al. (2005) reported a *cpc* operon from *Arthrospira platensis* FAHCB341 which contained six parts: the 427 bp upstream sequence (*ussB*), 519 bp *cpcB* gene, 111 bp intergenic spacer between *cpcB* and *cpcA* genes, 489 bp *cpcA* gene, 184 bp upstream sequence (*ussH*) of *cpcH*, and partial sequence of *cpcH* gene (357 bp) which related to the synthesis of one of the rod-linker polypeptide. In all cyanobacteria studied till date, the *cpcA* gene coding for the α subunit of PC is located downstream from the *cpcB* gene encoding β subunit of PC. In some cyanobacteria, multiple gene sets for some types of phycobiliproteins and linker polypeptides have been found. For example, two *cpcBA* gene sets, viz., *cpcB1A1* and *cpcB2A2* occur in *Synechococcus* sp. strain PCC 6301 (Lind et al. 1987; Lau et al. 1987; Kalla et al. 1988) and *Pseudanabaena* sp. strain PCC 7409 (Dubbs and Bryant 1987, 1993), while three different *cpcBA* gene sets have been reported in *Calothrix* sp. strain PCC 7601 (Mazel et al. 1988; Capuano et al. 1988; Mazel and Marliere 1989). In *Synechococcus* sp. strain PCC 6301 and PCC 7942, the duplicated phycocyanin genes encode identical polypeptides and are arranged as a tandem repeat unit with the genes for the rod linkers between the *cpcB1A1* gene set and the downstream *cpcB2A2EF* gene set (Kalla et al. 1988, 1989).

The *cpeBA* operon consisting of the genes encoding for PE has the same order of genes, i.e., *cpeB* gene is located upstream to the *cpeA* gene (Mazel et al. 1986; Dubbs and Bryant 1987, 1993; Anderson and Grossman 1990; Bernard et al. 1992). In *Calothrix* sp. strain PCC 7601, the two genes, *cpeB* and *cpeA*, were found to be separated by 79 base pairs (Mazel et al. 1986). A similar orientation is also reported in *Pseudanabaena* sp. PCC 7409 where the two genes were separated by 74 base pairs (Dubbs

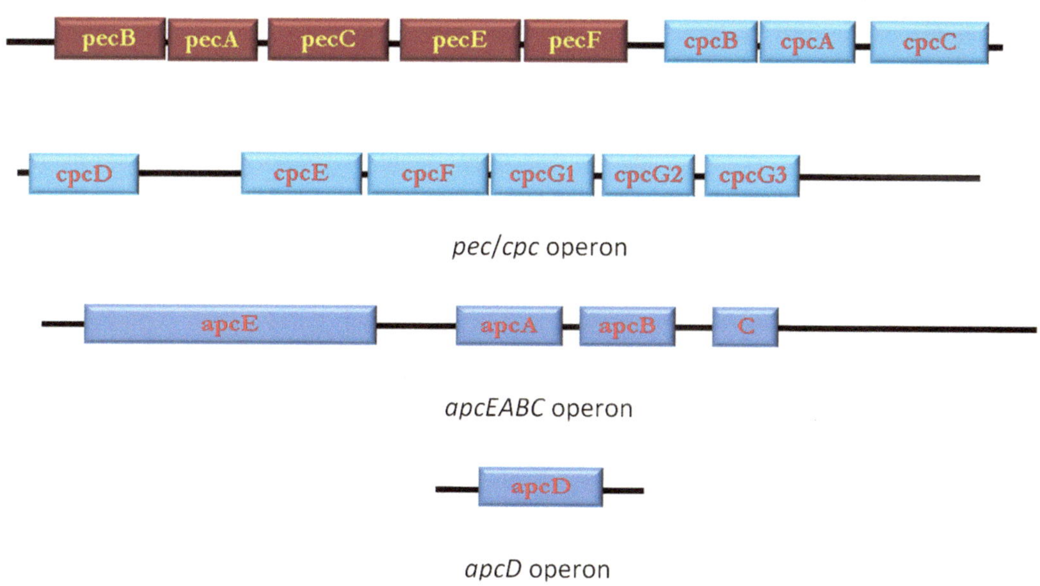

Fig. 4.3 Organization of phycobiliprotein genes in *Mastigocladus laminosus* (Adopted from Bhaya et al. (2000))

and Bryant 1987, 1993). However, none of these two cyanobacteria contained genes for linker polypeptides in *cpe*BA operon. In *Calothrix* sp. strain 7601, *cpeCDE* operon contained the genes for PE-associated rod-linker polypeptides that are only transcribed when cells are grown in green light (Federspiel and Grossman 1990; Federspiel and Scott 1992). DNA sequence analysis of *Synechococcus* sp. strain WH 8020 showed the genes for the α and β subunits of PE-I and PE-II as well as for the γ subunit of PE-II to be clustered in a large 15 kb region of the genome. Twelve open reading frames are included in this region (Wilbanks and Glazer 1993).

Unlike *cpc* and *cpe* operons, the *apc* operons have a reverse order of the genes: the *apcA* gene is located upstream to the *apcB* gene (Bryant 1988, 1991). The *apc* operons from several cyanobacteria like *Synechococcus* sp. strain PCC 6301 (Houmard et al. 1986), *Synechococcus* sp. PCC 7002 (Bryant 1988, 1991), *Calothrix* sp. strain 7601 (Houmard et al. 1988, 1990), *Anabaena variabilis* (Johnson et al. 1988), etc. The *apcC* gene encoding the small linker polypeptide $L_C^{8.9}$ lies downstream from the *apcB* gene. The *apcD* and *apcF* genes encode minor APC-related phycobiliproteins. The *apcE* encodes the large core membrane linker phycobiliprotein (L_{CM}). In *Synechococcus* sp. strain PCC 7002 (Bryant 1988, 1991), the *apcE* gene is transcribed as a separate unit, whereas in *Synechococcus* sp. strain PCC 6301 (Capuano et al. 1991), *Calothrix* sp. strain 7601 (Houmard et al. 1990), and *Mastigocladus laminosus*, it is located on a transcriptional unit with *apcA*, *apcB*, and *apcC*.

As cyanobacteria can be found in locations with varying chemical and physical parameters like nutrient availability, quality and quantity of light, temperature, water activity, etc., the expression of phycobiliprotein coding genes also modulate accordingly to adapt to particular conditions. Bryant (1981) studied the regulation of PC synthesis in response to growth in chromatic illumination in 69 cyanobacteria and reported controlled phycocyanin synthesis through photoregulated expression of the α and β subunits. For these strains, the expression of one pair of PC subunits was constitutive irrespective of light quality while expression of the other occurred particularly under red light. Oelmuller et al. (1988) showed that the levels of transcripts encoding the APC subunits, the core linker polypeptide, and the constitutive PC subunits were similar under red and green light conditions while very large increases PE and inducible PC mRNAs occurred rapidly after transferring *Fremyella diplosiphon* from red light to green light and green light to red light, respectively. The green light condition used in those experiments was effective in eliminating essentially all inducible PC mRNA synthesis while the red light did not eliminate all PE mRNA synthesis and the decrease in the phycoerythrin mRNA level in red light was much slower than the decline in the levels of the inducible phycocyanin transcripts in green light. Grossman et al. (1988) reported that mRNA of *cpc1* operon in *Calothrix* sp. strain PCC 7601 accumulated constitutively while the transcript of *cpc2* operon accumulated only when the cells are grown in red light. Similar results have been described for the group III chromatic adapter *Pseudanabaena* sp. strain PCC 7409 (Dubbs and Bryant 1993).

Fremyella diplosiphon UTEX481 (*Calothrix* sp. strain PCC 7601) is widely studied for changes in its PBP composition in response to varying light color. Studies have shown that the quality of light could modulate the composition of PBsome through a couple of two-component regulatory system. The regulator for complementary chromatic adaptation (Rca) activates *cpc2* and *pcyA* and represses *cpeC* expression in red light, while the control of green light induction (Cgi) system has no detectable effect on *cpc2* or *pcyA* expression and only represses *cpeC* expression in red light (Sobczyk et al. 1994; Li et al.2008; Seib and Kehoe 2002; Alvey et al. 2003; Li and Kehoe 2005; Kehoe and Gutu 2006). The final gene in the *cpeC* operon, *cpeR*, encodes an activator required for *cpeBA* and *pebAB* expression (Cobley et al. 2002; Seib and Kehoe 2002). Both the Rca and Cgi systems regulate these two operons by controlling *cpeC* expression.

Light intensity is the most significant environmental factor influencing the light harvesting complexes (phycobilisomes) in cyanobacteria. The influence of light intensity on PBP content has been reported in different organisms (Yamanaka and Glazer 1980; Lonneborg et al. 1985; Muller et al. 1993; Gamier et al. 1994). de Lorimier et al. (1992) reported 1.8-fold decrease in PC:APC ratio in *Agmenellum quadruplicatum* PR-6 during growth in a light intensity of 1260 μE as compared to that in 20 μE. The relative level of RNAs encoding phycocyanin and allophycocyanin was found to vary with light intensity in parallel with the phycobiliprotein ratio. Nomsawai et al. (1999) reported that rapid changes in light intensity from high to low light caused a dramatic increase in the accumulation of PC, APC, and Chl *a*. *Spirulina platensis* Cl underwent a significant change in pigmentation mainly due to the reduction of PC and Chl *a* synthesis under high light intensity. Shifting cells from high-light to low-light conditions resulted from an increase in transcription levels of the *cpc* and *apc* operons ultimately increasing the content of PBPs. When cells were shifted from low to high light, the 3.5 kb transcript species corresponding to the *cpcBAHID* operon disappeared in concomitant with the disappearance of 33 kDa linker polypeptide within 24 h. However, the other two rod-linker polypeptides, 34.5 and 15 kDa, were still associated with the intact phycobilisomes.

Hihara et al. (2001) reported downregulation of *cpc* genes in *Synechocystis* sp. PCC 6803 grown under high light intensity (300 μmole photons/m^2/s) and *cpc* genes were downregulated more strictly than *apc* genes because phycocyanin is the primary target for the reduction of antenna size. It was interesting to note that *cpcD*, which is part of the *cpcBACD* operon, apparently was not downregulated as much as the other *cpc* genes. Singh et al. (2008) also reported downregulation of phycocyanin-encoding genes in *Synechocystis*.

Upon transfer to high light intensity conditions, both the amounts of photosystems and PBsomes are downregulated to avoid absorption of excess light energy. The capacity for CO_2 fixation and other cellular metabolism is upregulated to increase energy consumption. Protection mechanisms to prevent photodamage have been developed to cope with the increased production of reactive oxygen species. However, the molecular mechanisms that enable the cells to perceive and acclimate to the changing light environment were poorly understood at that time (Muramatsu and Hihara 2012).

As mentioned earlier, phycobiliprotein synthesis as well expression of the phycobiliprotein-encoding genes may also be modulated by nutrients like nitrogen, sulfur, and iron. Iron-, nitrogen-, and sulfur-deficient cells contain less than half of the normal complement of thylakoid membranes (Sherman and Sherman 1983; Wanner et al. 1986). The remaining membranes are disorganized and interspersed with large deposits of glycogen. A dramatic response of cyanobacteria to nutrient limitation is a decrease in the abundance of pigment molecules in the cell. Cultures deprived of nitrogen for 30 h exhibited no detectable PC and sulfur-deprived cultures showed a similar rapid decline in the absolute levels of PC and APC (Collier and Grossman 1992; Wanner et al. 1986). During sulfur and nitrogen starvation, there was a rapid and near complete degradation of the PBsome. PBsome degradation could provide nitrogen-limited cells with amino acids used for the synthesis of proteins important for the acclimation process. It is more difficult to understand why PBsome would be degraded in cells limited for other macronutrients, such as sulfur, since phycobiliproteins are a poor source of sulfur amino acids. However, the PBsome was degraded in the same ordered manner in both sulfur- and nitrogen-starved cells (Collier and Grossman 1992; Yamanaka and Glazer 1980). The levels of mRNAs encoding the phycobiliproteins have also been reported to decline during nutrient-limited growth. de Lorimier et al. (1984) reported that the level of PC mRNA decreased to nearly zero in nitrogen-starved *Synechococcus* sp. strain PCC 7002. Further analyses had shown that 3–5 h after the initiation of nitrogen deprivation, the transcripts from the *cpcBA* operon were essentially undetectable (Bryant 1991). The results obtained by Gasparich et al. (1987) using *cpcB-lacZ* fusion construct suggested that nitrogen deprivation resulted in a marked decrease in the transcription of *cpcBA*. In *Synechococcus* sp. strain PCC 7942, the mRNAs encoding both PC and APC declined rapidly during nitrogen or sulfur limitation and less rapidly during phosphorus limitation. However, levels of all of these mRNA species remained at 5–10 % that of nutrient-replete cells (Collier and Grossman 1992), even 48 h after the cells were transferred to medium lacking nitrogen or sulfur. This change in the steady-state levels of phycobiliprotein mRNAs may be a consequence of both altered rates of transcription and mRNA turnover. These results suggested that the production of phycobiliproteins during nutrient-limited growth might be blocked by both transcriptional and posttranscriptional events.

Similarly, iron has been shown to modulate PBP production and regulation. Hardie et al. (1983) reported that C-PC content in *Agmenellum quadruplicatum* started decreasing after 16 h and decreased up to 200 h until iron was replenished in the medium. Troxler et al. (1989) showed that phycobiliprotein mRNA, which was absent in dark-grown cultures of *Cyanidium caldarium*, appeared when the culture was grown in presence of heme (Fe-protoporphyrin IX) under dark conditions. This was the first report to show that heme is regulatory factor specifically involved in transcriptional regulation of phycobiliprotein genes. Singh et al. (2003) reported downregulation of *cpc* genes in *Synechocystis* sp. strain PCC 6803 when the cyanobacteria was grown in iron-deficient BG-11 (with nitrogen) medium. Hemalata and Fatma (2009) observed that phycobiliprotein content in *Anabaena* NCCU 9 decreased significantly when grown in iron-deficient medium. Chakdar (2012) reported that ferric ammonium citrate (1.2 ppm) was optimal iron source for maximal phycocyanin production in three heterocystous cyanobacteria. In the same study, modulation of *cpcB* gene was also reported by concentration and source of iron which was probably modulated through altered synthesis of heme or phycocyanobilins (Fig. 4.4).

The composition and function of PBPs in cyanobacteria have been reported to change in response to stress conditions (Grossman et al. 1993), but among various stresses, acclimation to high salt concentrations is of high importance for basic as well as applied research, since a high percentage of irrigated land suffers from increasing levels of salts. Marin et al. (2004) reported that *cpcBACD* genes were downregulated in *Synechocystis* sp. strain 6803 when the cyanobacteria was grown with 684 mM NaCl. Chakdar et al. (2012) also reported modulation of cpcB gene under salinity. It was observed that cpcB gene expression and phycocyanin production was upregulated when *Anabaena* (CCC162 & CCC421) and *Nostoc* (CCC391) were grown with 10 mM Nacl. Rapid entry of sodium ions might result in detachment of phycobilisomes from the thylakoid membranes, resulting in decreased photosynthesis (Rafiqul et al. 2003), energy transfers from phycobiliproteins to PS II reaction center (Schubert et al. 1993; Verma and Mohanty 2000), and uptake of other mineral nutrients, such as K^+, Ca^{2+}, and Mn^{2+} (Hasegawa et al. 2000).

4.3 Environmental Stress Conditions for Enhancing Phycobiliprotein Production

As cyanobacteria can be found in locations which exhibit widely fluctuating chemical and physical parameters like nutrient availability, light intensity, light wavelength, temperature, water activity, etc., synthesis of phycobiliprotein is also accordingly modulated to adapt to a particular condition. The composition and function of phycobiliproteins in cyanobacteria have also been reported to change under stress conditions (Grossman et al. 1993). Among various conditions which influence the phycobiliprotein production, light, temperature, and pH are most important. Appropriate alteration of such parameters can render enhanced production of phycobiliproteins.

4.3.1 Light

Both color and intensity of light can influence phycobiliprotein synthesis in cyanobacteria. Complementary chromatic adaptation (CCA) has been a well-studied phenomenon in cyanobacteria where the relative amount of PC and PE

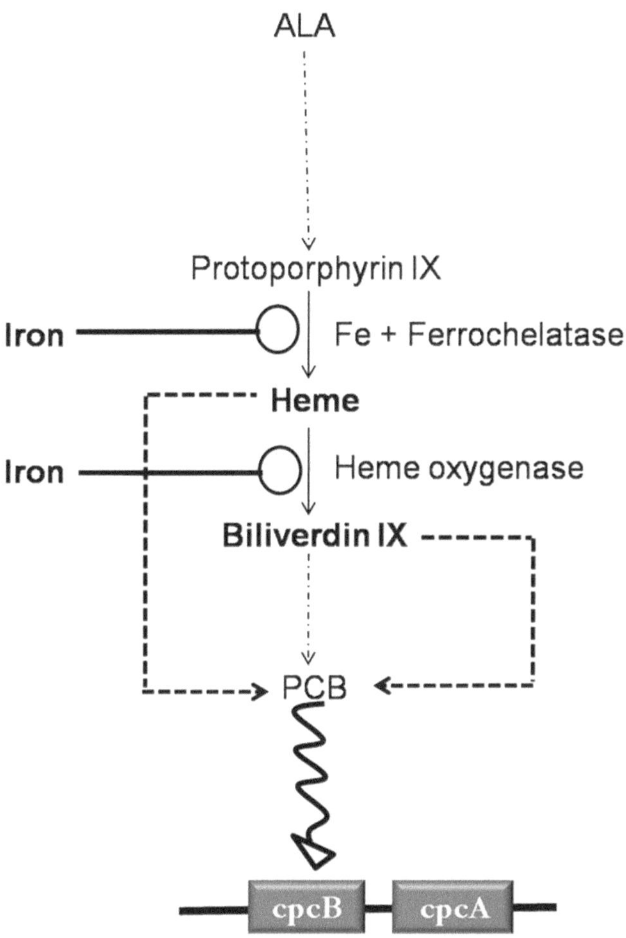

Fig. 4.4 Probable mode of regulation of cpcB gene through iron- A hypothetical view ——○ indicates steps where probably the key enzymes in PCB biosynthesis are regulated by iron which ultimately influenced the synthesis of Heme or Biliverdin IX; ------ indicates the regulation by the key products like heme or biliverdin IX which influences the synthesis of PCB. This could be the possible way through which iron actually regulates the expression of phycocyanin beta subunit gene (Adapted from Chakdar 2012)

changes depending on light color. CCA has been most extensively studied in the cyanobacterium *Calothrix* sp. strain 7601 (also known as *Fremyella diplosiphon*), although the phenomenon has been observed in a number of other cyanobacteria (Tandeau de Marsac 1977; Bryant and Cohen-Bazire 1981; Bryant 1981, 1982). Stowe et al. (2011) showed that *F. diplosiphon* exhibited a 3.3-fold induction of PC in red light and a 5.77-fold induction of PE in green light. In the same study, they reported that, in *Gloeotrichia* UTEX 583, PE was synthesized in green light and not synthesized in red light, but PC is highly abundant in both red and green light. So, chromatic adaptation property of cyanobacteria can be exploited for commercial production of high-value pigments like phycocyanin and phycoerythrin.

Light intensity is the most significant environmental factor influencing the light-harvesting complexes (phycobilisomes) in cyanobacteria. As such, light intensity is of particular importance to the outdoor cultivation since it is largely variable in natural environment. The influence of light intensity on PBP content has been reported in different organisms (Yamanaka and Glazer

1980; Lonneborg et al. 1985; Muller et al. 1993; Gamier et al. 1994).

de Lorimier et al. (1992) reported 1.8-fold decrease in PC:APC ratio in *Agmenellum quadruplicatum* PR-6 during growth in a light intensity of 1260 μE as compared to that in 20 μE. Hemalata and Fatma (2009) reported that 25 μmol photons/m^2/s to be the most suitable light intensity for maximum production (124.95 mg/g dry wt.) of phycocyanin in *Anabaena* NCCU9. 25 μmol photons/m^2/s was also reported to be optimum light intensity for *Synechococcus* NKBG 042902 (Takano et al. 1995), *Spirulina subsalsa*, *S. maxima* (Tomasseli et al. 1995, 1997), and *Synechocystis* (Hong and Lee 2008), while 12.5 μmol photons/m^2/s was found optimum for *Nostoc* UAM 206 (Poza-Carrion et al. 2001) and *N. muscorum* (Ranjitha and Kaushik 2005a). On the other hand, 150 μmol photons/m^2/s was optimal light intensity for phycobiliprotein production in *Arthronema africanum* (Chaneva et al. 2007). However, a number of studies have reported reduction in phycobiliprotein gene expression and synthesis (Hihara et al. 2001; Tu et al. 2004). Upon transfer to high light conditions, both the amounts of photosystems and the light-harvesting antenna complex, i.e., phycobilisome (PBsome), are downregulated to avoid absorption of excess light energy (Muramatsu and Hihara 2012). In fact, when the amount of light energy absorbed by light-harvesting complexes exceeds the limit of energy requirements by various cellular processes, the excess energy produces reactive oxygen species (ROS) leading to severe photodamage to cellular components. Nonetheless, photosynthetic organisms have evolved different acclimatory responses to high light conditions to maintain a balance between supply and consumption of energy and to protect the photosynthetic apparatus from damage. Cyanobacteria prefer low light intensities and stimulate phycobiliprotein synthesis due to their low specific maintenance energy and their pigment composition. Not only this, low irradiances actually broadens the overall light absorption band in such a way that the balance of light energy distribution between the two photosystems is maintained that optimizes the rate of light energy conversion.

4.3.2 Temperature

Like any other organism, all metabolic processes in cyanobacteria are also influenced by temperature. The optimal growth temperature and tolerance to the extreme values vary from strain to strain. Phycobiliprotein synthesis in cyanobacteria is also regulated by temperature, and depending on the strain, the optimum temperature for maximum production of phycobiliprotein production varies. Anderson et al. (1983) reported that PE to PC ratio in *Nostoc* sp. strain MAC increased when the cyanobacteria was grown at 39 °C (PE:PC = 2.3:1) as compared to growth at 30 °C (PE:PC = 1:1) under white light and light intensity of 10^{-2} μE/cm^{2}s. Like light quality and quantity, temperature can also regulate the synthesis of phycobiliproteins and change the relative amount of PC and PE according to the conditions. Hemalata and Fatma (2009) showed that a temperature of 30 °C is optimum for maximum production (127.02 mg/g of dry wt.) of phycobiliproteins in *Anabaena* NCCU9, and phycobiliprotein production reduced significantly at higher or lower temperature (23.6 % reduction at 20 °C and 38 % reduction at 40 °C). Chaneva et al. (2007) reported that phycobiliprotein content in *Arthronema africanum* increased up to 36 °C under a light intensity of 150 μmol photons/m^2/s and correlated to the changes in growth curve. But further increase in temperature resulted in decrease in phycobiliprotein content by the tune of 10–20 %. Sakamoto and Bryant (1998) reported a temperature of 36 °C to be most suitable for maximum production of phycobiliproteins in *Synechococcus*. In *Plectonema boryanum* UTEX 485, phycobiliprotein pigments reduced significantly at 15 °C as compared to 29 °C temperature under light intensity of 150 μmol photons/m^2/s (Misckiewicz et al. 2002). Most of these studies (Chaneva et al. 2007; Misckiewicz et al. 2000) were carried out to assess the combined effect of light intensity and temperature on phycobiliprotein production in different cyanobacteria, and it was clear from these studies that influence of light intensity and temperature in combination is more pronounced than the individual effects of these two factors. So, optimization of both the parameters rather than one of the factors will be the appropriate intervention for enhancing the production of phycobiliprotein production in cyanobacteria.

4.3.3 pH

In general, cyanobacteria are alkalophiles capable of maintaining an internal, constant pH of 7.1–7.5 in the range of external pH from 5 to 10 (Ritchie 1991). Changes in pH affect the solubility and bioavailability of nutrients, transport of substances across the cytoplasmic membranes, and the activity of intra- and extracellular enzymes, as well as photosynthetic electron transport and the osmotic potential of the cytoplasm (Walsby 1982). It was reported that phycobiliprotein concentration in *Gloeotrichia natans* changed significantly with respect to changes in the pH of the medium; however, growth rate of the cyanobacterium remained unaffected (Boussiba 1991). At a pH of 7.00, phycobiliprotein constituted up to 10 % of the total protein, but at pH 9.0, the amount was 28 % of the total protein. In *Nostoc* sp. UAM206, allophycocyanin content increased significantly at pH 9.0 as compared to pH 7.0 (Poza-Carrion et al. 2001). *Nostoc* sp. UAM206 exhibited an increase in PC and PE with the increase in pH under limited availability of inorganic carbon; however, when inorganic carbon was made available, the effect of pH was nullified (Poza-Carrion et al. 2001). Hence, the effect of pH on photosynthetic pigments is more pronounced under nutrient-limited conditions. Desmukh and Puranik (2012) also reported pH 10.0 to be the optimum pH for phycobiliprotein production in *Synechocystis* sp., while Hong and Lee (2008) showed pH 8.0 to be best for phycobiliprotein production in *Synechocystis* sp. PCC 6701. *Anabaena* NCCU9 also resulted in maximum phycobiliprotein production (102.24 mg/g dry wt.) when grown at pH 8.0 but the culture became white at pH 2 and 12 (Hemalata and Fatma 2009). For a particular culture, it is always essential to determine the optimum pH of the medium in which the culture can perform the best in terms of phycobiliprotein production.

4.3.4 Salt Stress

Among various stress conditions, salt stress is of utmost importance for basic as well as applied research, since a high percentage of irrigated land suffers from increasing levels of salts. Salt stress mainly results in decrease of phycocyanin content and thereby interrupts the energy transfer from phycobiliproteins to PSII reaction center (Schubert and Hagemann 1990; Schubert et al. 1993; Lu et al. 1999; Lu and Vonshak 2002). However, there has been no study to point out if there is any specific effect of salt stress on phycocyanin production as well as the expression of phycocyanin-encoding genes. Marin et al. (2004) in order to study genome-wide response of gene expression under salt stress, reported that *cpcBACD* genes were downregulated in *Synechocystis* sp. strain 6803 when the cyanobacteria was grown with 684 mM NaCl. Phycocyanin production as well as relative expression of *cpcB* gene increased in *Nostoc commune* (CCC391) and *Anabaena variabilis* (CCC421) when grown with 10 mM NaCl (Chakdar 2012). Hemalata and Fatma (2009) also reported enhancement of phycobiliprotein production in *Anabaena* NCCU9 with 10 mM sodium chloride. Although, all cyanobacteria do not respond in a similar fashion to salt stress, it can be used as a stimulus for salt-tolerant cyanobacteria to produce higher amount of phycobiliprotein

4.4 Extraction and Purification Strategies for Phycocyanin and Phycoerythrin

Commercial exploitation of phycobiliproteins requires rapid and efficient extraction followed by effective separation from undesirable proteins using appropriate purification strategy. But, extraction of phycobiliproteins from cyanobacteria is notoriously difficult because of their extremely resistant multilayered cell wall and considerably small size (Stewart and Farmer 1984; Wyman 1992). A number of methods have been employed to extract and quantify phycobiliproteins from different cyanobacteria; however, there is no standard technique for quantitative extraction of pigments from these organisms (Jeffrey and Mantoura 1997; Wiltshire et al. 2000). Methods like freezing and thawing (Abalde et al. 1998; Minkova et al. 2003), use of sonicators (Abalde et al. 1998) and French pressure cells (Alberte et al. 1984), use of enzymes like lysozyme (Boussiba and Richmond 1979), combination of EDTA and lysozyme (Stewart and Farmer 1984; Kilpatrick 1985; Vernet et al. 1990), etc., have been employed for disruption of the cells in order to release phycobiliproteins. Viskari and Colyer (2003) described a method for extracting phycobiliproteins from a *Synechococcus* CCMP 833 cyanobacteria culture that utilizes 3 % 3-[(3-cholamidopropyl) dimethylammonio] propanesulfonic acid (CHAPS) 0.3 % asolectin combined with nitrogen cavitation and achieved extraction efficiencies of greater than 85 %. Lawrenz et al. (2011) reported that disruption of cells by freezing–thawing and sonication both resulted in significantly higher extraction efficiencies than disruption with a tissue grinder. It has been reported by other workers (Abalde et al. 1998; Doke 2005) also that freeze/thaw cycles was the most efficient and economical way to extract C-PC from wet cyanobacterial biomass. Different buffers were generally used to extract phycobiliproteins with reliance on spectrometric detection and quantification. Phosphate buffer is mostly used for such extraction (Sarada et al. 1999; Doke 2005; Ranjitha and Kaushik 2005b; Benedetti et al. 2006); however, some people have also used Tris buffer (Bhaskar et al. 2005), HEPES buffer (Gupta and Sainis 2009), etc.

A number of processes are available for purification of PC and PE from cyanobacteria. Most of these procedures involve steps like ammonium sulfate precipitation, dialysis, gel filtration chromatography, anion exchange chromatography, hydroxy apatite chromatography, etc. A few procedures have also been reported which involve some additional steps like membrane filtration or use of organic solvents. These purification strategies involving a different combination of the abovementioned steps have been employed to purify PC and PE from a number of cyanobacteria. Some of the strategies used for purification of PC from cyanobacteria are presented in Table 4.1 (modified from Eriksen 2008). Abalde et al. (1998) reported 83.4 % recovery with a purity of 3.2 from *Synechococcus* sp. IO920 after hydrophobic interaction chromatography, while Santiago-Santos et al. (2004) obtained 80 % recovery with a purity ratio of 3.5 using similar procedure. Abalde et al. (1998) reported a final purity of 4.85 with 76.56 % recovery following a combination of hydrophobic interaction chromatography (using butyl sepharose) and ion exchange chromatography (using Q Sepharose), while Boussiba and Richmond (1979) attained a purity of 4.15 in *S. platensis* following hydroxyapatite and ion exchange chromatography. Following a novel method using rivanol treatment, 45.7 % PC with a purity of 4.30 was recovered from *S. fusiformis* (Minkova et al. 2003). In 2006, Benedetti et al. reported a purity of 4.78 from *Aphanizomenon flos-aquae* after hydroxyapatite chromatography. Soni et al. (2006) used ammonium sulfate precipitation, gel filtration chromatography, and DEAE-cellulose column chromatography to obtain phycocyanin with a purity of 3.31 from *Oscillatoria quadripunctulata* while they further attained PC with a purity of 4.52 from *Phormidium fragile* using a novel single-step hydrophobic interaction chromatography (Soni et al. 2008). Recently, Yan et al. (2011) attained a purity of 5.59 with a final PC recovery of 67.04 % from *Spirulina platensis* following ammonium sulfate precipitation and anion exchange chromatography using DEAE sepharose fast flow column. Patil et al. (2006) studied purification of C-PC by aqueous two-phase extraction (ATPE) combination with ion exchange chromatography, and the results showed that ATPE purification yielded high-purity C-PC with 73 % recovery. This not only enabled an increase in product purity without the

Table 4.1 Strategies used for purification of phycocyanin from different cyanobacteria

Numbers and types of central unit operations	Cyanobacterial species	Reference
Chromatography based methods		
1. Ammonium sulfate precipitation 2. Hydroxyapatite chromatography 3. Ion exchange chromatography	*Spirulina platensis*	Boussiba and Richmond (1979)
1. Hydrophobic interaction chromatography 2. Ion exchange chromatography	*Synechococcus* sp.	Abalde et al. (1998)
1. Ammonium sulfate fractionation 2. Ion exchange chromatography 3. Gel filtration	*S. platensis*	Zhang and Chen (1999)
1. Rivanol treatment 2. Ammonium sulfate precipitation 3. Gel filtration 4. Ammonium sulfate precipitation	*S. fusiformis*	Minkova et al. (2003)
1. Ammonium sulfate precipitation 2. Size exclusion chromatography 3. Anion exchange chromatography	*Oscillatoria quadripunctulata*	Soni et al. (2006)
1. Ammonium sulfate precipitation 2. Hydroxyapatite chromatography	*Aphanizomenon flos-aquae*	Benedetti et al. (2006)
1. Expanded bed adsorption chromatography 2. Anion exchange chromatography	*S. platensis*	Niu et al. (2007)
1. Ammonium sulfate fractionation 2. Hydrophobic interaction chromatography	*Phormidium fragile*	Soni et al. (2008)
1. Ammonium sulfate precipitation 2. Dialysis 3. Anion exchange chromatography	*S. platensis*	Yan et al. (2011)
1. Ammonium sulfate precipitation 2. Dialysis 3. Anion exchange chromatography	*Nostoc commune, Anabaena oryzae, and A. variabilis*	Chakdar (2012)
Two phase aqueous extraction based methods		
1. Two-phase aqueous extraction 2. Ultrafiltration 3. Ammonium sulfate precipitation	*S. maxima*	Rito-Palomares et al. (2001)
1. Chitosan adsorption 2. Two-phase aqueous extraction	*S. platensis*	Patil et al. (2006)
1. Chitosan adsorption 2. Two-phase aqueous extraction 3. Ion exchange chromatography	*S. platensis*	Patil et al. (2006)
1. Repeated two-phase aqueous extraction 2. Ultrafiltration	*S. platensis*	Patil and Raghavarao (2007)

need of multiple steps but also reduced the volume of the contaminant proteins.

The basic strategies for purification of PE are almost same as for PC. Although most of the PE purification studies have been carried out in eukaryotic alga, some reports are also available for cyanobacteria. Tchernov et al. (1999) achieved a purity of more than 5 from *Nostoc* sp. following rivanol treatment, gel filtration chromatography, aminohexyl-sepharose column chromatography, and ammonium sulfate precipitation. Reis et al. (1998) obtained a purity of 5 from *Nostoc* sp. after ammonium sulfate precipitation, gel filtration chromatography, and ion exchange chromatography using Q-sepharose fast flow column. Eighty-five percent recovery of phycoerythrin with a purity of 2.89 after 55 % ammonium sulfate precipitation from *Nostoc muscorum* has been reported (Ranjitha and Kaushik 2005b), while 72 % PE was recovered with a purity of 8.12 following anion exchange chromatography. Tripathi et al. (2007) also reported a purity of 5.25 from *Lyngbya arboricola* using a procedure involving acetone precipitation, gel filtration in addition to ammonium sulfate precipitation, and DEAE-cellulose column chromatography. Eighty percent recovery of PE content with a purity ratio of around 1.5 for young and old cultures from three cyanobacteria, viz., *Phormidium* sp. A27DM, *Lyngbya* sp. A09DM, and *Halomicronema* sp. A32DM, has been reported after the treatment of crude extract with 70 % ammonium sulfate (Parmar et al.2011). Fractionation of this PE-enriched 70 % ammonium sulfate precipitate of the crude extract with Sephadex G-150 further increased the purity ratio of PE to 3.9 and 3.6 for young and old culture of *Phormidium* sp. A27DM, 3.7 and 3.6 for young and old culture of *Lyngbya* sp. A09DM, and 4.0 for both the young and old culture of *Halomicronema* sp. A32DM respectively (Parmar et al.2011). The yield of PE obtained by the described extraction procedure was 62.6 %, 64.9 %, and 66.2 % for young cultures of *Phormidium* sp. A27DM, *Lyngbya* sp. A09DM, and *Halomicronema* sp. A32DM, respectively, and 68.2 %, 74.0 %, and 68.2 % for old cultures of *Phormidium* sp. A27DM, *Lyngbya* sp. A09DM, and *Halomicronema* sp. A32DM, respectively (Parmar et al. 2011). Chakdar and Pabbi (2012) developed a simple protocol for purifying phycoerythrin (PE) from *Anabaena variabilis* (CCC421) involving ammonium sulfate precipitation and dialysis followed by a single-step anion exchange chromatography using DEAE-cellulose-11 and acetate buffer. Precipitation of phycobiliproteins with 65 % ammonium sulfate resulted in 85.81 % recovery of PE with a purity of 2.81, while 62.5 % PE was recovered after chromatographic separation with a purity of 4.95.

4.5 Applications of Phycobiliproteins

Phycobiliproteins (PBPs) as natural colorants are gaining importance over synthetic colors as they are environment friendly, nontoxic, and noncarcinogenic. Dainippon Ink and Chemicals (Sakura, Japan) has developed a product called "Lina blue" (PC extract from *S. platensis*) which is used in chewing gum, ice sherbets, popsicles, candies, soft drinks, dairy products, and wasabi. Despite its lower stability to heat and light, phycocyanin is considered more versatile than gardenia and indigo, showing a bright blue color in jelly gum and coated soft candies (Lone et al. 2005). Besides this, there are number of other companies commercializing different products based on phycocyanins like C-Phycocyain from Cyanotech; PhycoLink® Biotinylated C-Phycocyanin from PROzyme; PhycoPro™ C-Phycocyanin from Europa Bioproducts Ltd.; C-Phycocyanin from Sigma Aldrich, C-Phycocyani from Fisher Scientific; etc. (Chakdar et al.2012). Use of phycobilins in cosmetics like lipstick, eyeliners, etc., is also gaining momentum. Properties like high molar absorbance coefficients, high fluorescence quantum yield, large Stokes shift, high oligomer stability, and high photostability make PBPs particularly PC and PE very powerful and highly sensitive fluorescent reagents. Purified native phycobiliproteins and their subunits fluoresce strongly and have been widely used as external labels for cell sorting and analysis and a wide range of other fluorescence based assays (Glazer and Stryer 1984). A number of commercial products based on C-phycoerythrin are also available in the market like Lightning Link® C-PE from Innova Biosciences Ltd., C-Phycoerythrin and Lightning Link C-PE antibody labeling kit from Novus Biologicals, Monoclonal Anti-mouse-Junctional Adhesion Molecule-C-Phycoerythrin from R & D Systems Inc., C-Phycoerythrin from Hash Biotech Ltd., Biotin Cr-PE (C-Phycoerythrin) from AssayPro, Lightning Link C-Phycoerythrin from Gentaur Molecular Products, and Stretavidin C-Phycoerythrin from Sigma Aldrich. The stabilized phycobilisomes designated PBXL-3 L was accessed as a fluorochrome for flow cytometric immuno-detection of surface antigens on immune cells (Telford et al. 2001). PE is also a very important reagent in proteomics and genomics and form the basis of the detection system in Affymetrix chips (DNA microarrays). Phycoerythrin-labeled streptavidin is added after complete binding and produces a strong signal from array elements containing the biotin-labeled DNA or protein probes (De Rosa et al. 2003). In vivo fluorescence from PC has been used for online monitoring of growth in cyanobacterial cultures (Sode et al. 1991), detection of toxic cyanobacteria in drinking water (Izydorczyk et al. 2005), and remote sensing of cyanobacteria in natural waters (Simis et al. 2005). A large number of patents on fluorescence-based applications of phycobiliproteins reflect their growing demands in industry (Sekar and Chandramohan 2008). The price of phycobiliproteins or its components like phycocyanin may go up to thousands of US $ per milligram depending on their purity and intended use (Sekar and Chandramohan 2008).

A number of investigations have also shown the health-promoting properties and broad range of pharmaceutical applications of phycobiliproteins. The pharmacological properties like antioxidant, anti-inflammatory, and neuroprotective and hepatoprotective activity have been attributed to phycocyanin (Benedetti et al. 2004; Cherng et al. 2007). Phycocyanin has also been reported to reduce the levels of tumor necrosis factor (TNF-α) in the blood serum of mice-treated with endotoxin. Phycocyanin from *Aphanizomenon flos-aquae* has been described as a strong antioxidant (Bhat and Madyastha 2000; Romay et al. 2003), and its protective nature against oxidative damage has also been demonstrated in vitro (Benedetti et al. 2004).

As allophycocyanin (APC) is greater than 100 times more sensitive than conventional organic fluorophores, it is routinely used for flow cytometry, live cell staining, and multicolor immunofluorescent staining. Allophycocyanin (Molecular Probes®) from Invitrogen, PhycoPro™ PB25 Cross-Linked APC and Prozyme xl APC from Prozyme, Lightning-Link® Allophycocyanin (APC) from Innova Biosciences Ltd., Streptavidin cross-linked allophycocyanin conjugated from AnaSpec, etc., are some of the commercially available allophycocyanin-based products.

References

Abalde J, Betancourt L, Torres E et al (1998) Purification and characterization of phycocyanin from the marine cyanobacterium *Synechococcus* sp. IO9201. Plant Sci 136:109–120

Adir N (2005) Elucidation of the molecular structures of components of the phycobilisome: reconstructing a giant. Photosynth Res 85:15–32

Alberte RS, Wood MA, Kursar TA et al (1984) Novel phycoerythrins in marine *Synechococcus* spp. Plant Physiol 75:732–739

Alvey RM, Karty JA, Roos E et al (2003) Lesions in phycoerythrin chromophore biosynthesis in *Fremyella diplosiphon* reveal coordinated light regulation of apoprotein and pigment biosynthetic enzyme gene expression. Plant Cell 15:2448–2463

Anderson LK, Grossman AR (1990) Structure and light-regulated expression of phycoerythrin genes in wild-type and phycobilisome assembly mutants of *Synechocystis* sp. Strain PCC 6701. J Bacteriol 172(3):1297–1305

Anderson LK, Rayner MC, Sweet RM et al (1983) Regulation of *Nostoc* sp. phycobilisome structure by light and temperature. J Bacteriol 155(3):1407–1416

Arciero DM, Bryant DA, Glazer AN (1988) *In vitro* attachment of bilins to apophycocyanin. I. Specific covalent adduct formation at cysteinyl residues involved in phycocyanobilin binding in c-phycocyanin. J Biol Chem 263:18343–18349

Bald D, Kruip J, Rogner M (1996) Supramolecular architecture of cyanobacterial thylakoid membranes: how is the phycobilisomes connected with photosystems? Photosynth Res 49:103–118

Beale SI, Cornejo J (1984) Enzymatic transformation of biliverdin to phycocyanobilin by extracts of the unicellular red alga *Cyanidium caldarium*. Plant Physiol 76:7–15

Beale SI, Cornejo J (1991a) Biosynthesis of phycobilins: 15,16-dihydrobiliverdin IXα is a partially reduced intermediate in the formation of phycobilins from biliverdin IXα. J Biol Chem 266:22341–22345

Beale SI, Cornejo J (1991b) Biosynthesis of phycobilins. Ferredoxin-mediated reduction of biliverdin catalyzed by extracts of *Cyanidium caldarium*. J Biol Chem 266:22328–22332

Beale SI, Weinstein JD (1990) Tetrapyrrole metabolism in photosynthetic organisms. In: Dailey HA (ed) Biosynthesis of heme and chlorophylls. McGraw-Hill, New York, pp 287–391

Benedetti S, Benvenutti F, Pagliarani S, Francogli S et al (2004) Antioxidant properties of a novel Phycocyanin extract from the blue-green alga *Aphanizomenon flos-aquae*. Life Sci 75:2353–2362

Benedetti S, Rinalducci S, Benvenuti F et al (2006) Purification and characterization of phycocyanin from blue-green alga *Aphanizomenon flos-aquae*. J Chromatogr B 833:12–18

Bernard C, Thomas JC, Mazel D et al (1992) Characterization of the genes encoding phycoerythrin in the red alga *Rhodella violacea*: evidence for a splitting of the *rpeB* gene by an intron. Proc Natl Acad Sci U S A 89(20):9564–9568

Bhaskar US, Gopalswamy G, Raghu R (2005) A simple method for efficient extraction and purification of C-Phycocyanin from *Spirulina platensis* Geitler. Ind J Exp Biol 43:277–279

Bhat VB, Madyastha KM (2000) C-phycocyanin: a potent peroxyl radical scavenger *in vivo* and *in vitro*. Biochem Biophys Res Commun 275:20–25

Bhaya D, Scwarz R, Grossman AR (2000) Molecular responses to environmental stress. In: Whitton BA, Potts M (eds) The ecology of cyanobacteria. Kluwer, Dordrecht, pp 523–561

Bogorad L (1975) PBPs and complementary chromatic adaptation. Annu Rev Plant Physiol 26:369–401

Boussiba S (1991) Nitrogen fixing cyanobacteria potential uses. Plant Soil 137:177–180

Boussiba S, Richmond AE (1979) Isolation and characterization of phycocyanins from the blue-green alga *Spirulina platensis*. Arch Microbiol 12:155–159

Brejc K, Ficner R, Huber R et al (1995) Isolation, crystallization, crystal structure analysis and refinement of allophycocyanin from the cyanobacterium *Spirulina platensis* at 2.3 Å resolution. J Mol Biol 249:424–440

Brown SB, Holroyd A, Vernon DI et al (1984) Ferrochelatase activity in the photosynthetic alga *Cyanidium caldarium*. Biochem J 220:861–863

Bryant DA (1981) The photoregulated expression of multiple phycocyanin species: general mechanism and control of phycocyanin synthesis in chromatically adapting cyanobacteria. Eur J Biochem 119:425–429

Bryant DA (1982) Phycoerythrocyanin and phycoerythrin: properties and occurrence in cyanobacteria. J Gen Microbiol 128:835–844

Bryant DA (1988) Genetic analysis of phycobilisomes biosynthesis assembly structure and function in the cyanobacterium *Synechococcus* sp. PCC 7002. In: Stevens SE Jr, Bryant DA (eds) Light energy transduction in photosynthesis: higher plants and bacterial models. American Society of Plant Physiologists, Rockville, pp 62–90

Bryant DA (1991) Cyanobacterial phycobilisomes: progress toward complete structural and functional analysis via molecular genetics. In: Bogorad L, Vasil IL (eds) Cell culture and somatic cell genetics of plants, vol 7B, The photosynthetic apparatus: molecular biology and operation. Academic Press, San Diego, pp 257–300

Bryant DA, Cohen-Bazire G (1981) Effects of chromatic illumination on cyanobacterial phycobilisomes: evidence for the specific induction of a second pair of phycocyanin subunits in *Pseudanabaena* 7409 grown in red light. Eur J Biochem 119:415–424

Bryant DA, Gugliemi G, Tandeau de Marsac N et al (1979) The structure of cyanobacterial phycobilisomes: a model. Arch Microbiol 123:113–127

Capuano V, Mazel D, Tandeau de Marsac N et al (1988) Complete nucleotide sequence of the red light specific set of phycocyanin genes from the cyanobacterium *Calothrix*. Nucleic Acids Res 16:1626–1630

Capuano V, Braux A, Tandeau de Marsac N, Houmard J (1991) The "Anchor Polypeptide" of Cyanobacterial Phycobilisomes: Molecular characterization of the Synechococcus sp. PCC 6301 apcE gene. J Biol Chem 266(11): 7239–7247

Chakdar H (2012) Isolation, purification and regulation of c-phycocyanin from cyanobacterial species. PhD thesis, Indian Agricultural Research Institute, New Delhi

Chakdar H, Pabbi S (2012) Extraction and purification of phycoerythrin from *Anabaena variabilis* (CCC421). Phykos 42(1):25–31

Chakdar H, Jadhav SD, Dhar DW et al (2012) Potential applications of blue green algae. J Sci Ind Res 71(1):13–20

Chaneva G, Furnadzhieva S, Minkova K et al (2007) Effect of light and temperature on the cyanobacterium *Arthronema africanum*: a prospective phycobiliprotein producing strain. J Appl Phycol 19:537–544

Cherng SC, Cheng SN, Tarn A et al (2007) Anti-inflammatory activity of c-phycocyanin in lipopolysaccharide-stimulated RAW 264.7 macrophages. Life Sci 81:1431–1435

Cobley JG, Clark AC, Weerasurya S et al (2002) CpeR is an activator required for expression of the phycoerythrin operon (*cpeBA*) in the cyanobacterium *Fremyella diplosiphon* and is encoded in the phycoerythrin linker-polypeptide operon (*cpeCDESTR*). Mol Microbiol 44:1517–1531

Collier JL, Grossman AR (1992) Chlorosis induced by nutrient deprivation in *Synechococcus* sp. strain PCC 7942: not all bleaching is the same. J Bacteriol 174:4718–4726

de Lorimier R, Bryant DA, Porter RD et al (1984) Genes for α and β phycocyanin. Proc Natl Acad Sci U S A 81:7946–7950

de Lorimier R, Smith RL, Stevens SE Jr (1992) Regulation of phycobilisome structure and gene expression by light intensity. Plant Physiol 98:1003–1010

De Rosa SC, Brenchley JM, Roederer M (2003) Beyond six colours: a new era in flow cytometry. Nat Med 9:112–117

Debreczeny MP, Sauer K, Zhou J et al (1995a) Comparison of calculated and experimentally resolved rate constants for excitation energy transfer in c-phycocyanin: I. Monomers. J Phys Chem 99:8412–8419

Debreczeny MP, Sauer K, Zhou J et al (1995b) Comparison of calculated and experimentally resolved rate constants for excitation energy transfer in C-phycocyanin: II. Trimers. J Phys Chem 99:8420–8431

Desmukh DV, Puranik PR (2012) Statistical evaluation of nutritional components impacting phycocyanin production in *Synechocystis* sp. Braz J Microbiol 43:348–355, http://dx.doi.org/10.1590/S1517-83822012000100041

Doke MJ Jr (2005) An improved and efficient method for the extraction of phycocyanin from *Spirulina* sp. Int J Food Eng 1(5):1–11

Dubbs JM, Bryant DA (1987) Organization of genes encoding phycoerythrin and the two differently expressed phycocyanins in the cyanobacterium *Pseudanabaena* PCC 7409. In: Bigins J (ed) Progress in photosynthesis research, vol IV. Martinus-Nijhoff, Dordrecht, pp 765–768

Dubbs JM, Bryant DA (1993) Organization and transcription of the genes encoding two differentially expressed phycocyanins in the cyanobacterium *Pseudanabaena* sp. PCC 7409. Photosynth Res 36:169–183

Ducret A, Sidler W, Wehrli E et al (1996) Isolation, characterization and electron microscopy analysis of a hemidiscoidal phycobilisomes type from the cyanobacterium *Anabaena sp.* PCC 7120. Eur J Biochem 236:1010–1024

Ducret A, Muller SA, Goldie KN et al (1998) Reconstitution, characterisation and mass analysis of the pentacylindrical allphycocyanin core complex from the cyanobacterium *Anabaena sp.* PCC 7120. J Mol Biol 278:369–388

Eisele LE, Bakhru SH, Liu X et al (2000) Studies on c-phycocyanin from *Cyanidium caldarium*, a eukaryote at the extremes of habitat. Biochim Biophys Acta 1456:99–107

Eriksen NT (2008) Production of phycocyanin—a pigment with applications in biology, biotechnology, foods and medicine. Appl Microbiol Biotechnol 80:1–14

Everroad C, Six C, Partensky F, Thomas JC, Holtzendorff J, Wood AM (2006) Biochemical bases of type IV chromatic adaptation in marine *Synechococcus* spp. J Bacteriol 188:3345–3356

Fairchild CD, Glazer AN (1994a) Nonenzymatic bilin addition to the gamma subunit of an apo-phycoerythrin. J Biol Chem 269:28988–28996

Fairchild CD, Glazer AN (1994b) Oligomeric structure, enzyme kinetics, and substrate specificity of the phycocyanin alpha subunit phycocyanobilin lyase. J Biol Chem 269:8686–8694

Fairchild CD, Zhao J, Zhou J et al (1992) Phycocyanin α-subunit phycocyanobilin lyase. Proc Natl Acad Sci U S A 89:7017–7021

Federspiel NA, Grossman AR (1990) Characterization of the light-regulated operon encoding the phycoerythrin-associated linker proteins from the cyanobacterium *Fremyella diplosiphon*. J Bacteriol 172:4072–4081

Federspiel NA, Scott L (1992) Characterization of a light-regulated gene encoding a new phycoerythrin-associated linker protein from the cyanobacterium *Fremyella diplosiphon*. J Bacteriol 174(18):5994–5998

Gamier F, Dubacq JP, Thomas JT (1994) Evidence for the transient association of new proteins with the *Spirulina maxima* phycobilisome in relation to light intensity. Plant Physiol 160:747–754

Gantt E (1975) Phycobilisomes: light harvesting pigment complexes. BioSci 25:781–788

Gantt E (1980) Structure and function of phycobilisomes: light harvesting pigment complexes in red and blue-green algae. Int Rev Cytol 66:45–80

Gasparich GE, Buzby JS, Bryant DA et al (1987) The effects of light intensity and nitrogen starvation on the phycocyanin promoter in the cyanobacterium *Synechococcus* sp. PCC 7002. In: Biggins J (ed) Progress in photosynthesis research, vol IV. Martinus-Nijhoff, Dordrecht, pp 761–764

Glazer AN (1985) Light harvesting by phycobilisomes. Annu Rev Biophys Biophys Chem 14:47–77

Glazer AN (1986) Phycobilisomes: relationship of structure to energy flow dynamics. In: Youvan DC, Daldal F (eds) Microbial energy transduction. Genetics, structure, and function of membrane proteins. Cold Spring Harbor Laboratory, New York, pp 31–36

Glazer AN (1989) Directional energy transfer in a photosynthetic antenna. J Biol Chem 5:1–4

Glazer AN (1994) Phycobiliproteins—a family of valuable, widely used fluorophores. J Appl Phycol 6:105–112

Glazer AN, Stryer L (1984) Phycofluor probes. Trends Biochem Sci 9:423–427

Gossauer A, Nydeggaer F, Benedikt E et al (1989) Synthesis of bile pigments. Part 16. Synthesis of a vinyl-substituted 2.3-Dihydrobilinedione: possible role of this new class of bile pigments in phycobilin biosynthesis. Helv Chim Acta 72(3):518–529

Grossman AR (2003) A molecular understanding of complementary chromatic adaptation. Photosynth Res 76:207–215

Grossman AR, Lemaux PG, Conley PB et al (1988) Characterization of phycobiliprotein and linker polypeptide genes in *Fremyella diplosiphon* and their regulated expression during complementary chromatic adaptation. Photosynth Res 17:32–56

Grossman AR, Schaefer MR, Chiang GG et al (1993) The phycobilisomes, a light harvesting complex responsive to environmental conditions. Microbiol Rev 57:725–749

Grossman A, Bhaya D, Apt K et al (1995) Light-harvesting complexes in oxygenic photosynthesis: diversity, control and evolution. Annu Rev Genet 29:231–288

Grossman AR, Bhaya D, He Q (2001) Tracking the light environment by cyanobacteria and the dynamic nature of light harvesting. J Biol Chem 276:11449–11452

Gupta A, Sainis JK (2009) Isolation of c-phycocyanin from *Synechococcus* sp. (*Anacystis nidulans* BD1). J Appl Phycol 22(3):231–233

Hardie LP, Balkwill DL, Stevens SE (1983) Effects of iron starvation on the physiology of the cyanobacterium *Agmenellum quadruplicatum*. Appl Environ Microbiol 45:999–1006

Hasegawa PW, Bressan RA, Zhu JK et al (2000) Plant cellular and molecular responses to high salinity. Annu Rev Plant Physiol Plant Mol Biol 51:463–499

Hemalata, Fatma T (2009) Screening of cyanobacteria for phycobiliproteins and effect of different environmental stress on its yield. Bull Environ Contam Toxicol 83:509–515

Hihara Y, Kamei A, Kanehisa M et al (2001) DNA microarray analysis of cyanobacterial gene expression during acclimation to high light. Plant Cell 13:793–806

Hong SJ, Lee CG (2008) Statistical optimization of culture media for production of phycobiliprotein by *Synechocystis* sp. PCC 6701. Biotechnol Bioprocess Eng 13:491–498

Houmard J, Mazel D, Moguet C et al (1986) Organization of nucleotide sequence of genes encoding core components of the phycobilisomes from *Synechococcus* sp., PCC 6301. Mol Gen Genet 205:404–410

Houmard J, Capuano V, Coursin T et al (1988) Genes encoding core components of the phycobilisomes in the cyanobacterium *Calothrix* sp. strain PCC 7601: occurrence of a multigene family. J Bacteriol 170:5512–5521

Houmard J, Capuano V, Colombano MV et al (1990) Molecular characterization of the terminal energy acceptor of cyanobacterial phycobilisomes. Proc Natl Acad Sci U S A 87:2152–2156

Izydorczyk K, Tarczynska M, Jurczak T et al (2005) Measurement of phycocyanin fluorescence as an online early warning system for cyanobacteria in reservoir intake water. Environ Toxicol 20:425–430

Jeffrey SW, Mantoura RFC (1997) Phytoplankton pigments in oceanography: guidelines to modern methods. UNESCO, Paris

Johnson TR, Hyanes JI, Wealand JL et al (1988) Structure and regulation of genes encoding phycocyanin and allophycocyanin from *Anabaena variabilis* ATCC 29413. J Bacteriol 174(4):1858–1865

Kalla SR, Lind LK, Lidholm J et al (1988) Transcriptional organization of the phycocyanin subunit gene clusters of the cyanobacterium *Anacystis nidulans* UTEX 625. J Bacteriol 170:2961–2970

Kalla SR, Lind LK, Gustafsson P (1989) Genetic analysis of phycobilisomes mutants in the cyanobacterium *Synechococcus* sp. PCC6301. Mol Microbiol 3:339–347

Kehoe DM, Gutu A (2006) Responding to color: the regulation of complementary chromatic adaptation. Annu Rev Plant Biol 57:127–150

Kikuchi H, Wako H, Yura K et al (2000) Significance of a two domain structure in subunits of phycobiliproteins revealed by the normal mode analysis. Biophys J 79:1587–1600

Kilpatrick KA (1985) The development of a method for measurement of marine cyanobacterial phycoerythrin extracted in solvents. Dissertation, Texas A M University, Texas

Kufer W, Högner A, Eberlein M et al (1991) Structure and molecular evolution of the gene cluster encoding proteins of the rod substructure of the phycobilisome from the cyanobacterium *Mastigocladus laminosus*. Gene Bank: M75599

Lau PCK, Condie A, Alvarado-Urbina G et al (1987) Nucleotide sequence of phycocyanin β subunit gene of cyanobacterium *Anacystis nidulans* strain R2. Nucleic Acids Res 15:2394–2399

Lawrenz E, Fedewa EJ, Richardson TL (2011) Extraction protocols for the quantification of phycobilins in aqueous phytoplankton extracts. J Appl Phycol 23:865–871

Li L, Kehoe DM (2005) *In vivo* analysis of the roles of conserved aspartate and histidine residues within a complex response regulator. Mol Microbiol 55:1538–1552

Li L, Alvey RM, Bezy RP et al (2008) Inverse transcriptional activities during complementary chromatic adaptation are controlled by the response regulator RcaC binding to red and green light-responsive promoters. Mol Microbiol 68:286–297

Lind LK, Kalla SR, Lonneborg A et al (1987) Organization of phycocyanin gene clusters in *Anacystis nidulans*. Acta Chem Scand B 41:112–115

Liu J, Jiang T, Zhang J et al (1999) Crystal structure of Allophycocyanin from red algae *Porphyra yezoensis* at 2.2 Å resolution. J Biol Chem 274:16945–16952

Liu LN, Chen XN, Zhang YZ et al (2005) Characterization, structure and function of linker polypeptides in phycobilisomes of cyanobacteria and red algae: an overview. Biochim Biophysa Acta 1708:133–142

Lone J, Lene DS, Karsten O et al (2005) Heat and light stability of three natural blue colourants for use in confectionery and beverages. Eur Food Res Technol 220(3–4):261–266

Lonneborg A, Lind LK, Kalla SR et al (1985) Acclimation process in the light harvesting system of the cyanobacterium *Anacystis nidulans* following a light shift from white to red. Plant Physiol 78:110–114

Loos D, Cotlet M, Schryer FD et al (2004) Single-molecule spectroscopy selectively probes donor and acceptor chromophores in the phycobiliprotein allophycocyanin. Biophys J 87:2598–2608

Lu C, Vonshak A (2002) Effects of salinity on photosystem II function in cyanobacterial *Spirulina platensis* cells. Physiol Plant 114:405–413

Lu C, Torzilo G, Vonshak A (1999) Kinetic response of photosystem II photochemistry in cyanobacterium *Spirulina platensis* to high salinity is characterized by two distinct phases. Aust J Plant Physiol 26: 283–292

MacColl R (1998) Cyanobacterial phycobilisomes. J Strl Biol 124:311–334

MacColl R (2004) Allophycocyanin and energy transfer. Biochim Biophys Acta 1657:73–81

MacColl R, Guard-Friar D (1987) Phycobiliproteins. CRC Press, Boca Raton

Marin K, Kanesaki Y, Los DA et al (2004) Gene expression profiling reflects physiological processes in salt acclimation of *Synechocystis* sp. strain PCC 68031. Plant Physiol 136:3290–3300

Mazel D, Marliere P (1989) Adaptive eradication of methionine and cysteine from cyanobacterial light harvesting proteins. Nature 341:245–248

Mazel D, Guglielmi G, Houmard J et al (1986) Green light induces transcription of the phycoerythrin operon in the cyanobacterium *Calothrix* 7601. Nucleic Acids Res 14:8279–8290

Mazel D, Houmard J, Tandeau de Marsac N (1988) A multigene family in *Calothrix* sp. PCC 7601 encodes phycocyanin, the major component of the cyanobacterial light-harvesting antenna. Mol Gen Genet 211(2):296–304

Minkova KM, Tchernov AA, Tchorbadjieva MI et al (2003) Purification of c-phycocyanin from *Spirulina (Arthrospira) fusiformis*. J Biotechnol 102:55–59

Miskiewicz E, Ivanov AG, Huner NPA (2002) Stoichiometry of the photosynthetic apparatus and phycobilisome structure of the cyanobacterium *Plectonema boryanum* UTEX 485 are regulated by both light and temperature. Plant Physiol 130:1414–1425

Muller C, Reuter W, Wehrmeyer W et al (1993) Adaptation of the photosynthetic apparatus of *Anacystis nidulans* to irradiance and CO2-concentration. Bot Acta 106:480–487

Muramatsu M, Hihara Y (2012) Acclimation to high-light conditions in cyanobacteria: from gene expression to physiological responses. J Plant Res 125:11–39

Niu JF, Wang GC, Lin XZ, Zhou BC (2007) Large-scale recovery of C-phycocyanin from *Spirulina platensis* using expanded bed adsorption chromatography. J Chromatogr B 850: 267–276

Nomsawai P, Tandeau de Marsac N, Thomas JC et al (1999) Light regulation of phycocyanin and gene expression in *Spirulina platensis* C1 (*Arthrospira* sp. PCC9438). Plant Cell Physiol 40:1194–1202

Oelmuller R, Conley PB, Federspiel N et al (1988) Changes in accumulation and synthesis of transcripts encoding phycobilisome components during acclimation of *Fremyella diplosiphon* to different light qualities. Plant Physiol 88:1077–1083

Ohki K, Gantt E, Lipschultz CA et al (1985) Constant phycobilisomes size in chromatically adapted cells of the cyanobacterium *Tolypothrix tenuis*, and variation in *Nostoc sp*. Plant Physiol 79:943–948

Olsen GJ, Woese CR, Overbeek R (1994) The winds of (evolutionary) change: breathing new life into microbiology. J Bacteriol 176:1–6

Ong LJ, Glazer AN (1991) Phycoerythrins of marine unicellular cyanobacteria: I Bilin types and locations and energy transfer pathways in *Synechococcus* spp. J Biol Chem 266:9515–9527

Palenik B (2001) Chromatic adaptation in marine *Synechococcus* strains. Appl Environ Microbiol 67:991–994

Parmar A, Singh NK, Kaushal A et al (2011) Purification, characterization and comparison of phycoerythrins from three different marine cyanobacterial cultures. Bioresour Technol 102:1795–1802

Pastore A, Lesk AM (1990) Comparisons of the structures of globins and phycocyanins. Evidence for evolutionary relationship. Proteins 8:133–155

Patil G, Raghavarao KSMS (2007) Aqueous two phase extraction for purification of c-phycocyanin. Biochem Eng J 34:156–164

Patil G, Chethana S, Sridevi AS et al (2006) Method to obtain c-phycocyanin of high purity. J Chromatogr A 1127:76–81

Pizarro SA, Sauer K (2001) Spectroscopic study of the light-harvesting protein C-phycocyanin associated with colorless linker peptides. Photochem Photobiol 73:556–563

Poza-Carrion C, Fernadez-Valiente E, Fernadez-Pinas F, Leganes F (2001) Acclimation of photosynthetic pigments and photosynthesis of the cyanobacterium Nostoc sp. Strain UAM 206 to combined fluctuations of irradiance, pH and inorganic carbon availability. J Plant Physiol 158: 1455–1461

Rafiqul IM, Hassan A, Sulebele G, Orosco CA, Roustaian P, Jalal KCA (2003) Salt stress culture of blue green algae *Spirulina fusiformis*. Pak J Biol Sci 6: 648–650

Ranjitha K, Kaushik BD (2005a) Purification of phycobiliproteins from *Nostoc muscorum*. J Sci Ind Res 64:372–375

Ranjitha K, Kaushik BD (2005b) Influence of environmental factors on accessory pigments of *Nostoc muscorum*. Indian J Microbiol 45:67–69

Reis A, Mendes A, Lobo-Fernandes H et al (1998) Production, extraction and purification of phycobiliproteins from *Nostoc* sp. Bioresour Technol 66:181–187

Reuter W, Muller C (1993) Adaptation of the photosynthetic apparatus of cyanobacteria to light and CO_2. J Photochem Photobiol B Biol 21:3–27

Reuter W, Nickel-Reuter C (1993) Molecular assembly of the phycobilisomes from the cyanobacterium *Mastigocladus laminosus*. J Photochem Photobiol B Biol 18:51–66

Ritchie RJ (1991) Membrane potential and pH control in the cyanobacterium *Synechococcus* R-2 PCC7242. J Plant Physiol 137:409–418

Rito-Palomares M, Nuñez L, Amador D (2001) Practical application of aqueous two-phase systems for the development of a prototype process for c-phycocyanin recovery from *Spirulina maxima*. J Chem Technol Biotechnol 76:1273–1280

Romay CH, Gonzalez R, Ledon N et al (2003) C-phycocyanin: a biliprotein with antioxidant, anti-inflammatory and neuroprotective effects. Curr Protein Pept Sci 4:207–216

Rowan KS (1989) Photosynthetic pigments of algae. Cambridge University Press, New York, pp p166–p211

Rüdiger W, Brandlmeier T, Bios I et al (1980) Isolation of the phytochrome chromophore. The cleavage reaction with hydrogen bromide. Z Naturforsch 35:763–769

Sakamoto T, Bryant DA (1998) Growth at low temperature causes nitrogen limitation in the cyanobacterium *Synechococcus* sp. PCC 7002. Arch Microbiol 169:10–19

Santiago-Santos MC, Noyalo PT, Olvera RR (2004) Extraction and purification of phycocyanin from *Calothrix* spp. Process Biochem 39:2047–2052

Sarada R, Pillai MG, Ravishankar A (1999) Phycocyanin from *Spirulina* sp. influence of processing of biomass an phycocyanin yield, analysis of efficacy of extraction methods and stability studies on phycocyanin. Process Biochem 24:795–801

Schirmer T, Bode W, Huber R et al (1985) X-ray crystallographic structure of the light-harvesting biliprotein c-phycocyanin from the thermophilic cyanobacterium *Mastigocladus laminosus* and its resemblance to globin structures. J Mol Biol 184:257–277

Schluchter WM, Glazer AN (1997) Characterization of cyanobacterial biliverdin reductase. Conversion of biliverdin to bilirubin is important for normal phycobiliprotein biosynthesis. J Biol Chem. 272(21): 13562–13569

Schluchter WM, Glazer AN (1999) Biosynthesis of phycobiliproteins in cyanobacteria. In: Peschek GA, Löffelhardt W, Schmetterer G (eds) The phototrophic prokaryotes. Kluwer/Plenum Press, New York, pp p83–p95

Schubert H, Hagemann M (1990) Salt effects on 77K fluorescence and photosynthesis in the cyanobacterium *Synechocystis* sp. PCC 6803. FEMS Microbiol Lett 71:169–172

Schubert H, Fulda S, Hagemann M (1993) Effects of adaptation to different salt concentrations on photosynthesis and pigmentation of the cyanobacterium *Synechocystis* sp. PCC 6083. J Plant Physiol 142:291–295

Seib LO, Kehoe DM (2002) A turquoise mutant genetically separates expression of genes encoding phycoerythrin and its associated linker peptides. J Bacteriol 184:962–970

Sekar S, Chandramohan M (2008) Phycobiliproteins as a commodity: trends in applied research, patents and commercialization. J Appl Phycol 20:113–136

Shen G, Saunee NA, Gallo E et al (2004) Identification of novel phycobiliprotein lyases in cyanobacteria. In: Niederman RA, Blankenship RE, Frank H et al (eds) Light-harvesting systems workshop. International Society of Photosynthesis Research, Saint Adele, pp 14–15

Sherman DM, Sherman LA (1983) Effect of iron deficiency and iron restoration on ultrastructure of *Anacystis nidulans*. J Bacteriol 156:393–401

Sidler WA (1994) Phycobilisomes and phycobiliprotein structures. In: Bryant DA (ed) The molecular biology of cyanobacteria. Kluwer, Dordrecht, pp p139–p216

Simis SGH, Peters SWM, Gons HJ (2005) Remote sensing of the cyanobacterial pigment phycocyanin in turbid inland water. Limnol Oceanogr 50:237–245

Singh AK, McIntyre LM, Sherman LA (2003) Microarray analysis of the genome-wide response to iron deficiency and iron reconstitution in the cyanobacterium *Synechocystis* sp. PCC 6803. Plant Physiol 132:1825–1839

Singh AK, Elvitigala T, Bhattacharyya-Pakrasi M et al (2008) Integration of carbon and nitrogen metabolism with energy production is crucial to light acclimation in the cyanobacterium *Synechocystis*. Plant Physiol 148:467–478

Sobczyk A, Bely A, Tandeau de Marsac N et al (1994) A phosphorylated DNA-binding protein is specific for the red-light signal during complementary chromatic adaptation in cyanobacteria. Mol Microbiol 13:875–885

Sode KJ, Horikoshi K, Takeyama J et al (1991) On line monitoring of marine cyanobacterial cultivation based on phycocyanin fluorescence. J Biotechnol 21:209–218

Soni B, Kalavadia B, Trivedi U et al (2006) Extraction, purification and characterization of phycocyanin from *Oscillatoria quadripunctulata*—isolated from the rocky shores of Bet-Dwarka, Gujarat, India. Process Biochem 41:2017–2023

Soni B, Trivedi U, Madamwar D (2008) A novel method of single step hydrophobic interaction chromatography for the purification of phycocyanin from *Phormidium fragile* and its characterization for antioxidant property. Bioresour Technol 99:188–194

Stewart DE, Farmer FH (1984) Extraction, identification and quantitation of phycobiliprotein pigments from phototrophic plankton. Limnol Oceanogr 29: 392–397

Stowe WC, Brodie-Kommit J, Stowe-Evans E (2011) Characterization of complementary chromatic adaptation in *Gloeotrichia* UTEX 583 and identification of a transposon-like insertion in the cpeBA operon. Plant Cell Physiol 52(3):553

Swanson RV, de Lorimier R, Glazer AN (1992) Genes encoding the phycobilisomes rod substructures are clustered on the *Anabaena* chromosome. Characterization of the phycoerythrocyanin operons. J Bacteriol 174:2640–2647

Takano H, Arai T, Hirano M et al (1995) Effects of intensity and quality of light on phycocyanin production by a marine cyanobacterium *Synechococcus* sp. NKBG 042902. Appl Microbiol Biotechnol 43:1014–1018

Tandeau de Marsac N (1977) Occurrence and nature of chromatic adaptation in cyanobacteria. J Bacteriol 130:82–91

Tandeau de Marsac N (2003) Phycobiliproteins and phycobilisomes: the early observations. Photosynth Res 76:197–205

Tchernov AA, Minkova KM, Houbavenska NB et al (1999) Purification of phycobiliproteins from *Nostoc* sp. by aminohexyl-sepharose chromatography. J Biotechnol 69:69–73

Telford WG, Moss MW, Moreseman JP et al (2001) Cyanobacterial stabilized phycobilisomes as fluorochromes for extracellular antigen detection by flow cytometry. J Immunol Methods 254:13–30

Toledo G, Palenik B, Brahamsha B (1999) Swimming marine *Synechococcus* strains with widely different photosynthetic pigment ratios form a monophyletic group. Appl Environ Microbiol 65:5247–5251

Tomasseli L, Margheri MC, Saachi A (1995) Effects of light on pigments and photosynthetic activity in a phycoerythrin rich strain of *Spirulina subsalsa*. Aquat Microb Ecol 9:27–31

Tomasseli L, Boldrini G, Margheri MC (1997) Physiological behaviour of *Arthrospira* (*Spirulina*) *maxima* during acclimation to changes in irradiance. J Appl Phycol 9:37–43

Tripathi SN, Kapoor S, Shrivastave A (2007) Extraction and purification of an unusual phycoerythrin in a terrestrial desiccation tolerant cyanobacterium *Lyngbya arboricola*. J Appl Phycol 19:441–447

Troxler RF, Greenwald LS, Zilinskas BA (1980) Allophycocyanin from *Nostoc* sp. phycobilisomes: properties and amino acid sequence at the NH_2 terminus of the α and β subunits of allophycocyanins I, II, and III. J Biol Chem 255:9380–9387

Troxler RF, Lin S, Offner GD (1989) Heme regulates expression of phycobiliprotein photogenes in the unicellular rhodophyte, *Cyanidium caldarium*. J Biol Chem 264:20596–22060

Tu CJ, Shrager J, Burnap RL et al (2004) Consequences of a deletion in *dspA* on transcript accumulation in *Synechocystis* sp. strain PCC6803. J Bacteriol 186:3889–3902

Verma K, Mohanty P (2000) Alterations in the structure of phycobilisomes of the cyanobacterium, *Spirulina platensis* in response to enhanced Na^+ level. World J Microbiol Biotechnol 16:795–798

Vernet M, Mitchell EG, Holm-Hansen O (1990) Adaptation of *Synechococcus in situ* determined by variability in intracellular phycoerythrin-543 at a coastal station of the Southern California coast, USA. Mar Ecol Prog Ser 63:9–16

Viskari PJ, Colyer CL (2003) Rapid extraction of phycobiliproteins from cultured cyanobacteria samples. Anal Biochem 319:263–271

Walsby AE (1982) Cell-water and cell-solute relations. In: Carr NG, Whitton BA (eds) The biology of cyanobacteria. Blackwell Science Publications, Oxford, pp 237–262

Wanner G, Henkelmann G, Schmidt A et al (1986) Nitrogen and sulfur starvation of the cyanobacterium *Synechococcus* 6301. An ultrastructural, morphometrical, and biochemical comparison. Z Naturforsch 41:741–750

Weller JP, Gossauer A (1980) Synthese und Photoisomerisierung des racem. Phytochromobilin-dimethylesters. Chem Ber 113:1603–1611

Wilbanks SM, Glazer AN (1993) Rod structure of a phycoerythrin II containing phycobilisome. I. Organization and sequence of the gene cluster encoding the major phycobiliprotein rod components in the genome of marine *Synechococcus* sp. WH8020. J Biol Chem 268:1226–1235

Wiltshire KH, Boersma M, Moller A et al (2000) Extraction of pigments and fatty acids from the green alga *Scenedesmus obliquus* (Chlorophyceae). Aquat Ecol 34:119–126

Woese CR (1987) Bacterial evolution. Microbiol Rev 51:221–271

Wu SH, Lagarias JC (2000) Defining the bilin lyase domain: lessons from the extended phytochrome superfamily. Biochemistry 39:13487–13495

Wyman M (1992) An *in vivo* method for the estimation of phycoerythrin concentration in marine cyanobacteria (*Synechococcus* spp.). Limnol Oceanogr 37:1300–1306

Yamanaka G, Glazer AN (1980) Dynamic aspects of phycobilisome structure. Phycobilisome turnover during nitrogen starvation in Synechococcus sp. Arch Microbiol 124:39–47

Yan S, Zhu L, Su H, Zhang X et al (2011) Single step chromatography for simultaneous purification of c-phycocyanin and allophycocyanin with high purity and recovery from *Spirulina* (*Arthrospira*) *platensis*. J Appl Phycol 23:1–6

Zhang YM, Chen F (1999) A simple method for efficient separation and purification of c-phycocyanin and allophycocyanin from *Spirulina platensis*. Biotechnol Techniques 13:601–603

Zhao KH, Wu D, Wang L et al (2002) Characterization of phycoviolobilin phycoerythrocyanin- α-84-cystein-lyase-(isomerizing) from *Mastigocladus laminosus*. Eur J Biochem 269:4542–4550

Zhao KH, Su P, Böhm S et al (2005) Reconstitution of phycobilisome core-membrane linker, Lcm, by autocatalytic chromophore binding to ApcE. Biochim Biophys Acta 1706:81–87

Zhao KH, Su P, Tu JM et al (2007) Phycobilin: cystein-84 biliprotein lyase, a near-universal lyase for cysteine-84-binding sites in cyanobacterial phycobiliproteins. Proc Natl Acad Sci U S A 104: 14300–14305

Zhou J, Gasparich GE, Stirewalt VL et al (1992) The *cpcE* and *cpcF* genes of *Synechococcus sp.* PCC 7002. Construction and phenotypic characterization of interposon mutants. J Biol Chem 267: 16138–16145

Zilinskas BA (1982) Isolation and characterization of the central component of the phycobilisome core of *Nostoc* sp. Plant Physiol 70:1060–1065

Zilinskas BA, Greenwald LS (1986) Phycobilisome structure and function. Photosynth Res 10:7–36

Good Manufacturing Practices (GMP): "Planning for Quality and Control in Microbiology"

5

Upendra Krishnalal Jani

Abstract

Good manufacturing practices with comprehensive design and proper implementation of the quality system infrastructure helps manufacturer to achieve quality objectives. Quality assurance covers various aspects of manufacturing and quality control process which ensures that products are manufactured consistently as per the quality standards. Quality assurance integrates cGMP and quality control along with environmental monitoring and occupational health and safety hazards. To implement quality management system effectively, manufacturer must have instructions, processes, and adequate resources relevant to the need of product. Monitoring and controlling the QMS not only measures the effectiveness, but helps for the continuous improvement. cGMP gives consumers a confidence that products are meeting required standards for release and safe for intended use. Compliance with basic requirements of cGMP elements ensures/assures that products have achieved quality attributes during manufacturing cycle. Quality assurance is the responsibility of all personnel engaged for the manufacturing and control within the organization.

Keywords

Good Manufacturing Practices • Quality Assurance • Quality Control • Environmental Monitoring Quality Management System • Proactive Compliance Approach

5.1 Introduction

During aseptic manufacturing of biologics, sound process, equipment, and proper facility design is used to minimize/eliminate contaminants. Monitoring such important means and infrastructure, one can measure the effectiveness of

U.K. Jani (✉)
Hester Biosciences Limited,
Ahmedabad, Gujarat, India
e-mail: upendra.jani@hester.in

P. Shukla (ed.), *Frontier Discoveries and Innovations in Interdisciplinary Microbiology*,
DOI 10.1007/978-81-322-2610-9_5

contamination control. With trend analysis, identification of threats to the purity of the product can be taken care to build quality products. cGMP is a set of scientifically proven sound methods, practices, or principles that are implemented and documented during research and development of the product and during commercial manufacturing. cGMP regulations establish minimum requirements for methods, controls, and protocols. Its implementation during manufacturing process religiously assures product quality attributes for their intended use when marketed.

5.2 Basics of cGMP

5.2.1 Quality Assurance

QA usually (a) reviews and approves all standard operating procedures (*SOPs*) related to production, quality control, and maintenance; (b) reviews batch manufacturing and test records; (c) performs self-audits; (d) performs annual product quality reviews (*APQR*); (e) evaluates trend analysis; and (f) ensures system control, consistency, and validation. QA ensures that products are made in accordance with manufacturing standards and meets their predetermined specifications (World Health Organization 2013a). Good manufacturing practices (GMP) is based on science- and risk-based approaches (World Health Organization 2010, 2011; Lynn 2013). Quality assurance along with quality control is a part of the broader concept of quality management. QA is often explained by the "plan, do, check, act" (PDCA) cycle.

5.2.2 Production and Process Controls

Written standard operating procedures for manufacturing and quality control shall be followed and documented while performing the activity. Any change/variation from these procedures shall be recorded and justified with proper explanation. Maintenance is coordinated for the preventive/planned maintenance. Calibrations and validations are followed as per the validation master plan (*VMP*) for the critical equipment, systems, and processes (Process Validation 2012; U.S. Department of Health and Human Services et al. 2011).

5.2.3 Quality Control

QC usually (a) assesses that incoming raw materials, containers, closures, labels, packing materials, in-process materials, and the finished products are suitable for use; (b) evaluates process performance against set standards and limits; and (c) determines acceptance criteria of batch prior to distribution. Quality control is a reactive process which ensures the quality of materials and products, checks for and detects failures, and identifies and corrects the defects. Both QC and production areas should be independent of each other. Dedicated areas required for chemical, biological, and microbiological analysis/testing. Separate instrument room with adequate area with sophisticated instruments should be provided for analysis. Enough storage space with controls is a must to store test samples, retained samples, laboratory reagents, and documents. According to WHO, quality control is the sum of all procedures undertaken to ensure the identity and purity of a finished product. QC ensures the safety and efficacy of a product before it is released for marketing.

5.2.4 Vendor Audit

Audit vendors on a regular basis for the consistent supply of the quality raw materials and packaging materials. The purpose of this audit is also to determine whether the supplier is capable to supply quality material with consistency. It is the responsibility and authority of the quality assurance department to define the relationship and expectations between organization and its supplier for the continuous improvement (Wingate 2014).

5.2.5 Buildings and Facilities

The location, design, and construction should facilitate proper cleaning, maintenance, and operations for the product being manufactured. Adequate space is required for the man-material movement avoiding mix-ups and cross-contamination. The utilities like water, steam, gas, compressed air, and systems such as HVAC should be maintained, qualified, and validated as per the schedule before use (Witcher et al. 2012). Adequate lighting and work environment should be provided. Proper drainage and sewage treatment plant should be provided to meet the regulatory norms. The written procedures for proper sanitation of the facility with sanitation program, where responsibility, schedule, methods, equipment, etc. are mentioned, are a must.

5.2.6 In-Process Testing

One must have written procedures for (a) testing of product while being manufactured and (b) monitoring aseptic environment to assure batch sterility, uniformity, and integrity.

5.2.7 Laboratory Controls

For each batch produced, QC tests the product for the compliance with the finished product specifications as guided by the regulatory requirements before release (Sandle 2010). A final release approval is given by quality assurance for the distribution in the market.

5.2.8 Expiry Date

Results of the stability testing of the product at defined time, temperature conditions mentioned in the pharmacopoeias, determine the expiry date. It assures that product meets applicable standards as per the label claim.

5.2.9 Packaging and Labeling

Correct and released labels and packaging materials should be used as specified in the written procedure. Reconciliation of labels and packaging material is a must. It helps to control mix-ups and recalls.

5.2.10 Stability Testing

A well-designed "stability testing program" shall be followed to determine storage conditions and expiry date (CFR 211.22). Trending stability test results help to monitor/control acceptance criteria and consistency of production. It's a key element and essential part for new drug development program in R&D. Stability test results help to decide storage conditions and shelf life.

5.2.11 Production Record Review

Batch manufacturing records and QC test records shall be reviewed and approved against established standards/acceptance criteria of master formula record (*MFR*) for the compliance before release/distribution in the market.

5.2.12 Deviation Investigations

Sometimes certain unplanned departures from QA written specifications may occur. If such deviations do not affect the quality and safety of the product adversely and one decides to accept such material/product under deviation, a written procedure should be established. This procedure is applicable to all raw materials, packaging materials, and semifinished and finished products. No deviation is permitted from pharmacopoeial and statutory requirements. Some examples of deviations include: production, EHS, quality improvement, audit, customer service, technical, material complaint, and system routing deviations (US Food and Drug Administration 2006).

5.2.13 Complaints

A written procedure for handling of all written and oral complaints is required. Complaints are handled through proper route – from field to manufacturer via quality assurance department. Quality control checks/tests their retained samples with respective batch. Results are conveyed to the complainant in the form of report with findings/justification. Reports are retained for 2 years after expiry of the said batch.

5.2.14 Records and Reports

Documentation of various records and reports is essential for the review and trend analysis. The review of validation reports, out of specification (*OOS*) results, FDA inspection report, and deviation reports is a must. Customer feedbacks shall be analyzed for quality improvement. Robust effective "recall" procedure is required.

5.2.15 Self-Inspection/Self-Audit

The company shall carry out self-inspection with the help of experienced team members within the organization on a regular time interval. Members with the adequate resources and authority should focus on globally harmonized regulatory standards (Harmonized Tripartite Guideline 2008, 2009). Nonconformities are discussed among the senior officials, and corrective/preventive actions are taken. The audits shall be conducted periodically, the frequency and the depth of audit being influenced by factors such as urgency and focus at a given point and a given situation. Audit checklist shall be drawn by quality assurance, in consultation with, but not necessarily under directions from, persons responsible for specific activities which the audit intends to cover.

5.2.16 Training

The deficiency-related training includes new employee orientation, process redesigns, SOP revision, and technical training (Welty 2009). The company shall train all the personnel (production, quality control, and maintenance) as per the "annual training plan." Training updates the knowledge of the employees regarding cGMP concepts and revisions/amendments. Training need is identified and adequate resources either internal or external should be provided. There are two basic needs for training in pharmaceutical companies:

- Extend quality training to all departments
- Perform specific departmental training in each functional area

5.3 Environmental Monitoring

A *clean room* is a controlled environment where products are manufactured aseptically (International Organization for Standardization 1999). It has a controlled level of contamination specified by the numbers of particles per cubic meter with a defined particle size. It must be designed, qualified, and operated according to the international regulatory standards. The layouts/plans with personnel and material flows, air handling systems (HVAC), and utilities (water and gas system) are approved by FDA prior to the commercial manufacturing (WHO 2012). *Contaminants* are generated from *five* basic sources which include: people, facilities, tools, fluids, and product being manufactured. *Contamination* is a process that causes materials or surfaces to be soiled with contaminants. Elimination of such airborne contamination using advanced equipments and technologies required to maintain aseptic environment. People are the major source of contamination in the clean room. Contamination can lead to expensive downtime with increased manufacturing costs. The personnel engaged in manufacturing in aseptic environment should be trained for "clean room behaviors." Environmental monitoring should be conducted using "risk-based approach" (World Health Organization 2013b; Commission Europeenne 2004). The risk associated with cleaning, disinfection, and change over validation and personal health/hygiene activities is assessed to comply legal requirements mentioned in the cGMP guidelines. A thorough and unbiased investigation which rules out impact on quality, purity, or safety of the product is essential. *W. Edward Deming* has rightly said, "If you can't describe what you are doing as a process, you don't know what you are doing." The precisely monitored and audited "clean room management program" assures that documented procedures and defined protocols are well understood and effectively implemented at all levels within aseptic manufacturing process.

5.4 Quality Management System

Implementing an effective quality management system allows biological manufacturers to meet their ethical and regulatory obligations (Landerville 2014). Adopting a proactive approach to quality management is essential to achieve the step change in quality performance. The critical utilities, equipment, processes, and test methods all need to be validated and controlled appropriately as per validation master plan. Recently, current GMP norms have integrated quality system and risk management approaches. The objective behind such integration was to encourage the industry to adopt modern and innovative technologies. However, after the major revision in GMP regulations (1978), people observed advancements in the manufacturing science, technology, and understanding regarding quality systems. The ultimate aim of the cGMP regulation is to provide flexibility. It depends upon the type of business; however, they are very similar and share essential elements.

5.5 The Concepts of Modern QMS

The concepts of modern QMS focus on the quality product. Quality in totality comprises established strength, identity, and purity. The manufacturer must develop the product in such a manner that the finished product with quality attributes is produced with consistency.

5.5.1 Quality by Design (QbD)

QbD provides sound framework for transferring the technology from development stage in R&D to scale-up level for the commercial production. This knowledge and process understanding helps to resolve post-development changes and optimization, if any.

5.5.2 Quality Risk Management

It is an essential part of QMS. It helps to establish specifications, process parameters, and acceptance criteria. Assessment helps to mitigate the changes to process or specifications and determines the corrective and preventive actions after successful investigations (Harmonized Tripartite Guideline 2005).

5.5.3 Risk Management

Quality is the science of identifying and controlling variations. Risk management is the future of quality management. It is an ongoing cyclical process. It should enable control or elimination of significant risks as well as the identification of any new risks and processes. Risk management aims to be proactive approach once embedded in an organization. Risk reduction measures will be preventive actions rather than corrections or corrective actions.

5.5.4 CAPA

It is an integral part of QMS with other system-like deviation management, handling of out of specification (OOS), and change control procedure. The successfully implemented CAPA procedure helps to handle failures, nonconformances, and deviations.

Some advantages are:

- Prevents recurrence if applied effectively.
- Provides a structured plan to address identified issues.

- Provides for continuous improvement and effective use may result in proactive actions being taken.
- Provides a record.

Some disadvantages are:

- Retrospective for correction and corrective actions.
- Not a standalone tool.
- Training and knowledge are required to apply effectively and understand the differences between correction and corrective and preventive action.
- Requires established standards and controls for a baseline to be set.

5.5.5 Change Control

It is another cGMP tool, which focuses on how to manage and control unintended issues during aseptic manufacturing of biologics. The changes that alter specifications, or critical attributes, require regulatory approvals. It creates an environment which encourages continual improvement following change control.

5.5.6 Six-System Inspection Model

It is a system-based approach, where quality system is integrated with production, facilities and equipment, laboratory controls, materials, and packaging and labeling to help manufacturer to keep systems under control to meet the regulatory compliances.

5.6 Proactive Compliance Approach

It was rightly said by *John Ruskin*, "Quality is never an accident; it is always the result of intelligent effort." A collective challenge facing the industry is to achieve *proactive compliance.* This involves effective management and control of the manufacturing environment to avoid problems rather than just responding to problems after they have happened. The fundamentals of quality and compliance must never be compromised. Today's modern businesses are becoming more proactive and less reactive. Quality assurance involves taking a proactive approach to ensure drug products are made in accordance with manufacturing standards and met their predefined product specifications (Wright 2012a). The aim is for quality and compliance to be achieved *right the first time* rather than depend on detecting problems. The aim is to continually improve manufacturing standards, eliminating errors along the way. A holistic approach to quality assurance is needed. The internal control framework needs to cover governance, systems, and processes, as well as distinct activities that encourage supportive mindset and organizational behavior. The holistic approach to quality assurance needs to promote transparency in support of performance improvement. An open and trusting relationship must be maintained so that production problems are raised as they occur for rapid resolution. A learning culture needs to replace a *mistakes-are-punished* or a *someone-is-to-blame* approach to quality issues (Wright 2012b). Quality can be better managed when it is recognized and understood that the control of variability and prevention of waste are imperative to achieve a cost-effective business. Ideally, we strive to keep quality, cost, and supply in harmony, but when we need to prioritize, it is only possible to achieve two, and quality must always be preserved. Quality management when structured with quality assurance using cost analysis as a business driver reaps the cost benefits of a proactive approach. Shared beliefs, values, attitudes, and behavior patterns are pieces of the jigsaw that must come together. The energy and motivation for quality comes from the top. The management must acknowledge the challenge of change in their organizations and stay vigilant. A culture of quality will empower teams to continually improve and solve the problems (Gallant 2014). We must remember that the person at the end of our supply chain is depending on us to provide safe and effective products. Better real-time data presents more meaningful information, contributing to a better knowledge and understanding of the process, an increase in product quality, and a safer product. Improved yields, cost savings, and increase profits will be the result. A knowledge update on clean rooms, associated air management elements, and other accessories is the need of the era.

References

CFR 211.22(a), Responsibilities of quality control unit, Sub part B – organization and personnel, Ch. 1, vol 4, pp 144

Commission Europeenne (2004) Good manufacturing practice, annex 1: manufacture of sterile medicinal products. Bruxelles, bio contamination control (ISO 14698). International Organization for Standardization, Geneva

Gallant J (2014) The road to a culture of quality [Article]. Pharmaceutical Online, 19 Sept 2014

Guideline on Process Validation, 29 Mar 2012 EMA/CHMP/CVMP/QP/70278, Rev 1

ICH Harmonized Tripartite Guideline (2005) Quality risk management, Q9, Current step 4 version dated 9 Nov 2005

ICH Harmonized Tripartite Guideline (2008) Pharmaceutical quality system, Q10, Current step 4 version dated 4 June 2008

ICH Harmonized Tripartite Guideline (2009) Pharmaceutical development Q8 (R2), Current step 4 version dated Aug 2009

International Organization for Standardization (1999) Clean rooms and associated controlled environments (ISO 14644). International Organization for Standardization, Geneva

Landerville T (2014) Quality management systems guide. Isocert Solutions, Santa Rosa

Lynn S (2013) Product manufacturing and manufacturing practices (CGMPs). USFDA, 13 Oct 2013

Sandle T (2010) Best practices in microbiology laboratory training. In: Hodges N, Hanlon G (eds) Industrial pharmaceutical microbiology: standards and controls. Euromed, Basingstoke, pp S11.1–S11.16, Supplement 11

U.S. Department of Health and Human Services, Food and Drug Administration, Center for Drug Evaluation and Research (CDER), Center for Biologics Evaluation and Research (CBER), Center for Veterinary Medicines (CVM) (2011) Guidance for industry, process validation: general principles and practices. Current Good Manufacturing Practices (CGMP) revision 1, Jan 2011

US Food and Drug Administration (2006) Guidance for an industry: quality systems approach to pharmaceutical CGMP regulations. US Department of Health and Human Services, Rockville

Welty G (2009) Developing a continuing CGMP training program. J GXP Compliance 13(4):86–96

WHO (2012) Environmental monitoring of clean rooms in vaccine manufacturing facilities- points to consider for manufacturers of human vaccines. WHO, Geneva

Wingate G (2014) Moving from quality control to quality assurance. Pharm Eng 34:2

Witcher MF et al (2012) Biopharmaceutical manufacturing in the twenty-first century: the next generation manufacturing facility. Pharm Eng 32(2):10–22

World Health Organization (2010) Good manufacturing practices for sterile pharmaceutical preparations. In: WHO expert committee on specifications for pharmaceuticals preparations. Forty-fourth reports, vol 957, WHO technical report series. World Health Organization, Geneva, Annex 4

World Health Organization (2011) WHO good manufacturing practices: main principles for pharmaceutical products. In: WHO expert committee on specifications for pharmaceutical preparations. Forty-fifth report, vol 961, WHO technical report series. World Health Organization, Geneva, Annex 3

World Health Organization (2013a) Quality assurance of pharmaceuticals WHO guidelines, related guidance and GxP training materials. World Health Organization, Geneva

World Health Organization (2013b) WHO guidelines on quality risk management. In: WHO expert committee on specifications for pharmaceutical preparations. Forty-seventh report, vol 981, WHO technical report series. World Health Organization, Geneva, Annex 2

Wright R (2012a) How to take charge of your manufacturing process [Magazine Article]. Pharmaceutical Online, 29 Nov 2012

Wright R (2012b) How to create an innovative culture in a pharma company [Magazine Article]. Pharmaceutical Online, 1 June 2012

Advances in Molecular Mechanism Toward Understanding Plant-Microbe Interaction: A Study of *M. oryzae* Versus Rice

6

Jahangir Imam, Nimai Prasad Mandal, Mukund Variar, and Pratyoosh Shukla

Abstract

Rice blast, caused by the fungus *Magnaporthe oryzae*, is the most devastating disease of rice which causes considerable economic loss worldwide. The interaction between rice and *M. oryzae* is an important model system for studying host-pathogen interactions. Since genomes of both species are sequenced, research is more focused by exploiting modern genetics, genomics, proteomics and bioinformatics. Recent research on functional genomics and candidate gene identification has helped to elucidate the role of resistance (*R*) and avirulence (*Avr*) genes and their interactions. Over the years, many avirulence genes (Avr) involved in pathogenicity and resistance genes (R) of rice involved in pathogen signaling molecules and defence response have been identified and characterized. Recent development in cloning of *Avr* genes (13) and *R* genes (22) and identification of many quantitative trait loci (QTL) has improved our understanding of rice-*M. oryzae* interaction at molecular level. This chapter is focusses on the current *R* and *Avr* genes cloned and characterized their structure, function and co-evolution, and the future research directions to study and understand the molecular mechanism of rice-*M. oryzae* interactions for better targeting and exploitation of host plant resistance towards management of this disease.

J. Imam
Biotechnology Laboratory, Central Rainfed Upland Rice Research Station (CRRI), Hazaribagh 825301, Jharkhand, India

Enzyme Technology and Protein Bioinformatics Laboratory, Department of Microbiology, Maharshi Dayanand University, Rohtak 124001, Haryana, India

N.P. Mandal • M. Variar
Biotechnology Laboratory, Central Rainfed Upland Rice Research Station (CRRI), Hazaribagh 825301, Jharkhand, India

P. Shukla (✉)
Enzyme Technology and Protein Bioinformatics Laboratory, Department of Microbiology, Maharshi Dayanand University, Rohtak 124001, Haryana, India
e-mail: pratyoosh.shukla@gmail.com

P. Shukla (ed.), *Frontier Discoveries and Innovations in Interdisciplinary Microbiology*,
DOI 10.1007/978-81-322-2610-9_6

Keywords

R genes • *Avr* genes • Blast disease • Gene interaction • *Magnaporthe oryzae* • Rice

6.1 Introduction

Rice feed almost half of the world and is the most staple food having major contribution toward world food security. During the past four decades, green revolution has brought manyfold increase in rice production through hybrid rice and semi-dwarf cultivar cultivation (Khush and Jena 2009; Liu et al. 2010) and high-input crop culture. These accentuated several biotic and abiotic stresses in different production systems, rice blast being one such major constraint depressing yield. Rice blast causes loss to the crop depending upon the rice production zones, physical environment, crop management, and pathogen population dynamics; the loss of the crop varies depending on the above mentioned factors (Variar et al. 2009). During monsoons followed by cold season the crop in South India, the bunds, plateau and uplands rice growing areas are the most severely infected sub-systems. Traditional varieties are more affected by blast than the high-yielding varieties as in the case of the traditional Basmati types grown in the North. In the present scenario, unavailability of suitable blast-resistant variety adds to the increase in disease incidence and epidemics. Use of high doses of nitrogen for Basmati and hybrid rice leads to severe blast infection and loss up to 50 %. According to International Rice Research Institute, rice blast is estimated to cause economic losses of up to USD 60 million annually in South and Southeast Asia (Babujee and Gnanamanickam 2000). In Japan, frequent breakdown of resistance to blast results in significant yield losses (20–100 %). The loss of resistance of cultivars against blast fungus is because of the instability of *Magnaporthe oryzae* genome and rapid development of virulent strains when challenged with new resistant variety.

Chemical control of blast disease is the most effective and yet most undesirable as it possess serious environmental threats. Therefore, to overcome yield losses due to this disease, resistant cultivars have emerged as the best substitute (Babujee and Gnanamanickam 2000). Currently, blast occurs in more than 80 countries, especially in Asia where more than 90 % of the world's rice are grown and have the highest consumption. This necessitates continuous efforts to develop broad-spectrum resistant varieties against prevalent races in quick succession. This is possible through strong and combined efforts of conventional and molecular breeding for developing *R* genes exploiting resistance governed by QTLs.

6.2 Development of Near-Isogenic Lines (NILs) of Rice for Blast Resistance

Resistance to blast disease is an important objective for most rice breeding programs. During 1990s, the genetics and molecular studies of resistance were complicated tasks because of the variability of the blast pathogen and unavailability of rice genotypes with single resistance genes (Mackill and Bonman 1992). Near-isogenic lines (NILs), monogenic to resistance (*R*) genes, have proved to be a proficient method to illustrate the pathogenicity of blast isolates when used as differentials and as a source of resistance for breeding programs. A set of monogenic and NILs also offers great advantages for allelism test in NILs which has become much easier compared to donor cultivars where resistance is often conferred by two or more major genes. Mackill and Bonman (1992) developed a set of NILs in the background of CO39, a highly susceptible and short-duration cultivar (about 95 days to maturity), by backcrossing method. Initially, Yamada et al. (1976) and Kiyosawa et al. (1981) selected 12 differential varieties carrying known *R* genes in Japan. But the genetic background of each variety was not uniform and also contained other unknown *R* genes. Under the Japan-IRRI collaborative program, two sets of monogenic differentials (NILs), one in the background of a highly susceptible Chinese cultivar Lijiangxintuanheigu (LTH) and another in the background of a universal susceptible variety (US-2), were developed targeting 24 blast-resistance genes (Fukuta et al. 2004). Backcross breeding method was employed to transfer the genes singly in the recurrent parent, LTH. In both, LTH and US-2, no resistance gene has been identified so far (Tsunematsu et al. 2000). CO39 genetic background NILs, a representative of *indica*, were also developed through a recurrent cycle of backcrossing (BC_6F_{10}) for 23 resistance genes except *Pia* resistance gene as CO39 harbors *Pia* gene. This was the first *indica-type* differential variety set for a large number of resistance genes (Tsunematsu et al. 2000; Kobayashi et al. 2007). US-2 can be cultivated easily under both tropical and temperate condition unlike LTH and CO39. Therefore, US-2 is now the suitable choice for the development of NILs for blast resistance.

The near-isogenic lines for many blast-resistance genes in different genetic backgrounds have taken a major part in the rice breeding program for rice resistance variety development against blast. NILs has been proved to be instrumental in studying the rice-*M. oryzae* interaction at molecular level by using qRT-PCR, cDNA microarray, whole-genome transcriptional analysis, etc. Microarray-based transcriptomic analysis of early response genes to *M. oryzae* on NILs, IRBL18 (carrying *Pi1*), and IRBL22 (carrying *Pi9*) was studied which deciphered the incompatible interaction. The results clearly showed the upregulation and downregulation of some important genes which leads to disease response and identified OsWRKYs as an important regulator in rice blast resistance (Wei et al. 2013). Similarly, the rice cultivar Zhong156 carrying *Pi-24(t)* blast-resistance gene was used to characterize the differentially expressed genes when inoculated with virulent and avirulent *M. oryzae* isolates (Haiyan et al. 2012). The gene expression profiles can provide powerful information for the identification of functionally related new genes. The exploitation of NILs resulted in the rapid accumulation of microarray data which has allowed the creation of co-expression network to study rice-*M. oryzae* interaction and their gene expression analysis at different experimental conditions (Cao et al. 2012).

6.3 Rice Blast-Resistance (*R*) Genes and Quantitative Trait Loci (QTLs)

The rice-*M oryzae* interaction follows the classical gene-for-gene interaction. The product of *R* gene interacts with the effector of *Avr* gene (Flor 1971). The two ways to classify resistance i.e qualitative (complete) and quantitative (partial) depends on the way the gene(s) affect pathogen reproduction. The strong selection by the rapid evolution of the *M. oryzae* fungus leads to the breakdown of qualitative resistance in rice (Kou and Wang 2010). The partial-resistance gene is race nonspecific, incomplete, and usually polygenic and is characterized by a moderate infection type as suggested by the genetic and molecular evidences. Multiple genes or quantitative trait loci (QTLs) mainly control the partial resistance against blast fungus (Fukuoka and Okuno 2001). But when a major gene is effective against a part of the pathogen population, it will also appear as partial resistance. Till now, 35 QTLs have been documented, mainly *indica* and *japonica* rice cultivars.

6.3.1 Dynamics of Blast-Resistance (R) Genes

Most of the blast resistance encodes nucleotide-binding sites-leucine-rich repeats (NBS-LRR) proteins as evidenced by several researchers, except *Pi-d2* (Chen et al. 2006). The phylogenetic analysis of the 22 cloned NBS-LRR resistance proteins were performed (Fig. 6.1). The cloning and phylogenetic analysis of NBS-LRR regions of *R* genes clearly revealed that *Pi2*, *Piz-t* , and *Pi9* are present at the same locus, indeed

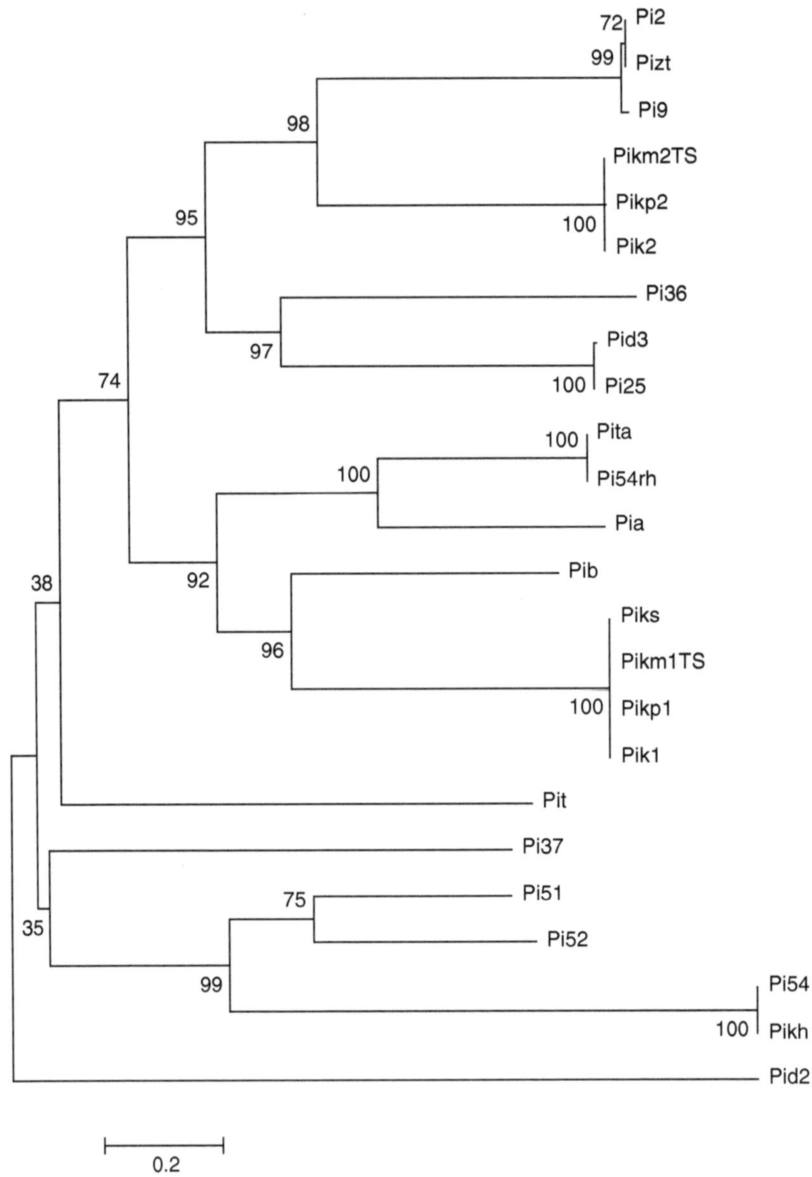

Fig. 6.1 Phylogenetic tree of selected cloned blast *R* genes. This tree was constructed using MEGA4 using NBS-LRR amino acid sequences collected from the National Center for Biotechnology Information (NCBI)

allelic to each other and clustered in the same clade. This is obvious and expected as the three genes are tightly linked to each other (Zhou et al. 2006, 2007). The *Pi2*/*Pi9* locus, other than tightly linked *Piz-t*, also harbors *Piz*, *Pigm(t)*, *Pi26*, and *Pi40* genes (Liu et al. 2002; Hayashi et al. 2004; Deng et al. 2006; Jeung et al. 2007). The cloning and sequence analysis of *Pi2*/*Piz-t* revealed that both these genes are true alleles to each other and have eight amino acids that are different and showed similar expression pattern (Zhou et al. 2006). Expression study results for the three genes (*Pi2*, *Piz-t*, and *Pi9*) confirmed that their expressions are constitutive and not induced by blast infection. One striking point is that the three *Pi2*, *Piz-t*, and *Pi9* genes are identified in different species. *Pi2* was introgressed from *indica* cultivar 5173, *Piz-t* was from *indica* rice cultivar TKM1, and *Pi9* from *Oryza minuta*, a tetraploid wild species of *Oryza* genus (Mackill and Bonman 1992; Nagai et al. 1970; Amante-Bordeos et al. 1992). The clustering of these related NBS-LRR genes at a single locus may result in the generation of orthologs and paralogs with new *R* alleles of more resistance and specificities which may be able to combat the blast disease.

The clusters of *Pik*, *Pikp*, *Pikh*, *Piks*, and *Pikm* alleles are allelic and tightly linked blast-resistance genes (Xu et al. 2008; Kiyosawa 1972; McCouch et al. 1994). The phylogenetic analysis showed that *Pikh* gene is far from *Pikp*, *Piks*, and *Pikm* alleles which are clustered in two clades near to each other. Two adjacent NBS-LRR class genes are essential for complete resistance to blast for *Pik*, *Pikp*, and *Pikm*. The *Pikm* locus has two NBS-LRR genes: *Pikm1-TS* and *Pikm2-TS*. Complementary test showed that neither of the transgenic plants carrying either of the NBS-LRR genes confers *Pikm* resistance specificity (Costanzo and Jia 2010). The evolutionary results reveal that *Pikp* allele found in wild relatives of rice can be considered as the ancestral allele and the other *Pik* alleles have evolved after domestication of rice (RoyChowdhury et al. 2012).

Pi5 gene similar to *Pikm* requires two candidate NBS-LRR genes *Pi5-1* and *Pi5-2* for conferring resistance (Lee et al. 2009). The presence of unique C-terminus and more number of introns (four and five, respectively) makes this Pi5 (Pi5-1 and Pi5-2) proteins distinct from other R proteins (Chen et al. 2006). These distinct properties of *Pi5-1* and *Pi5-2* genes put them in a single clade different from other NBS-LRR genes. The molecular basis of these two R proteins is still unclear and the expression analysis work indicates that the *Pi5-1* transcript level is high during pathogen attack, in contrast *Pi5-2* gene is expressed constitutively (Jeon et al. 2003). The *Pi37* gene of rice cultivar St. No. 1 exhibits partial and complete resistance to Japanese and Chinese *M. oryzae* isolates (Ezuka et al. 1969; Yunoki et al. 1970; Chen et al. 2005). A 1290-amino acid long NBS-LRR protein product is encoded by *Pi37* gene, and its resistance phenotype is because of the presence of V239A and 1247M in NBS region (Lin et al. 2007).

Pita, *Pi36*, and *Pid2* are single-copy *R* genes in which resistance specificity is determined by single amino acid difference. Both *Pita* and *Pi36* are NBS-LRR-encoding *R* genes but *Pid2* is the only non-NBS-LRR-encoding *R* gene, which encodes a receptor-like kinase. All cultivars carrying *Pita-2* also possess *Pita* gene and are mostly inseparable. The reason for this can be the recombination suppression observed near the centromere region (Bryan et al. 2000). The *indica* rice variety Kasalath carries the gene *Pi36*, conferring resistance to Chinese *M. oryzae* isolates, and is located on chromosome 8. The *Pi36* gene encodes a 1056-amino acid NBS-LRR protein. A single amino acid substitution event (Asp to Ser) at position 590 of *Pi36* controls expression of resistance or susceptibility. *Pi36* is more closely related to the barley powdery mildew resistance genes *Mla1* and *Mla6* and is constitutively expressed similar to other blast-resistance genes (Liu et al. 2007). *Pi-d2* is the only non-NBS-LRR-encoding *R* gene (Chen et al. 2006). A single amino acid difference at position 441 of *Pi-d2* determines the expression of resistance or susceptibility. Due to the presence of novel extracellular domain, *Pi-d2* symbolizes a new class of plant resistance genes (Chen et al. 2006).

The allele mining of *Pid3* reveals an ortholog *Pid3-A4* in the common wild rice A4 (*Oryza rufipogon*) (Q et al. 2013). The *Pi25* was mapped on chromosome 6 at the same locus of *Pi2*/*Pi9*, which is the locus of resistance genes. The *Pi25* confers neck blast resistance to Chinese isolate 92-183 (race ZC15). Since the locus has cluster of genes, it may be possible that presence of two or more genes helps in conferring neck blast resistance (Jian-li et al. 2004). The phylogenetic analysis revealed the similarity between the *Pid3* and *Pi25* genes and thus occupied the same clade.

The protein produced by *Pi54rh* gene is extracellularly localized and may play a role in signal transduction process. This is in contrast to other blast-resistance genes that are predicted to code for intracellular NBS-LRR-type resistance proteins. Pi54rh protein has a pathogen-inducible promoter and hence upregulated after pathogen infection (Das et al. 2012). The structural relatedness of *Pi54rh* is clearly analyzed in phylogenetic tree. Both *Pita* and *Pi54rh* are put in a same clade and this shows their evolutionary relationship.

The *Pi54* gene has been cloned from *indica* rice cultivar, Tetep, on chromosome 11. It is slightly away from the *Pik* locus, and its sequence similarity clearly reveals that it is very similar to *Pikh* gene (Sharma et al. 2005a, 2010). In addition to NBS-LRR domain, the Pi54 protein contains a small zinc finger domain. Against geographically diverse strains of *M. oryzae* from different parts of India, the *Pi54* gene showed a stable and high level of resistance (Sharma et al. 2005a; Rai et al. 2011). After infection, in both resistant and susceptible plants, there is constitutive basal level expression of *Pi54* gene. Its expression is induced by *M. oryzae* infection, and a more than twofold higher induction was observed at 72 h of post-inoculation in the resistant plant (Rai et al. 2011).

Pia and *Pish* are weak *R* genes which are brought forward the indigenous Japanese cultivars. However, *Pia* consists of a cluster of four NBS-LRR genes, located on the short arm of rice chromosome 11. Out of four, only *Pia-3* is functional and is the better candidate gene of *Pia*. The *Pia* shows a narrow-spectrum resistance to blast isolates (Kiyosawa 1974; Hittalmani et al. 2000; Yamasaki and Kiyosawa 1996; Fukuta et al. 2007). Similarly, *Pish* is a cluster of NBS-LRR genes, arranged as tandem repeats, and is highly similar to each other. An *in silico* analysis reveals that the *Pish* locus is hot spot for Tos17 insertions. The tos17 inserts are most frequent in the functional genes within hot spot regions (Takahashi et al. 2010). Both *Pia* and *Pish* are weaker genes and if present alone, they are not able to confer resistance. These *R* genes could have important roles if used in combination with other *R* genes (Kiyosawa 1974; Hittalmani et al. 2000; Liu et al. 2003).

6.3.2 Cloning, Characterization and Molecular Basis of Blast QTLs

Quantitative resistance often called as "field resistance" is controlled by multiple genes or quantitative trait loci (QTL). The field resistance consists of cumulative effects of multiple QTLs, with each QTL providing a partial level of resistance. Resistance offered collectively by QTLs is characterized by its durability and broad spectrum for fungal pathogen (Kou and Wang 2010), since the QTLs have low breakdown chances and QTL resistance can remain effective for longer period of time in a cultivar, even though the environment is favorable for blast disease spread (Krattinger et al. 2009). Therefore, there is an urgent need of breeding program focused on QTLs or "field resistance." Although the partial resistance provided by QTLs is thought to be general, there are reports characterizing it as specific partial resistance (Bonman et al. 1989; Fujii et al. 1995; Fukuoka and Okuno 1997; Lopez-Gerena 2006; Roumen 1992; Talukder et al. 2004). Ballini et al. 2008 stressed the need for meta-analysis approach for the identification of QTLs. A total of 49 metaQTLs as partial-resistance QTLs is identified by this approach. Assembling these 49 metaQTLs provides an opportunity to test the specificity of these QTLs as well as co-localization with the candidate genes (Ballini et al. 2008).

Many QTLs for blast have been identified by means of the facilitation of molecular markers (Table 6.1). Till now, two major blast-resistance QTLs *pi21* and *Pb1* have been cloned by map-based cloning strategy, which is the best choice to clone these minor QTLs (Hayashi and Yoshida 2009; Hayashi et al. 2010). Many of these QTLs have small effects, lees that 10 % on the phenotypic resistance against blast.

The major blast-resistance QTL, *pi21*, was mapped on chromosome 4 as a single recessive gene and confers durable resistance against blast fungus (Fukuoka and Okuno 2001). A proline-rich protein is encoded by this gene which is made up of a putative protein-protein interaction motif and putative heavy metal-binding domain.

Table 6.1 The characteristics of the cloned blast-resistance genes and quantitative trait loci (QTLs) in rice

S. No.	R gene	Chr. No.	Cognate Avr gene	Encoding protein	Expression	Cloning strategy	References
1	*Pib*	2	NA	NBS-LRR	Circadian, inducible by stress	Map based	Wang et al. (1999)
2	*Pita*	12	Yes	NBS-LRR	Constitutive	Map based	Bryan et al. (2000)
3	*Pi9*	6	NA	NBS-LRR	Constitutive	Map based	Qu et al. (2006)
4	*Pi2*	6	NA	NBS-LRR	Constitutive	Map based	Zhou et al. (2006)
5	*Piz-t*	6	Yes	NBS-LRR	Constitutive	Map based	Zhou et al. (2006)
6	*Pid2*	6	NA	Receptor kinase	Constitutive	Map based	Chen et al. (2006)
7	*Pi36*	8	NA	CC-NBS-LRR	Constitutive	Map based	Liu et al. (2007)
8	*Pi37*	1	NA	NBS-LRR	Constitutive	Map based in silico	Lin et al. (2007)
9	*Pikm*	11	Yes	NBS-LRR	Constitutive	Map based	Ashikawa et al. (2008)
10	*Pit*	1	NA	CC-NBS-LRR	Transcriptionally inactive	Map based	Hayashi and Yoshida (2009)
11	*Pi5*	9	NA	CC-NBS-LRR	Pi5-1 is pathogen dependent Pi5-2 is constitutive	Map based	Lee et al. (2009)
12	*Pid3*	6	NA	NBS-LRR	Constitutive	In silico homology based	Shang et al. (2009)
13	*pi21*[a]	4	NA	Proline-containing protein	Inducible by stress	Map based	Fukuoka et al. (2012)
14	*Pb1*[a]	11	NA	CC-NBS-LRR	Transcriptionally inactive	Map based	Hayashi et al. (2010)
15	*Pish*	1		CC-NBS-LRR	Constitutive	Mutant screening	Takahashi et al. (2010)
16	*Pik*	11	Yes	CC-NBS-LRR	Constitutive	Map based	Zhai et al. (2011)
17	*Pikp*	11	Yes	CC-NBS-LRR	Constitutive	Map based in silico	Yuan et al. (2011)
18	*Pi54* (*Pikh*)	11	NA	NBS-LRR	Constitutive	Map based	Sharma et al. (2005b)
19	*Pia*	11	Yes	CC-NBS-LRR	Constitutive	Multifaceted genomics approach	Okuyama et al. (2011)
20	*NLS1*	11	NA	CC-NBS-LRR	Constitutive	Map based	Tang et al. (2011)
21	*Pi25*	12	NA	NBS-LRR	Constitutive	Map based	Chen et al. (2011)
22	*Pi54rh*	11	NA	CC-NBS-LRR	Constitutive	Allele mining approach	Das et al. (2012)

NA not available, *NBS-LRR* nucleotide-binding site and leucine-rich repeats, *CC-NBS-LRR* coiled-coil nucleotide-binding site and leucine-rich repeats

[a]Cloned blast-resistance QTLs

It is tightly linked to genes of poor eating quality and thus, there is a strong association between blast resistance and undesirable eating character of rice (Fukuoka et al. 2012). The durability of a plant resistance may be contributed by the delayed induction of defense. The *pi21* triggers defense mechanism leading to pathogen recognition by activating different signaling pathway even though there is no direct gene-to-gene interaction (Jones and Dangl 2006).

The *Pb1* comes under a field-resistance gene because it is durable and shows quantitative resistance against *M. oryzae* (Hayashi et al. 2010). *Pb1* gene encodes a CC-NBS-LRR protein. However, Pb1 protein differs from other R protein in its NBS domain, in which the P-loop which is conserved in *R* genes is apparently absent along with degeneration of and some other motifs. The structure of the Pb1 also plays a crucial role in its slow and delayed action, commonly termed as transcriptionally inactive or "sleeping" response. The promoter sequence at 5′ position and gene duplication at the 60 kb locus are very critical for the activation of *Pb1* gene and its expression (Hayashi et al. 2010). Cultivars with *Pb1* show a characteristic "adult resistance" (Fujii and Hayano-Saito 2007).

The molecular basis of blast-resistance QTLs still remains poorly understood. Till now, only two QTLs, *pi21* and *Pb1*, have been cloned and characterized. The co-localization of blast-resistance QTLs confers resistance against races of *M. oryzae*. The broad-spectrum quantitative resistance mediates in two ways, either by functioning in overlapping signaling pathways or by regulation of the defense pathways.

6.4 Avirulence Genes

The gene-for-gene hypothesis proposed by Flor, while describing the relationship between *R* gene and *Avr* gene, states that for every *R* gene in the host plant, there is a complementary *Avr* gene in the pathogen, and their interaction leads to activation of host defense response which leads to plant resistance (Flor 1971). This type of *R-Avr* interaction is well documented in rice-Magnaporthe *oryzae*. Avirulence (*Avr*) genes encode molecules during normal growth and pathogenicity of the pathogen. The host range of pathogens is mainly determined by *Avr* genes, which determines the capability of *Avr* gene carrier to cause disease based on *R* genes. The natural population of *M. oryzae* has the common and interesting phenomenon of modification or shedding of avirulence genes which gives them the advantage of shifts in the host range (Sharma et al. 2012).

So far, more than 40 *Avr* genes of *M. oryzae* have been identified, out of which cloning and characterization of 12 are accomplished (Table 6.2), and two, *Avr-Pi15* and *Avr-Pi7*, have been mapped (Ma et al. 2006; Shujie et al. 2007). Out of 12, map-based cloning approach has been used to clone nine *Avr* genes while genome-wide association mapping is used for three. The first *Avr* gene cloned was *PWL2*, the member of *PWL* gene family consisting of four genes, *PWL2*, *PWL1*, *PWL3*, and *PWL4*. The *PWL1*, *PWL3*, and *PWL4* are the homologs of *PWL2* isolated from *M. grisea*. The amino acid sequences of PWL2 protein is 75, 51, and 57 % identical to predicted PWL1, PWL3, and PWL4 proteins (Kang et al. 1995; Sweigard et al. 1995).

Multiple translocation across the genome is contributed by retrotransposon which flanked the *Avr-Pita* gene, while *Avr-Pita3* lacks avirulence activity, is stable, and is located on chromosome 7 in most of the isolates (Chuma et al. 2011). The *AVR-Pita* avirulence function is easily abolished by the point mutation in the protease consensus sequence (Khang et al. 2008; Orbach et al. 2000). Pot3 transposon insertion in *Avr-Pita* inactivates the *Avr-Pita* interaction, which leads to the loss of rice blast resistance in *Pita* containing cultivars (Zhou et al. 2007). *Avr-Pita* is a race-specific elicitor that triggers *Pita*-mediated resistance since the *Avr-Pita* protein directly binds to the *Pita* leucine-rich domain (LRD) region inside the plant cell to initiate a *Pita*-mediated defense response (Jia et al. 2000).

The *AVR* gene *Avirulence Conferring Enzyme1* (*ACE1*) has been identified by the cross between Guy11 and ML25 *M. grisea* isolates (Silue et al. 1992). Map-based cloning strategy has been used to isolate *ACE1* gene and it encodes a putative hybrid protein which makes it a unique *Avr* gene (Bohnert et al. 2004). *ACE1* is expressed specifically during penetration without being secreted out of the cell. It is a large protein of 4035 amino acids and the fungus could not be recognized by the plant with a single amino acid change and this suggests that virulence biosynthetic activity of *ACE1* is important (Bohnert et al. 2004).

The *Avr1-CO39* gene controls avirulence to the rice cultivar CO39. It is mapped on chromo-

Table 6.2 Summary of the corresponding *R* gene and predicted product of cloned *Avr* gene *Magnaporthe oryzae*

S. No.	Avr gene	Corresponding R gene	Encoding protein	Cloning strategy	References
1	*PWL1*	NA	Glycine-rich, hydrophilic protein, secreted protein	Map based	Kang et al. (1995)
2	*PWL2*	NA	Glycine-rich, hydrophilic protein, secreted protein	Map based	Sweigard et al. (1995)
3	*PWL3*	NA	Glycine-rich, hydrophilic protein, secreted protein	Map based	Sweigard et al. (1995)
4	*PWL4*	NA	Glycine-rich, hydrophilic protein, secreted protein	Map based	Sweigard et al. (1995)
5	*Avr1-CO39*	*Pi-CO39*	Secreted protein	Map based	Farman and Leong (1998)
6	*Avr-Pita*	*Pita*	Secreted protein	Map based	Orbach et al. (2000)
7	*ACE1*	*Pi33*	Polyketide synthase/ peptide synthase	Map based	Bohnert et al. (2004)
8	*Avr-Piz-t*	*Piz-t*	Secreted protein	Map based	Li et al. (2009)
9	*Avr-Pia*	*Pia*	Secreted protein	Association genetics approach	Yoshida et al. (2009)
10	*Avr-Pii*	*Pii*	Secreted protein	Association genetics approach	Yoshida et al. (2009)
11	*Avr-Pik/Km/Kp*	*Pik/Km/Kp*	Secreted protein	Association genetics approach	Yoshida et al. (2009)
12	*Avr-Pi54*	*Pi54*	Secreted protein	Map based	

some 1 and cloned by chromosome walking from linked RFLP markers (Farman and Leong 1998). The *Avr1-CO39* gene came from an *M. grisea* isolate from weeping love grass and was not present in the rice-infecting *M. grisea* isolates. Molecular analysis at the *Avr1-CO39* locus revealed the presence of truncated left and right border in rice-infecting *M. grisea* isolates and this is why most rice-infecting isolates are virulent on CO39 (Farman et al. 1998; Farman et al. 2002).

Recently, a map-based cloning strategy has been utilized to clone *Avr-Piz-t* gene from an avirulent field isolate 81278ZB15 (Li et al. 2009). A 108-amino acid protein is synthesized by *Avr-Piz-t* gene and the encoded protein elicits resistance reaction in rice mediated by *Piz-t*. *Avr-Piz-t* helps in the suppression of pathogen-associated molecular pattern (PAMP)-triggered immunity (Park et al. 2012). No homolog of this Avr-Piz-t protein is found in *M. oryzae* or in other sequenced fungi. The abundance of active transposable elements surrounding the *Avr-Piz-t* genomic sequence determines the avirulence and also provides an ability to rapidly change and thus overcome the *R* genes (Li et al. 2012).

Avr-Pii and *Avr-pik* have developed paralogs while *Avr-Pia* does not and deletion of functional paralogs of *Avr-pii* and *Avr-Pik* seems to be common mechanism for *M. oryzae* isolates to overcome *Pii* and *Pik* genes (Yoshida et al. 2009). Phylogenetic analysis showed that out of five alleles of *Avr-Pik* (*Avr-Pik-A*, *Avr-Pik-B*, *Avr-Pik-C* , *Avr-Pik-D*, and *Avr-Pik-E*), only the *Avr-Pik-D* allele is most likely the ancestral allele (Kanzaki et al. 2012).

The avirulent gene, *Avr-Pi54*, has been identified and cloned by whole-genome sequencing technique from *M. oryzae* strain 70–15 and from 37.878 Mb genome, 474 *Avr* candidate genes have been identified. *Avr-Pi54-1* and *Avr-Pi54-2* have been identified having potential of being the true candidate gene.

Recently, avirulence gene *AvrPi9* has been identified by pulldown assays and comparative proteomics approach by comparing avirulent and virulent isolates. Four putative AvrPi9 candidates were identified and it was confirmed by

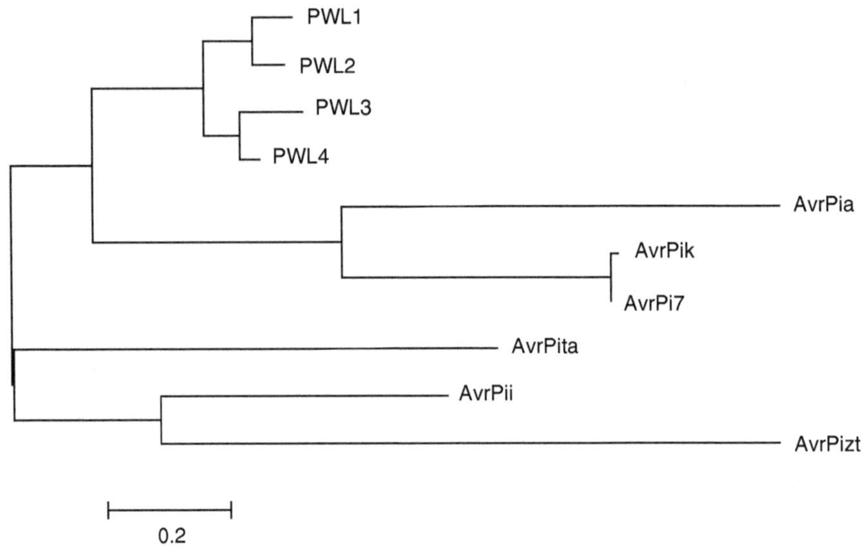

Fig. 6.2 Phylogenetic tree of selected cloned and mapped *Avr* genes. This tree was constructed using MEGA4 using coding region (amino acid sequence)

complementation and allele swapping that *R55* is the *AvrPi9*. R55 encodes a predicted secreted protein in avirulent strains. Insertion of Mg-SINE into the exon of *R55* or loss of *R55* caused the gain of virulence toward rice cultivar containing *Pi9*. Preliminary studies suggested that *R55* is expressed only during plant growth and localizes to BIC and host cells (Kou et al. 2013).

The phylogenetic analysis showed that *PWL1* and *PWL2* genes are clustered together while *PWL3* and *PWL4* are clustered in another clade near to each other. Similarly, the amino acid sequence of *Avr-Pik* and *Avr-Pi7* showed similarity and thus included in the same clade. *Avr-Pita* is altogether in a separate clade, while *Avr-Pii* and *Avr-Piz-t* have been included in the same clade as it showed similarity in their amino acid sequence (Fig. 6.2).

6.5 Role of Downstream Genes During Rice-*Magnaporthe oryzae* Interaction

The rice-*Magnaporthe oryzae* interaction has been studied for decades. Fungal spores germinate during 4–6 h after landing on the leaf surface depending on the availability of moisture. At this point, different downstream signaling pathways trigger which contributes to resistance and/or susceptibility (Jones and Dangl 2006; Dodds and Rathjen 2010; Nimchuk et al. 2003). The phenylpropanoid biosynthesis pathway is very important in this process. Downregulation of phenylpropanoid biosynthesis genes leads to plant resistance to fungal pathogens (Naoumkina et al. 2010; Maher et al. 1994; Kruger et al. 2002; Bhuiyan et al. 2009). Wei et al. (2013) found that the upregulation of enzymes like many enzymes occurs in the phenylpropanoid and phenylalanine biosynthesis pathways, which reflects the transcriptional reprogramming in metabolism during rice-*Magnaporthe oryzae* interaction. The identification of the genes governing these downstream responses is imperative for the information of pathway. With the help of Yeast two-hybrid system (YTH) has been employed to identify an Avr-Pita interacting protein, named AV13.

6.6 Coevolution of *R* and *Avr* Genes

Avirulence loss to *R* genes is because of the highly diverse genetic structure of the blast pathogen over the world (Skamnioti and Gurr

2009). The short life of many newly released cultivars results from the generation of new virulent races or pathotypes (Tharreau et al. 2009; Liu et al. 2010). New virulent races knock down major *R* genes due to multiple genetic mutation events, like partial and complete deletion, point mutations, frame shift mutation, and transposon insertions (Liu et al. 2010). In *Ace1*, *Avr-Pita*, and *Avr-Piz-t*, avirulence factor is lost due to the insertion of a retrotransposon hence activating virulence genes in the isolates (Zhou et al. 2007; Li et al. 2009; Fudal et al. 2005). The phylogenetic analysis of *Avr-Pik* alleles suggests that *Avr-Pik-D* is the ancestral allele. Surprisingly, most of the *Pik* alleles recognize *AVR-Pik-D* allele (Kanzaki et al. 2012). The other *Avr-Pik* alleles evolve with one or few changes in the amino acid and that is how it evades the recognition by the *Pik* genes. The finding suggests that the observed variation patterns between the rice *Pik* gene and *M. oryzae Avr-Pik* genes are the results of coevolution between them. *Pita* engages in "trench warfare" with *AVR-Pita1* with both genes maintaining an array of strategies to prevent/cause disease. Positive selection pressure is behind the evolution of *AVR-Pita1* in nature. Inactivation of *Pita*-mediated resistance in commercial rice cultivars is majorly because of the variation at the *AVR-Pita1* locus (Dai et al. 2010).

The *R* genes are predicted to encode highly similar NBS-LRR proteins and predicted transposon-mediated evolution. The *Pit* and *Pita* genes were found to be activated by a transposon Pot3 in the promoter region and hence play a positive role in regulating blast *R* genes (Hayashi and Yoshida 2009; Lee et al. 2009). Genetic crosses and screening have identified a new locus, *Ptr(t)*, a critical gene that is required for both *Pita* and *Pita-2* gene. The absence of *Ptr(t)* resulted in disease susceptibility suggesting its critical roles as either upstream or downstream of *Pita-*/*Pita-2*-mediated disease resistance (Jia and Martin 2008). The synthesis of varied size functional proteins describes the importance of posttranslational modification processes like exon skipping and alternative splicing. Thus, both transposon and posttranslational modification processes play an important role in the evolution of a blast *R* gene. Recent studies on the rice germplasm and *M. oryzae* isolate collected from North East and Eastern India also demonstrate that the rice germplams harbor many *R* gene and the *M. oryzae* strain also possesses many corresponding *Avr* genes. The generation of virulent strains harboring the corresponding Avr gene clearly gives an impression that both *R* and *Avr* genes are evolving, comparatively *Avr* genes more frequently than *R* genes (Imam et al. 2013; 2014a, b, c, 2015).

6.7 Rice Blast-Resistance Gene Allele Mining

The improvement of high-yielding varieties of rice through plant breeding is mainly due to transfer of beneficial alleles found in the plant germplasm (Kumar et al. 2010). Natural mutation is the main reason behind the generation and evolution of new alleles. The evolution of new alleles occurs due to natural mutations like transition, transversion, point mutation, and insertion and deletion (InDels). Today, enormous sequence database information of rice genome is available in public databases. With the help of these databases, we can identify and retrieve the desired and superior alleles that are agronomically important. Allele mining approach has potential applications in rice improvement programs (Kumar et al. 2010). It helps in identification of new haplotypes and in tracing the evolution of alleles.

Two major approaches, TILLING (Targeting Induced Local Lesions in Genomes) and sequence-based allele mining, are available for the identification of sequence polymorphisms for a given gene in the naturally occurring populations. Till et al. (2003) discovered TILLING is for target gene-induced mutation (Till et al. 2003). TILLING has proven to be a practical, an efficient, and an effective approach for functional genomics studies in rice. For the identification of natural variation in selected genes in rice, a variation of TILLING technique, EcoTILLING, has been coined (Kumar et al. 2010). The presence of SNPs and InDels helps in analyzing individuals for haplotype structure and diversity. Sequence-based allele mining involves huge cost in sequencing. The cost and sophistication involved in these techniques lead to the introduction of a new allele mining approach called as next generation sequencing technologies in allele mining.

Initially, allele mining from wild rice accessions was initiated for the *R* genes like *Pita* + alleles for *Oryza rufipogon* (Yang et al. 2007; Geng et al. 2008), *Pita* from *O. rufipogon* (Huang et al. 2008), *Pid3* from wild rice accessions of *O. rufipogon* (Shang et al. 2009), *Pi54* from wild rice, and *Pi54rh* from *O. rhizomatis* (Das et al. 2012). Most of these genes were allele mined through sequence-based allele mining approach. Recently, allele mining of blast-resistance genes like *Pita*, *Pikh*, *Piz-t*, and *Pi54* has also been reported in cultivated rice (Sharma et al. 2009; Kumari et al. 2013; Kumar et al. 2010).

6.8 Rice Blast Molecular Breeding

Marker-assisted selection (MAS) is one of the most important and popular method for rice improvement and variety development. The deployment of single blast-resistance gene into the genetic background of popular variety is always subjected to breakdown of resistance because of the frequent variation in the fungal genome. Accessibility of different molecular markers like simple sequence repeats (SSR), restriction fragment length polymorphism (RFLP), single nucleotide polymorphism (SNP), and PCR-based allele-specific markers has fastened the procedure for gene introgression and gene pyramiding (Hayashi et al. 2006; Latif et al. 2011). Conventional rice breeding is a slow process for variety development and mostly influenced by environmental conditions. Conventional breeding requires repeated backcross (up to BC5 or BC6) to introduce genes into elite cultivars but also results in linkage drag (Brinkman and Frey 1977). MAS offers better selection strategies in rice breeding to minimize the linkage drag and also requires shorter period of time. The cross between *indica* and *japonica* rice variety linkage blocks (LB) is a common event, where a large set of genes were inherited together (Jia et al. 2012). MAS is more efficient, effective, and reliable than phenotype selection (Miah et al. 2013). Blast-resistance genes like *Pi1*, *Pi5*, *Piz-5*, *Pi54*, *Pita*, *Pi9*, *Pi2*, *Pid1*, and *Pib* (Hittalmani et al. 2000; Gouda et al. 2012; Singh et al. 2012; Narayanan et al. 2002; Wen and Gao 2011; Chen et al. 2004) have been introgressed in various elite rice genotypes using MAS.

6.9 Rice Innate Immunity and Signaling

Rice-*M. oryzae* interaction follows Flor's "gene-for-gene" hypothesis (Hogenhout et al. 2009). The fungal effector protein acts in two ways. First, by the recognition and activation of effector-triggered immunity (ETI) which results in hypersensitive response (HR) that halts fungal growth within 48 hours. Second, a limited pathogen-associated molecular patterns (PAMP)-triggered immunity (PTI) response in case that the fungal effectors are not recognized set in upon the recognition of fungal PAMPs (Kaku et al. 2006; Shimizu et al. 2010) and during this intermittent period of fungal hyphae continue to spread in plant tissues, resulting in disease symptoms in the initial few after infection (Kankanala et al. 2007; Chen and Ronald 2011). Thus, for the final outcome of the level of rice blast resistance, the early defense response immediately after *M. oryzae* invasion is critical (Xiang et al. 2008). In most cases, rapid production of reactive oxygen species (ROS) is triggered in defense responses along with hypersensitivity response (HR) in the infected cells (Hammond-Kosack and Jones 1996). Systemic-acquired resistance (SAR) is established in response to accumulation and also related to pathogenesis-related (PR) gene induction (Ryals et al. 1996; Dempsey et al. 1999; Durrant and Dong 2004). Defense response genes are those genes which function downstream of *R*- or HPRR-initiated defense signaling pathways. Out of 12 characterized genes contrib-

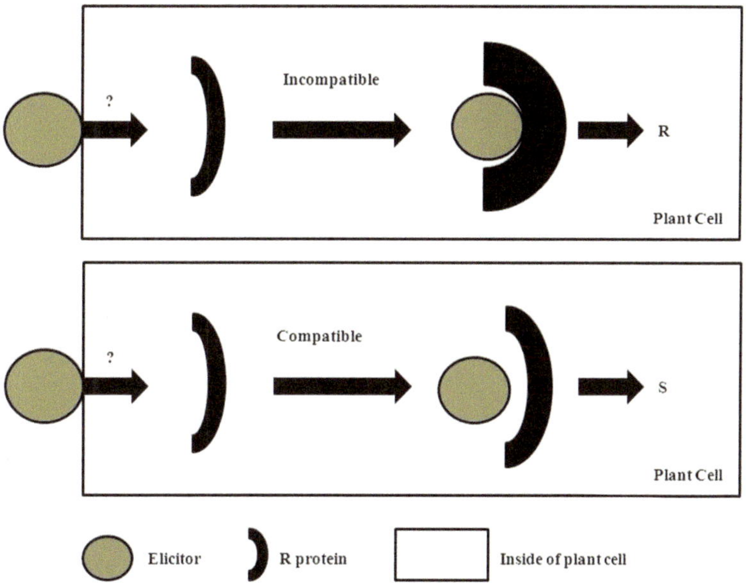

Fig. 6.3 Molecular mechanism of blast *R* gene-mediated responses in incompatible and compatible interactions. *R* indicates for resistance and *S* indicates for susceptible

uting to QTL resistance, 9 of them belong to the defense-responsive gene class which mediates *R* gene race-specific resistance (Fig. 6.3) (Kou and Wang 2010).

6.10 Future Perspectives

Development of resistant cultivars against blast and research on *R* and *Avr* gene interaction in order to combat the rapid changes in the virulence pattern of blast pathogen are continuous processes. Although significant progress has been made in this area in recent years, new challenges lie ahead which are listed below.

- Development of durable and broad-spectrum disease-resistant cultivars by transferring multiple gene combinations based on knowledge on the prevailing pathogen population.
- For the deployment of *R* genes, study of interaction of *R* and *Avr* genes is one of the most important and daunting tasks. Currently, yeast two-hybrid system (Y2H) and bimolecular fluorescence complementation (BiFC) analyses are employed to detect the interaction between *R* and *Avr* genes.
- Identification of downstream resistance genes by using new proteomic approaches will be also an important step in deciphering rice-*M. oryzae* interaction.
- In the absence of a circulating antibody system in plants similar to animals, efforts are to be made in understanding innate resistance mechanism in order to develop innate immunity in plants.

References

Amante-Bordeos A, Stich LA, Nelson R, Damacio RD, Oliva NP, Aswidinoor H, Leung H (1992) Transfer of bacterial blight and blast resistance from tetraploid wild rice *Oryza minuta* to cultivated rice, *Oryza sativa*. Theor Appl Genet 84:345–354

Ashikawa I, Hayashi N, Yamane H, Kanamori H, Wu J, Matsumoto T, Ono K, Yano M (2008) Two adjacent nucleotide-binding site-leucine-rich repeat class genes are required to confer *Pikm*-specific rice blast resistance. Genetics 180:2267–2276

Babujee L, Gnanamanickam SS (2000) Molecular tools for characterization of rice blast pathogen (*Magnaporthe grisea*) population and molecular marker-assisted breeding for disease resistance. Curr Sci 78:248–257

Ballini E, Morel JB, Droc G, Price A, Courtois B, Notteghem JL, Tharreau D (2008) A genome-wide meta-analysis of rice blast resistance genes and quantitative trait loci provides new insights into partial and complete resistance. Mol Plant Microbe Interact 21:859–868

Bhuiyan NH, Selvaraj G, Wei YD, King J (2009) Gene expression profiling and silencing reveal that monolignol biosynthesis plays a critical role in penetration defence in wheat against powdery mildew invasion. J Exp Bot 60:509–521

Bohnert HU, Fudal I, Dioh W, Tharreau D, Notteghem JL, Lebrun MH (2004) A putative polyketide synthase/peptide synthetase from *Magnaporthe grisea* signals pathogen attack to resistant rice. Plant Cell 16:2499–2513

Bonman JM, Bandong JM, Lee YH, Lee EJ (1989) Race-specific partial resistance to blast in temperate *japonica* rice cultivars. Plant Dis 73:496–499

Brinkman FN, Frey KJ (1977) Yield component analysis of oat isolines that produce different grain yields. Crop Sci 17:165–168

Bryan GT, Wu KS, Farrall L, Jia Y, Hershey HP, McAdams SA, Donaldson GK, Tarchini R, Valent B (2000) A single amino acid difference distinguishes resistant and susceptible alleles of the rice blast resistance gene *Pi-ta*. Plant Cell 12:2033–2046

Cao P, Jung KH, Choi D, Hwang D, Zhu J, Ronald PC (2012) The rice oligonucleotide array database: an atlas of rice gene expression. Rice 5:1–17

Chen X, Shang J, Chen D, Lei C, Zou Y, Zhai W, Liu G, Xu J, Ling Z, Cao G, Ma B, Wang Y, Zhao X, Li S, Zhu L (2006) A B-lectin receptor kinase gene conferring rice blast resistance. Plant J 46:794–804

Chen X, Ronald PC (2011) Innate immunity in rice. Trends Plant Sci 16:451–459

Chen XW, Li SG, Ma YQ, Li HY, Zhou KD, Zhu LH (2004) Marker-assisted selection and pyramiding for three blast resistance genes, *Pi-d(t)1*, *Pi-b*, *Pi-ta2*, in rice. Sheng Wu Gong Cheng Xue Bao Chin J Biotechnol 20:708–714

Chen S, Wang L, Que Z, Pan R (2005) Genetic and physical mapping of *Pi37(t)*, a new gene conferring resistance to rice blast in the famous cultivar St. No. 1. Theor Appl Genet 111:1563–1570

Chen J, Shi Y, Liu W, Chai R, Fu Y, Zhuang J, Wu J (2011) A *Pid3* allele from rice cultivar Gumei2 confers resistance to *Magnaporthe oryzae*. J Genet Genomics 38:209–216

Chuma I, Isobe C, Hotta Y, Ibaragi K, Futamata N et al (2011) Multiple translocation of the *Avr-pita* effector gene among chromosomes of the rice blast fungus *Magnaporthe oryzae* and related species. PLoS Pathog. doi:10.1371/journal.ppat.1002147

Costanzo S, Jia Y (2010) Sequence variation at the rice blast resistance gene *Pi-km* locus: implication for the development of allele specific markers. Plant Sci 178:523–530

Dai Y, Jia Y, Correll J, Wang X, Wang Y (2010) Diversification and evolution of the avirulence gene AVR-Pita1 in field isolates of Magnaporthe oryzae. Fungal Genet Biol 47:973–980

Das A, Soubam D, Singh PK, Thakur S, Singh NK, Sharma TR (2012) A novel blast resistance gene, *Pi54rh* cloned from wild species of rice, *Oryza rhizomatis* confers broad spectrum resistance to *Magnaporthe oryzae* Funct Integr. Genomics 12, 215–228

Dempsey DA, Shah J, Klessig DF (1999) Salicylic acid and disease resistance in plants. Crit Rev Plant Sci 18:547–575

Deng Y, Zhu X, Shen Y, He Z (2006) Genetic characterization and fine mapping of the blast resistance locus *Pigm(t)* tightly linked to *Pi2* and *Pi9* in a broad-spectrum resistant Chinese variety. Theor Appl Genet 113:705–713

Dodds PN, Rathjen JP (2010) Plant immunity: towards an integrated view of plant- pathogen interactions. Nat Rev Genet 11:539–548

Durrant WE, Dong X (2004) Systemic acquired resistance. Annu Rev Phytopathol 42:185–209

Ezuka A, Yunoki T, Sakurai Y, Shinoda H, Toriyama K (1969) Studies on the varietal resistance of rice to blast. II. Tests for field resistance in paddy fields and upland nursery beds. Bull Chugoku Natl Agric Exp Stn E4:32–53 (in Japanese, English summary)

Farman ML, Leong SA (1998) Chromosome walking to the *AVR1-CO39* avirulence gene of *Magnaporthe grisea*: discrepancy between the physical and genetic maps. Genetics 150:1049–1058

Farman ML, Eto Y, Nakao T, Tosa Y, Nakayashiki H, Mayama S, Leong SA (2002) Analysis of the structure of the *AVR1-CO39* avirulence locus in virulent rice-infecting isolates of *Magnaporthe grisea*. Mol Plant Microbe Interact 15:6–16

Flor HH (1971) Current status of gene-for-gene concept. Annu Rev Phytopathol 9:275–296

Fudal I, Böhnert HU, Tharreaub D, Lebrun MH (2005) Transposition of MINE, a composite retrotransposon, in the avirulence gene *ACE1* of the rice blast fungus *Magnaporthe grisea*. Fungal Genet Biol 42:761–772

Fujii K, Hayano-Saito Y (2007) Genetics of durable resistance to rice panicle blast derived from an *indica* rice variety Modan. Jpn J Plant Sci 1:69–76

Fujii K, Hayano-Saito Y, Shumiya A, Inoue M (1995) Genetical mapping based on the RFLP analysis for the panicle blast resistance derived from a rice parental line St. No. 1. Breed Sci 45:209 (in Japanese)

Fukuoka S, Okuno K (1997) QTL analysis for field resistance to rice blast using RFLP markers. Rice Genet Newsl 14:99

Fukuoka S, Okuno K (2001) QTL analysis and mapping of *pi21*, a recessive gene for field resistance to rice blast in Japanese upland rice. Theor Appl Genet 103:185–190

Fukuta Y, Mary JT, Yanoria H, Tsunematsu T, Imbe H, Kato LAE, Khush GS (2004) Development of international standard differential set for blast resistance in rice (Oryza sativa L.). In: Fisher RA New directions for a diverse planet. Proceedings of 4th international crop science congress, Brisbane, 26 Sep–1 Oct 2004

Fukuta Y, Eboron L, Kobayashi N (2007) Review: genetic and breeding analysis of blast resistance in elite *indica*-type rice (*Oryza sativa* L.) bred in international rice research institute. JAEQ 41:101–114

Fukuoka S, Saka N, Koga H, Ono K, Shimizu T, Ebana K, Hayashi N, Takahashi A, Hirochika H, Okuno K, Yano M (2012) Loss of function of a proline-containing protein confers durable disease resistance in rice. Science 325:998–1001

Geng XS, Yang MZ, Huang XQ, Cheng ZQ, Fu J, Sun T, Li J (2008) Cloning and analyzing of rice blast resistance gene *Pi-ta*+allele from Jinghong erect type of common wild rice (*Oryza rufipogon* Griff) in Yunnan. Yi Chuan 30:109–114

Gouda PK, Saikumar S, Varma CMK, Nagesh K, Thippeswamy S, Shenoy V, Ramesha MS, Shashidhar HE (2012) marker-assisted breeding of Pi-1 and Pi-5 genes imparting resistance to rice blast in PRR78, restorer line of PUSA RH-10 basmati rice hybrid. Plant Breed. doi:10.1111/pbr.12017

Haiyan H, Jieyun Z, Kangle Z, Mingjiu L (2012) Characterization of differentially expressed genes induced by virulent and avirulent Magnaporthe grisea strains in rice. Plant Omics J 5:542–546

Hammond-Kosack KE, Jones JD (1996) Resistance gene-dependent plant defense responses. Plant Cell 8:1773–1791

Hayashi K, Hashimoto N, Daigen M, Ashikawa I (2004) Development of PCR-based SNP markers for rice blast resistance genes at the *Piz* locus. Theor Appl Genet 108:212–220

Hayashi K, Yoshida H, Ashikawa I (2006) Development of PCR-based allele-specific and InDel marker sets for nine rice blast resistance genes. Theor Appl Genet 113:251–260

Hayashi K, Yoshida H (2009) Refunctionalization of the ancient rice blast disease resistance gene Pit by the recruitment of a retrotransposon as a promoter. Plant J 57:413–425

Hayashi N, Inoue H, Kato T et al (2010) Durable panicle blast-resistance gene *Pb1* encodes an atypical CC-NBS-LRR protein and was generated by acquiring a promoter through local genome duplication. Plant J 64:498–510

Hittalmani S, Parco A, Mew TV, Zeigler RS, Huang N (2000) Fine mapping and DNA marker-assisted pyramiding of the three major genes for blast resistance in rice. Theor Appl Genet 100:1121–1128

Hogenhout SA, Vander Hoorn RAL, Terauchi R, Kamoun S (2009) Emerging concepts in effector biology of plant-associated organisms. Mol Plant Microbe Interact 22:115–122

Huang CL, Hwang SY, Chiang YC, Lin TP (2008) Molecular evolution of the *Pi-ta* gene resistant to rice blast in wild rice (*Oryza rufipogon*). Genetics 179:1527–1538

Imam J, Alam S, Variar M, Shukla P (2013) Identification of rice blast resistance gene *Pi9* from Indian rice land races with STS marker and its verification by virulence analysis. Proc Natl Acad Sci India Sec B Biol Sci. doi:10.1007/s40011-013-0186-6

Imam J, Alam S, Mandal NP, Maiti D, Variar M, Shukla P (2014a) Molecular Diversity and Mating Type distribution of the rice blast pathogen *Magnaporthe oryzae* in North-east and Eastern India. Indian J Microbiol. doi:10.1007/s12088-014-0504-6

Imam J, Alam S, Mandal NP, Variar M, Shukla P (2014b) Molecular screening for identification of blast resistance genes in North East and Eastern Indian rice germplasm (*Oryza sativa* L.) with PCR based markers. Euphytica 196:199–211

Imam J, Mahto D, Mandal NP, Maiti D, Shukla P, Variar M (2014c) Molecular analysis of Indian rice germplasm accessions with resistance to blast pathogen. J Crop Improv 28:1–11

Imam J, Alam S, Mandal NP, Shukla P, Sharma TR, Variar M (2015) Molecular identification and virulence analysis of AVR genes in rice blast pathogen, Magnaporthe oryzae from Eastern India. Euphytica. doi:10.1007/s10681-015-1465-5

Jeon JS, Chen SD, Yi GH, Wang GL, Ronald PC (2003) Genetic and physical mapping of *Pi5(t)*, a locus associated with broad-spectrum resistance to rice blast. Mol Genet Genomics 269:280–289

Jeung JU, Kim BR, Cho YC, Han SS, Moon HP, Lee YT, Jena KK (2007) A novel gene, *Pi40(t)*, linked to the DNA markers derived from NBS-LRR motifs confers broad spectrum of blast resistance in rice. Theor Appl Genet 115:1163–1177

Jia Y, McAdams SA, Bryan GT, Hershey HP, Valent B (2000) Direct interaction of resistance gene and avirulence gene products confers rice blast resistance. Eur Mol Biol Org 19:4004–4014

Jian-li W, Ronh-yao C, Ye-yang F, De-bao L, Kang-le Z, Leung H, Jie-yun Z (2004) Clustering of major genes conferring blast resistance in a durable resistance rice cultivar Gumei 2. Rice Sci 11:161–164

Jia Y, Martin R (2008) Identification of a new locus, Ptr(t), required for rice blast resistance gene Pi-tamediated resistance. Mol Plant Microbe Interact 21:396–403

Jia Y, Jia MH, Wang X, Liu G (2012) Indica and Japonica crosses resulting in linkage block and recombination suppression on rice chromosome 12. PLoS One. doi:10.1371/journal.pone.0043066

Jones JDG, Dangl JL (2006) The plant immune system. Nature 444:323–329

Kaku H, Nishizawa Y, Ishii-Minami N, Akimoto-Tomiyama C, Dohmae N et al (2006) Plant cells recognize chitin fragments for defense signaling through a plasma membrane receptor. Proc Natl Acad Sci U S A 103:11086–11091

Kang S, Sweigard JA, Valent B (1995) The *PWL* host specificity gene family in the blast fungus *Magnaporthe grisea*. Mol Plant Microbe Interact 8:939–948

Kankanala P, Czymmek K, Valent B (2007) Roles for rice membrane dynamics and plasmodesmata during biotrophic invasion by the blast fungus. Plant Cell 19:706–724

Kanzaki H, Yoshida K, Saitoh H, Fujisaki K, Hirabuchi A, Alaux L, Fournier E, Tharreau D, Terauchi R (2012) Arms race co-evolution of *Magnaporthe oryzae AVR-Pik* and rice *Pik* genes driven by their physical interactions. Plant J. doi:10.1111/j.1365-313X.2012.05110.x

Khang C, Park SY, Lee YH, Valent B, Kang S (2008) Genome organization and evolution of the *Avr-Pita* avirulence gene family in the *Magnaporthe grisea* species complex. Mol Plant Microbe Interact 21:658–670

Khush GS, Jena KK (2009) Current status and future prospects for research on blast resistance in rice (*Oryza sativa* L.). In: Wang X, Valent B (eds) Advances in genetics, genomics and control of rice blast disease. Springer, Dordrecht, p 1–10

Kiyosawa S (1972) Genetics of blast resistance. In: International Rice Research Institute (eds) Rice breeding. International Rice Research Institute, Los Banos, pp 203–225

Kiyosawa S (1974) Studies on genetics and breeding of blast resistance in rice. Misc Publ Natl Inst Agric Sci 1:1–587 (in Japanese)

Kiyosawa S, Ikehashi S, Kato H, Ling ZZ (1981) Pathogenicity tests of Philippine isolates of blast fungus using two sets of rice varieties. Jpn J Breed 31:367–376

Kobayashi N, Yanoria MJT, Tsunematsu H, Hato H, Imbe T, Fukuta Y (2007) Development of new sets of International differential varieties for blast resistance in rice (Oryza sativa L.). JARQ 41:31–37

Kou Y, Wang S (2010) Broad-spectrum and durability: understanding of quantitative disease resistance. Curr Opin Plant Biol 13:181–195

Kou Y, Qu ZW, Zhou B, Naqvi NI (2013) A secreted effector cofers avirulence towards Pi9-mediated broadspectrum blast resistance. Abstract, IRBC, A-21

Krattinger SG, Lagudah ES, Spielmeyer W, Singh RP, Huerta-Espino J, McFadden H, Bossolini E, Selter LL, Keller B (2009) A putative ABC transporter confers durable resistance to multiple fungal pathogens in wheat. Science 323:1360–1363

Kruger WM, Carver TLW, Zeyen RJ (2002) Effects of inhibiting phenolic biosynthesis on penetration resistance of barley isolines containing seven powdery mildew resistance genes or alleles. Physiol Mol Plant Pathol 61:41–51

Kumar GR, Sakthivel K, Sundaram RM, Neeraja CN, Balachandran SM, Rani NS, Viraktamath BC, Madhav MS (2010) Allele mining in crops: prospects and potentials. Biotechnol Adv 18:451–461

Kumari A, Das A, Devanna BN, Thakur S, Singh PK, Singh NK, Sharma TR (2013) Mining of rice blast resistance gene *Pi54* shows effect of single nucleotide polymorphisms on the phenotype expression of the alleles. Eur J Plant Pathol. doi:10.1007/s10658-013-0216-5

Latif MA, Rafii MY, Rahman MM, Talukder MRB (2011) Microsatellite and minisatellite markers based DNA fingerprinting and genetic diversity of blast and ufra resistant genotypes. C R Biol 334:282–289

Lee S, Costanzo S, Jia Y, Olsen K, Caicedo A (2009) Evolutionary dynamics of the genomic region around the blast resistance gene *Pi-ta* in AA genome *Oryza* species. Genetics 183:1315–1325

Li W, Wang B, Wu J, Lu G, Hu Y, Zhang X, Zhang Z, Zhao Q, Feng Q, Zhang H, Wang Z, Wang G, Han B, Wang Z, Zhou B (2009) The *Magnaporthe oryzae* avirulence gene *AvrPiz-t* encodes a predicted secreted protein that triggers the immunity in rice mediated by the blast resistance gene *Piz-t*. Mol Plant Microbe Interact 22:411–420

Li P, Bai B, Zhang HY, Zhou H, Zhou B (2012) Genomic organization and sequence dynamics of the *AvrPiz-t* locus in *Magnaporthe oryzae*. J Zheijang Univ Sci B Biomed Biotechnol 13:452–464

Lin F, Chen S, Que Z, Wang L, Liu X, Pan Q (2007) The blast resistance gene *Pi37* encodes a nucleotide binding site-leucine-rich repeat protein and is a member of a resistance gene cluster on rice chromosome 1. Genetics 177:1871–1880

Liu G, Lu G, Zeng L, Wang L (2002) Two broad-spectrum blast resistance genes, *Pi9*(t) and *Pi2*(t), are physically linked on rice chromosome 6. Mol Genet Genomics 267:472–480

Liu S, Li X, Wang Z et al (2003) Gene pyramiding to increase the blast resistance in rice. Mol Plant Breed 1:22–26 (in Chinese)

Liu X, Lin F, Wang L, Pan Q (2007) The in silico map-based cloning of *Pi36*, a rice coiled-coil nucleotide-binding site leucine-rich repeat gene that confers race-specific resistance to the blast fungus. Genetics 176:2541–2549

Liu J, Wang X, Mitchell T, Hu Y, Liu X, Dai L, Wang GL (2010) Recent progress and understanding of the molecular mechanisms of the rice-*Magnaporthe oryzae* interaction. Mol Plant Pathol 11:419–427

Lopez-Gerena J (2006) Mapping QTL controlling durable resistance to rice blast in the cultivar Oryzica Llanos 5. Ph.D. thesis, Universidad del Valle, Plant Pathology College of Agriculture, Cali, Colombia and Kansas State University, Manhatten

Ma JH, Wang L, Feng SJ, Lin F, Xiao Y, Pan QH (2006) Identification and fine mapping of *AvrPi15*, a novel avirulence gene of *Magnaporthe grisea*. Theor Appl Genet 113:875–883

Mackill DJ, Bonman JM (1992) Inheritance of blast resistance in near-isogenic lines of rice. Phytopathology 82:746–749

Maher EA, Bate NJ, Ni WT, Elkind Y, Dixon RA et al (1994) Increased disease susceptibility of transgenic tobacco plants with suppressed levels of preformed phenylpropanoid products. Proc Natl Acad Sci U S A 91:7802–7806

McCouch SR, Nelson RJ, Tohme J, Zeigler RS (1994) Mapping of blast resistance genes in rice. In: Zeigler RS, Leong SA, Teng PS (eds) Rice blast disease. CAB International, Wallingford, pp 167–186

Miah G, Rafii MY, Ismail MR, Puteh AB, Rahim HA, Asfaliza R, Latif MA (2013) Blast resistance in rice: a review of conventional breeding to molecular approaches. Mol Biol Rep 40:2369–2388

Nagai K, Fujimake H, Yokoo M (1970) Breeding of a rice variety Tordie 1 with multi-racial resistance to leaf blast. Jpn J Breed 20:7–14

Naoumkina MA, Zhao QA, Gallego-Giraldo L, Dai XB, Zhao PX et al (2010) Genome-wide analysis of phenylpropanoid defence pathways. Mol Plant Pathol 11:829–846

Narayanan NN, Baisakh N, Vera Cruz CM, Gnanamanickam SS, Datta K, Datta SK (2002) Molecular breeding for the development of blast and bacterial blight resistance in rice cv. IR50. Crop Sci 42:2072–2079

Nimchuk Z, Eulgem T, Holt BE, Dangl JL (2003) Recognition and response in the plant immune system. Annu Rev Genet 37:579–609

Okuyama Y, Kanzaki H, Abe A et al (2011) A multifaceted genomics approach allows the isolation of rice *Pia* blast resistance gene consisting of two adjacent NBS-LRR protein genes. Plant J 66:467–479

Orbach MJ, Farrall L, Sweigard JA, Chumley FG, Valent B (2000) A telomeric avirulence gene determines efficacy for the rice blast resistance gene Pi-ta. Plant Cell 12:2019–2032

Park CH, Chen S, Shirsekar G, Zhou B, Khang CH, Songkumarn P, Afjal AJ, Ning Y, Wang R, Bellizi M, Valent B, Wang GL (2012) The *Magnaporthe oryzae* effector *AvrPiz-t* targets the RING E3 ubiquitin ligase AP1P6 to suppress pathogen-associated molecular pattern-triggered immunity in rice. Plant Cell 24:4748–4762

Q L, Xu X, Shang J, Jiang G, Pang Z, Zhou Z, Wang J, Liu Y, Li T, Li X, Xu J, Cheng Z, Zhao X, Li S, Zhu L (2013) Functional analysis of Pid3-A4, an ortholog of rice blast resistance gene Pid3 revealed by allele mining in common wild rice. Phytopahology 103:594–599

Qu S, Liu G, Zhou B, Bellizzi M, Zeng L, Dai L, Han B, Wang GL (2006) The broad spectrum blast resistance gene *Pi9* encodes a nucleotide-binding site leucine-rich repeat protein and is a member of a multigene family in rice. Genetics 172:1901–1914

Rai AK, Kumar SP, Gupta SK, Gautam N, Singh NK, Sharma TR (2011) Functional complementation of rice blast resistance gene *Pi-kh* (*Pi54*) conferring resistance to diverse strains of *Magnaporthe oryzae*. J Plant Biochem Biotechnol 20:55–65

Roumen EC (1992) Small differential interactions for partial resistance in rice cultivars to virulent isolates of the blast pathogen. Euphytica 64:143–148

RoyChowdhury M, Jia Y, Jia MH, Fjellstrom B, Cartwright R (2012) Identification of the rice blast resistance gene *Pi-b* in the national small grains collection. Phytopathology 102:700–706

Ryals JA, Neuenschwander UH, Willits MG, Molina A, Steiner HY, Hunt MD (1996) Systemic acquired resistance. Plant Cell 8:1809–1819

Shang J, Tao Y, Chen X, Zou Y, Lei C, Wang J, Li X, Zhao X, Zhang M, Lu Z, Xu J, Cheng Z, Wan J, Zhu L (2009) Identification of a new rice blast resistance gene, Pid3, by genome-wide comparison of paired nucleotide-binding site-leucine-rich repeat genes and their pseudogene alleles between the two sequenced rice genomes. Genetics 182:1303–1311

Sharma TR, Madhav MS, Singh BK, Shanker P, Jana TK, Dalal V, Pandit A, Singh A, Gaikwad K, Upreti HC, Singh NK (2005a) High-resolution mapping, cloning and molecular characterization of the *Pi-kh* gene of rice, which confers resistance to *Magnaporthe grisea*. Mol Genet Genomics 274:569–578

Sharma TR, Shanker P, Singh BK, Jana TK, Madhav MS, Gaikwad K, Singh NK, Plaha P, Rathour R (2005b) Molecular mapping of rice blast resistance gene *Pi-kh* in rice variety Tetep. J Plant Biochem Biotechnol 14:127–133

Sharma TR, Gupta YK, Thakur S, Singh PK, Upreti HC, Singh NK, Singh UD, Rathour R, Kapoor AS, Kaushik RP, Variar M, Maiti D, Mandal NP, Prashanti SK, Hanamaratti NG (2009) Allele mining for important blast resistance genes from Indian land races of rice. 6th International rice genetics symposium, Manila

Sharma TR, Rai AK, Gupta SK, Singh NK (2010) Broad spectrum blast resistance gene *Pi-kh* designated as *Pi-54*. J Plant Biochem Biotechnol 19:987–989

Sharma TR, Rai AK, Gupta SK, Vijayan J, Devanna BN, Ray S (2012) Rice blast management through host-plant resistance: retrospect and prospects. Agric Res 1:37–52

Shimizu T, Nakano T, Takamizawa D, Desaki Y, Ishii-Minami N et al (2010) Two LysM receptor molecules, CEBiP and OsCERK1, cooperatively regulate chitin elicitor signaling in rice. Plant J 64:204–214

Shujie F, Ling W, JunHong MA, Fei L, QuingHua P (2007) Genetic and physical mapping of *AvrPi7*, a novel avirulence gene of *Magnaporthe oryzae* using physical position-ready makers. Chin Sci Bull 52:903–911

Silue D, Tharreau D, Notteghem JL (1992) Identification of *Magnaporthe grisea* avirulence genes to seven rice cultivars. Phytopathology 82:1462–1467

Singh VK, Singh A, Singh SP, Ellur RK et al (2012) Incorporation of blast resistance into 'PRR78', an elite basmati rice restorer line, through marker assisted backcross breeding. Field Crop Res 128:8–16

Skamnioti P, Gurr SJ (2009) Against the grain: safeguarding rice from rice blast disease. Trends Biotechnol 27:141–150

Sweigard JA, Carroll AM, Kang S, Farrall L, Chumley FG, Valent B (1995) Identification, cloning, and characterization of *PWL2*, a gene for host species specificity in the rice blast fungus. Plant Cell 7:1221–1233

Takahashi A, Hayashi N, Miyao A, Hirochika H (2010) Unique features of the rice blast resistance *Pish* locus revealed by large scale retrotransposon-tagging. BMC Plant Biol 10:175

Talukder ZI, Tharreau D, Price AH (2004) Quantitative trait loci analysis suggests that partial resistance to rice blast is mostly determined by race-specific interactions. New Phytol 162:197–209

Tang J, Zhu X, Wang Y, Liu L, Xu B, Li F, Fang J, Chu C (2011) Semi-dominant mutations in the CC-NB-LRR-type R gene, NLS1, lead to constitutive activation of defense responses in rice. Plant J 66:996–1007

Tharreau D, Fudal I, Andriantsimialona D, Santoso-Utami D, Fournier E, Lebrun MH, Nott'eghem JL (2009) World population structure and migration of the rice blast fungus, *Magnaporthe oryzae*. In: Wang GL, Valent B (eds) Advances in genetics, genomics and control of rice blast disease. Springer Publication, Dordrecht, pp 209–215

Till BJ, Reynolds SH, Green EA, Codomo CA, Enns LC, Johnson JE (2003) Large-scale discovery of induced point mutations with high-throughput TILLING. Genome Res 13:524–530

Tsunematsu H, Yanoria MJT, Ebron LA, Hayashi N, Ando I, Kato H, Imbe T, Khush GS (2000) Development of monogenic lines of rice for rice blast resistance. Breed Sci 50:229–234

Variar M, Vera Cruz CM, Carrillo MG, Bhatt JC, Sangar RBS (2009) Rice blast in India and strategies to develop durably resistant cultivars. In: Xiaofan W, Valent B (eds) Advances in genetics, genomics and control of rice blast disease. Springer, New York, pp 359–374

Wang ZX, Yano M, Yamanouchi U, Iwamoto M, Monna L, Hayasaka H, Katayose Y, Sasaki T (1999) The *Pib* gene for rice blast resistance belongs to the nucleotide-binding and leucine-rich repeat class of plant disease resistance genes. Plant J 19:55–64

Wei T, Ou B, Li J, Zhao Y, Guo D et al (2013) Transcriptional profiling of rice early response to *Magnaporthe oryzae* identified OsWRKYs as important regulators in rice blast resistance. PLoS One. doi:10.1371/journal.pone.0059720

Wen S, Gao B (2011) Introgressing blast resistance gene *Pi-9(t)* into elite rice restorer Luhui17 by marker-assisted selection. Rice Genomics Genet 2:31–36

Xiang T, Zong N, Zou Y, Wu Y, Zhang J, Xing W, Li Y, Tang X, Zhu L, Chai J, Zhou JM (2008) *Pseudomonas syringae* effector *AvrPto* blocks innate immunity by targeting receptor kinases. Curr Biol 18:74–80

Xu X, Hayashi N, Wang CT, Kato H, Fujimura T, Kawasaki S (2008) Efficient authentic fine mapping of the rice blast resistance gene *Pik-h* in the *Pik* cluster, using new *Pik-h* differentiating isolates. Mol Breed 22:289–299

Yamada M, Kiyosawa S, Yamaguchi T, Hirano T, Kobayashi T, Kushibuchi K, Watanabe S (1976) Proposal of a new method of differentiating races of *Pyricularia oryzae* Cavara in Japan. Ann Phytopathol Soc Jpn 42:216–219

Yamasaki Y, Kiyosawa S (1996) Studies on inheritance of resistance of rice varieties to blast I. Inheritance of resistance of Japanese varieties to several strains of the fungus. Bull Natl Inst Agric Sci D 14:39–69 (in Japanese)

Yang MZ, Cheng ZQ, Chen SN, Qian J, Xu LL, Huang XQ (2007) A rice blast-resistance genetic resource from wild rice in Yunnan, China. Zhi Wu Sheng Li Yu Fen Zi Sheng Wu Xue Xue Bao 33:589–595

Yoshida K, Saitoh H, Fujisawa S, Kanzaki H, Matsumura H, Yoshida K, Tosa Y, Chuma I, Takano Y, Win J, Kamoun S, Terauchia R (2009) Association genetics reveals three novel avirulence genes from the rice blast fungal pathogen Magnaporthe oryzae. Plant Cell 21:1573–1591

Yuan B, Zhai C, Wang W, Zeng X, Xu X, Hu H, Lin F, Wang L, Pan Q (2011) The *Pik-p* resistance to *Magnaporthe oryzae* in rice is mediated by a pair of closely linked CC-NBS-LRR genes. Theor Appl Genet 122:1017–1028

Yunoki TA, Ezuka T, Morinaka Y, Sakurai H, Shinoda et al (1970) Studies on the varietal resistance of rice to blast. IV. Variation of field resistance due to fungus strains. Bull Chugoku Natl Agric Exp Stn E6 (in Japanese, English summary), pp 21–41

Zhai C, Lin F, Dong Z, He X, Yuan B, Zeng X, Wang L, Pan Q (2011) The isolation and characterization of *Pik*, a rice blast resistance gene which emerged after rice domestication. New Phytol 189:321–334

Zhou B, Qu S, Liu G, Dolan M, Sakai H, Lu G, Bellizzi M, Wang GL (2006) The eight amino-acid differences within three leucine-rich repeats between *Pi2* and *Piz-t* resistance proteins determine the resistance specificity to *Magnaporthe grisea*. Mol Plant Microbe Interact 19:1216–1228

Zhou E, Jia Y, Singh P, Correll JC, Lee FN (2007) Instability of the *Magnaporthe oryzae* avirulence gene *AVR-Pita* alters virulence. Fung Genet Biol 44:1024–1034

Plant Virus Detection and Diagnosis: Progress and Challenges

7

Neelam Yadav and S.M. Paul Khurana

Abstract

The production of healthy planting material requires robust diagnostic procedures. The advancements of the molecular virology and biotechnology have witnessed major breakthroughs in the recent years resulting in sensitive and effective technologies/methods.

PCR has been the major a breakthrough leading successful detection RNA/DNA viruses.

The development of qPCR assays helped to avoid the risk of post-PCR contamination and has benefits of high reproducibility and accuracy. The nonthermal methods, like HDA, RPA, NASBA, etc., have proved suitable for poorly resourced laboratories and on-site testing.

NGS technology involving massive parallel sequencing approach followed by bioinformatics analysis has revolutionized discovery of novel viruses. With the increasing competition for machinery and analytical reagents among companies, high-throughput multiplexing and virus detection techniques are becoming cheaper. However, the challenge remains in their use at the site ensuring to maintain highest crop health standards at affordable price.

Keywords

Detection of virus • Diagnostic techniques • ELISA • Immunological diagnosis • Isothermal amplification • Molecular diagnosis • Nucleic acid hybridization • PCR • RFLP

7.1 Introduction

Viruses are catastrophic despite being miniscule acellular parasites. The simplest viruses comprise small stretch of nucleic acid carrying the genetic information encapsidated by coat protein.

N. Yadav • S.M.P. Khurana (✉)
Amity Institute of Biotechnology, Amity University of Haryana, Manesar, Gurgaon 122413, Haryana, India
e-mail: reachtoneelam@gmail.com; smpkhurana@ggn.amity.edu

P. Shukla (ed.), *Frontier Discoveries and Innovations in Interdisciplinary Microbiology*,
DOI 10.1007/978-81-322-2610-9_7

Being obligate parasite, viruses exclusively utilize the host metabolism for its reproduction. Viruses are known to infect wide range of living organism, viz. animals, plants, fungi and bacteria. Most of our crops are under the threat of various viroids and plant viruses causing plants diseases. Symptoms by these pathogens ranges from mild to catastrophe resulting in loss of crops over large area. The plant viruses cause many economically significant plant diseases worldwide affecting crop yield and quality. Thus, virus infections result in altered physiology of plant that includes stunted growth, inferior quality, and reduced yield leading to severe economic losses. Plant viruses attribute to significant losses in crop production of cash crops, ornamentals, etc. (Table 7.1).

Table 7.1 List of economically significant plant viruses infecting food and horticultural crops

Crop plant	Important viruses reported
Fruits	
Apple	*Apple mosaic virus* (ApMV) *Apple stem grooving virus* (ASGV) *Apple stem pitting virus* (ASPV) *Apple chlorotic leaf spot virus* (ACLSV)
Banana	*Banana bunchy top virus* (BBTV) *Banana streak badnavirus* (BSV) *Banana bract mosaic virus* (BBrMV)
Citrus	*Citrus psorosis virus* (CPsV) *Citrus tristeza virus* (CTV) *Citrus ringspot virus* (CRSV)
Papaya	*Papaya ringspot virus* (PRSV) *Papaya leaf distortion mosaic virus* (PLDMV) *Papaya mosaic virus* (PMV) *Papaya leaf curl virus* (PLCV)
Watermelon	*Watermelon mosaic virus* (WMV) *Watermelon silver mottle virus* (WSMV) *Squash vein yellowing virus* (SqVYV)
Vegetables	
Okra	*Bhendi yellow vein mosaic virus* (BYVMV) *Okra mosaic virus* (OMV)
Brinjal	*Eggplant mottled crinkle tombusvirus*(EMCV) *Tomato bushy stunt virus* (TBSV) *Eggplant mottled dwarf virus* (EMDV)
Tomato	*Tomato spotted wilt virus* (TSWV) *Tomato leaf curl virus* (ToLCV) *Tomato yellow leaf curl virus* (TYLCV)
Potato	*Potato leaf roll virus* (PLRV) *Potato virus X* (PVX) *Potato potyvirus* (PVY)
Cucumber	*Cucumber mosaic virus* (CMV)
Cucurbits	*Cucurbit mild mosaic virus* (CuMMV)
Pumpkin/squash	Squash mosaic virus (SqMV)
Zucchini	*Zucchini yellow mosaic virus* (ZYMV) *Zucchini yellow fleck potyvirus* (ZYFV)
Cereals	
Barley	*Barley stripe mosaic virus* (BSMV) *Barley yellow dwarf virus* (BYDV)
Maize	*Maize dwarf mosaic virus* (MDMV) *Maize streak virus* (MSV)

(continued)

Table 7.1 (continued)

Crop plant	Important viruses reported
Rice	*Rice ragged stunt virus* (RRSV) *Rice tungro bacilliform virus* (RTBV) *Rice tungro spherical virus* (RTSV) *Rice yellow stunt virus* (RYSV)
Oat	*Oat blue dwarf virus* (OBDV) *Oat chlorotic stunt virus* (OCSV)
Wheat	*Wheat American striate mosaic virus* (WASMV) *Wheat* streak mosaic *virus* (WSMV) *Triticum mosaic virus* (TriMV)
Legumes	
Bean	*Bean common mosaic virus* (BCMV) *Bean common mosaic necrosis virus* (BCMNV) *Bean yellow mosaic virus* (BYMV)
Mungbean	*Mungbean yellow mosaic virus* (MYMV)
Soybean	*Soybean mosaic virus* (SMV) *Soybean dwarf virus* (SbDV) *Soybean vein necrotic virus* (SVNV)
Pea	*Pea green mottle virus* (PGMV) *Pea enation mosaic virus* (PEMV) *Pea streak virus* (PSV) *Pea seed-borne mosaic virus* (PSbMV)
Tubers	
Cassava	*Cassava common mosaic virus* (CsCMV) *Cassava virus X* (CsVX) *Cassava green mottle virus* (CGMV) *Sri Lankan cassava mosaic virus* (SLCMV) *African cassava* mosaic *virus* (ACMV)
Cacao/cocoa	*Cacao swollen shoot virus* (CSSV) *Cacao mottle leaf virus* (CMLV) *Cacao necrosis virus* (CNV) *Cacao yellow mosaic virus* (CYMV)
Oil crops	
Groundnut/*peanut*	*Peanut* clump *virus* (PCV) *Peanut yellow mosaic virus* (PeYMV) *Groundnut bud necrosis virus* (GBNV)
Sunflower	*Sunflower chlorotic mottle virus* (SuCMoV) *Sunflower mosaic virus* (SuMV) *Sunflower necrosis virus* (SFNV)
Sugar yielding crops	
Sugarcane	*Sugarcane mosaic virus* (SCMV) *Sugarcane streak mosaic virus* (SCSMV)
Sweet potato	*Sweet potato feathery mottle virus* (SPFMV) *Sweet potato latent virus* (SPLV) *Sweet potato mild speckling virus* (SPMSV)
Beetroot	*Beet ringspot virus* (BRV) *Beet* necrotic yellow vein virus (BNYVV)

Few examples to show the estimated annual economic losses include *Tomato spotted wilt virus* infecting various plant species such as tomato, tobacco, groundnuts, etc. (Sherwood et al. 2003), which led to loss of ~ \$1 billion worldwide (Pappu 1997; Hull 2002); *Cacao swollen shoot virus* causing \$28 million losses to cocoa beans in Africa (Bowers et al. 2001); and *Rice tungro virus* causing \$1.5 billion losses to rice in southeast Asia (Hull 2002).

Due to variability in populations in genotypes of phytopathogens including viruses, it is hard to manage the infection and control the disease. Hence, prevention of viral diseases and subsequent economic damage warrants the fast and accurate methods for detection and diagnosis of the causal organisms, severity of disease and mechanism of virulence. Such information can then be significantly used to minimize the disease by inhibiting its virulence mechanisms.

Precise identification of viral diseases is essential because environmental stresses or other plant pathogens also produce symptoms similar to viruses. The development of diagnostic methods and their application is influenced by multiple factors. Plant viruses constitute highly diverse group of life lacking rRNA sequences unlike other cellular pathogens. Due to increasing globalization of trades including major ornamentals as well as the change in climate, it has significantly contributed to increase in the viruses and their vector movement, thus making the diagnostics more challenging and warranting the urgent needs for the transformations in diagnostic tools and techniques.

The most widely used techniques for screening the propagation materials, seeds and other plant samples for the specific virus infection or its latent presence include electron microscopy and immunological detection like ELISA, PCR and microarray. The viral diseases result in the spread of viruses under favourable conditions by causing either global damages or severe local losses. The conventional agronomic practices for virus control include the use of virus free stock/propagule, roughing of diseased plants, control of alternative vector or hosts and implementation of quarantines. Molecular diagnosis is of significance for the diseases that are not easily diagnosed by symptoms or are symptomless or similar symptoms are produced by other viruses as well. The several techniques known for viral disease diagnosis have advantages and limitations.

7.2 Serological/Immunological Techniques

Immunological tests are used to test the viral antigens by cross-reaction with antisera containing antigen-specific antibodies. These antigen-specific antibodies are raised in a model animal by the introduction of foreign/viral proteins into the bloodstream of animal. These antibodies can be polyclonal or monoclonal antibodies which are available for economically significant viruses both at laboratories and commercially. Monoclonal antibodies have been extensively used for the detection of plant viruses such as potyviruses (Karyeija et al. 2000; Balamurlikrishnan et al. 2002; Desbiez et al. 2002; Ounouna et al. 2002; Villmor et al. 2003; Crosslin et al. 2005).

7.2.1 Production of Antibodies

There are three major ways for the production of antibodies for diagnostics.

Polyclonal antibodies are raised by introducing the pathogen extracts into an animal (usually rabbit) followed by the collection of blood after some time. The blood sample collected is allowed to clot and serum thus produced contains antibodies against the protein(s) in the extract injected. Such antibodies are directly used from isolated serum or can be further used for purification of IgGs (Dijkstra and DeJager 1998). The genus- and strain-specific antibodies enable to differentiate the virus at generic level as well as can differentiate the viruses of the same species at the strain level as used for *Potato virus Y* (Bhat et al. 1997), *Bean common mosaic virus* and *Blackeye cowpea mosaic virus* (Khan et al. 1990).

Plant viruses have been successfully detected using polyclonal antibodies, but their utility is compromised as sometimes they are cross-reactive, specificity is variable across batches, and they are produced in limited quantities. Viral proteins are also expressed in bacteria to raise the polyclonal antibodies (Bragard et al. 2000; Hema et al. 2003; Fajardo et al. 2007; Yadav and Khan 2009). Unlike conventional method of raising antibodies against structural protein, bacterial expression system provides a chance to raise the antibodies against nonstructural proteins (replicase, HC-Pro, etc.) instead of structural protein (capsid or coat protein) (Riedel et al. 1998; Joseph and Savithri 1999; Rodoni et al. 1999).

Monoclonal antibodies are produced by hybridoma technology. Antibody secreting plasma cells are isolated from the spleen of primed animal and fused with cultured myeloma cells. It generates multiple hybrid cell lines called hybridomas. Every hybridoma produces antibody against an epitope of the antigen called monoclonal antibodies. The individual cell lines are propagated to harvest single monoclonal antibodies. Therefore, hybridoma technology provides antibody more specific to antigen and maintains its unlimited supply. Major limitations of monoclonal antibodies are slow production, cost ineffectiveness and maintenance of hybridomas that may undergo apoptosis or fail to secrete antibody. However, despite its limitations, monoclonal antibodies are extensively used in the diagnosis of plant pathogens including viruses (Gonsalves 1979; Karande et al. 1998; Torrance 1995).

7.2.2 Enzyme-Linked Immunosorbent Assay (ELISA)

ELISA is the most favourite immunodiagnostic method because of its simplicity, accuracy and low cost. It is based on the specificity of antibodies to interact with proteins, mostly the capsid protein of target virus. The microtitre plate is coated with the extract from virus-infected plant which is then detected with primary antibodies. The target virus if present in the plant sap will interact with primary antibodies. Before adding secondary antibodies, any traces of the unbound primary antibody are washed off. The secondary antibodies are already conjugated with the reporter molecule, commonly an enzyme, to enable the detection of the virus by producing the chromogenic product which acts on substrate. The quantity of the chromogenic product is determined by the spectrophotometer corresponding indirectly to the viral load (Clark and Adams 1977). Therefore, ELISA enables both quantitative and qualitative analyses. On the basis of order of antigen and antibodies added in the detection reaction, the ELISA can be of following types: antigen-coated (Clark et al. 1988; Cooper and Edwards 1986; Van Regenmortel and Dubs 1993) and sandwich ELISA (Fig. 7.1).

The antigen-coated ELISA (Fig. 7.1a) uses viral antigens bound to the surface of well of microtitre plate which can be detected directly using enzyme-conjugated primary antibodies

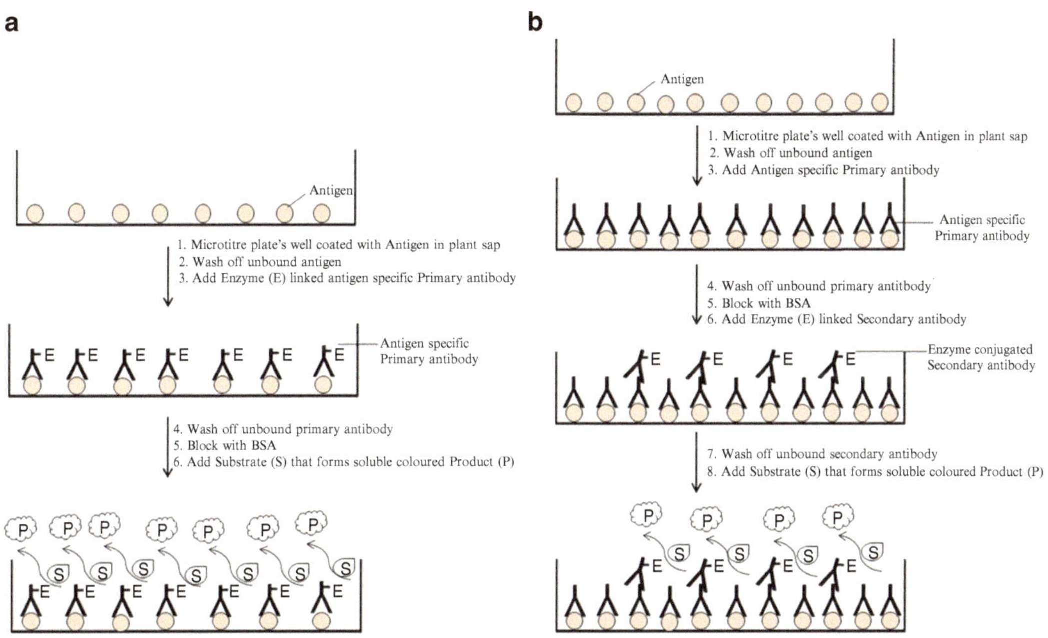

Fig. 7.1 Antigen-coated ELISA: (**a**) Direct ELISA and (**b**) indirect ELISA

followed by colour development by the addition of substrate; this is direct ELISA. It can also be done using indirect ELISA, where viral antigens coated on the surface of well of microtitre plate are initially detected by primary antibodies against target antigen, which is subsequently detected by enzyme-linked secondary antibodies followed by colour development by the addition of substrate. Secondary antibodies used are against the constant region of primary antibodies; e.g. for primary antibodies against viral antigen produced in rabbits, it will use enzyme-linked anti-rabbit IgGs raised in other mammals like bovine, horse and goat.

Another approach is sandwich or double-antibody sandwich (DAS)-ELISA where the viral antigen is sandwiched between two antibodies raised against uncommon epitopes (Fig. 7.2a). DAS-ELISA is highly strain-specific procedure that needs enzyme-conjugated antigen-specific antibody for detection. Another popular method is triple-antibody sandwich (TAS)-ELISA similar to DAS-ELISA, except that an additional step is involved before adding enzyme-conjugated secondary antibody specific to constant region of primary antibodies (Fig. 7.2b). Alkaline phosphatase and horseradish peroxidase are most common enzymes used as the reporter molecules in enzyme-conjugated antibody.

Utilizing multiple wells of the microtitre plate ELISA can enable testing different plant samples for the presence of a virus or single plant for multiple virus infections (Van den Heuvel and Peters 1989; Smith et al. 1991). ELISA tests are highly virus specific (Banttari and Khurana 1998). The accuracy of the test can be improved by performing the same reaction in duplicate or triplicate in the same microtitre plate. Such reactions can be performed with higher precision only if the polyclonal/monoclonal antibody is raised against the target viral protein under consideration and does not cross-react with host plant proteins. The presence of contaminant host protein(s) in purified viral protein to be used for antiserum generation results in antibodies that are cross-reactive as polyclonal antiserum is able to identify the host as well as viral pathogen epitopes. So in such cross-reactions, the antibodies should be cross adsorbed with healthy plant sap so that antibodies against the plant proteins are removed before using such polyclonal antisera for ELISA.

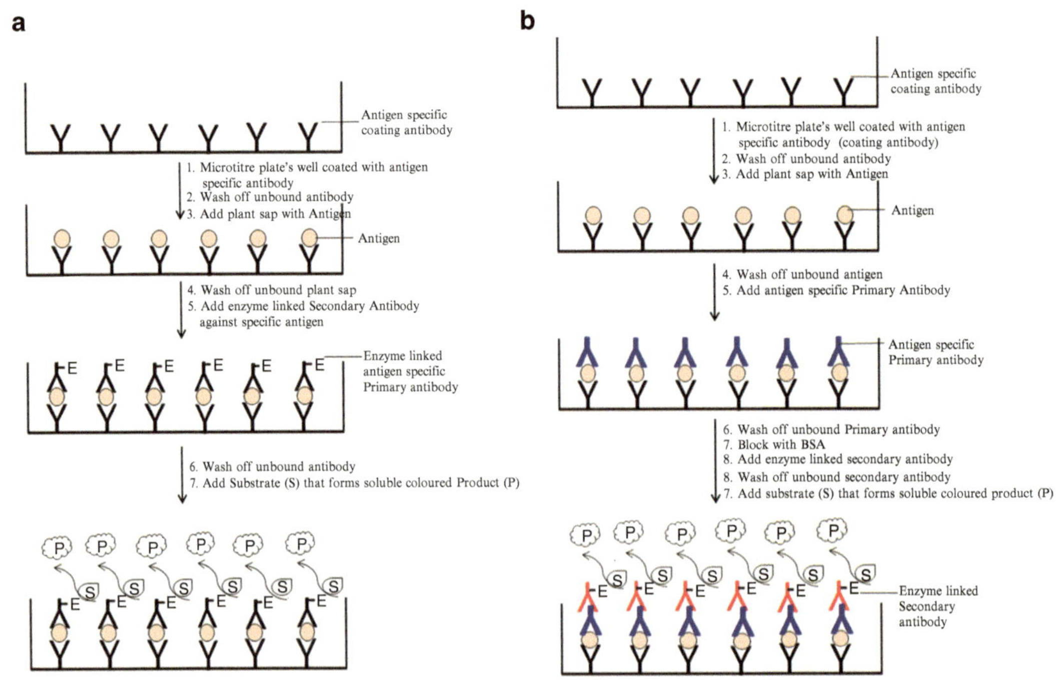

Fig. 7.2 Sandwich ELISA: (**a**) DAS ELISA and (**b**) TAS ELISA

Variant of ELISA, such as voltammetric enzyme immunoassay, is based on the detection of modulation in electrical conductivity of the substrate, instead of chromogenic detection. It is more sensitive than the conventional ELISA and has been used for the detection of CMV (Sun et al. 2001).

7.2.3 Phage Display

Phage display is a powerful technique to capture proteins/viral antigens that interact with phage surface-displayed antibody fragments. Mostly plant virus diagnostic assays make use of PAbs or MAbs. MAbs recognize single epitope of viral antigen and thus give less cross-reactions to discriminate between viral strains (Torrance 1995). MAbs generated against conserved epitopes can be used to group virus isolates for epidemiological and taxonomical studies.

Antibody fragments are made up of heavy and light chains. The antigen binding domains contain hyper-variable sequences that provide the specificity to the antibodies. Phage display is another method of producing antibodies for immunodiagnostic assays of plant pathogen. It uses libraries that express PCR-amplified functional single-chain variable fragment (scFv) or antigen binding fragment (Fab) of antibody molecules obtained from animals including human (Hoogenboom et al. 1991; McCafferty et al. 1990). Thus bacteriophage displaying functional antibody fragments provides a selection system with high fidelity to obtain specific MAbs from combinatorial phage-antibody libraries without the use of laboratory animals (Clackson et al. 1991).

This technique have been utilized for optimization of immunodetection assays for phytopathogens including plant viruses, such as *Blackcurrant reversion associated virus* (Susi et al. 1998), *Potato virus Y* (Boonham and Barker 1998), *Beet necrotic yellow vein virus* (Griep et al. 1999; Uhde et al. 2000), *Tomato spotted wilt virus* (Griep et al. 2000), *Rice black-streaked dwarf virus* (Bai et al. 2002) and *Banana streak virus* (Heng et al. 2007). Several antibodies raised against plant pathogens are used as capture and detection reagent in DAS-ELISA or in immunofluorescence (Fig. 7.3).

This technique enables to mimic the immune system by display of antibody fragments on filamentous bacteriophage on the minor coat protein to detect/isolate low concentrations of peptides/proteins through sequential cycles of phage growth and selection via binding to antibody (Kushwaha et al. 2013). For isolation of species-specific antibodies, phage display has advantages over PAbs/MAbs as it is a fast, cost-effective procedure with continued supply of antibodies (Ziegler and Torrance 2002).

7.2.4 Tissue Blot Immunoassay (TBIA)

TBIA is similar to ELISA as it requires antibodies raised against virus or viral antigens. Plant tissue is pressed onto the nitrocellulose or nylon membranes to make an impression blotting paper which is used to detect the virus using labelled probes usually chemiluminescent. The main advantages associated with the procedure over ELISA include less labour intensive. Also it is a simple (requiring no virus extraction), rapid, sensitive and inexpensive method. This enables the survey of thousands of plant samples in a day providing an option to collect the sample on blotting paper in field that can be processed later (Webster et al. 2004).

Commercial detection kits are available from the International Centre for Agriculture Research in the Dry Areas (ICARDA) for the detection of a variety of viruses infecting legume crops based on polyclonal antisera including *Chickpea chlorotic stunt virus* (CpCDV), *Cucumber mosaic virus* (CMV), *Alfalfa mosaic virus* (AMV), *Bean yellow mosaic virus* (BYMV) and *Pea seed-borne mosaic virus* (PSbMV) and few monoclonal antibody-based kits such as for *Beet western yellows virus* (BWYV), *Bean leafroll virus* (BLRV) and *Soybean dwarf virus* (SbDV) (Katul 1992; Kumari et al. 2006, 2008; Makkouk and Kumari 1996).

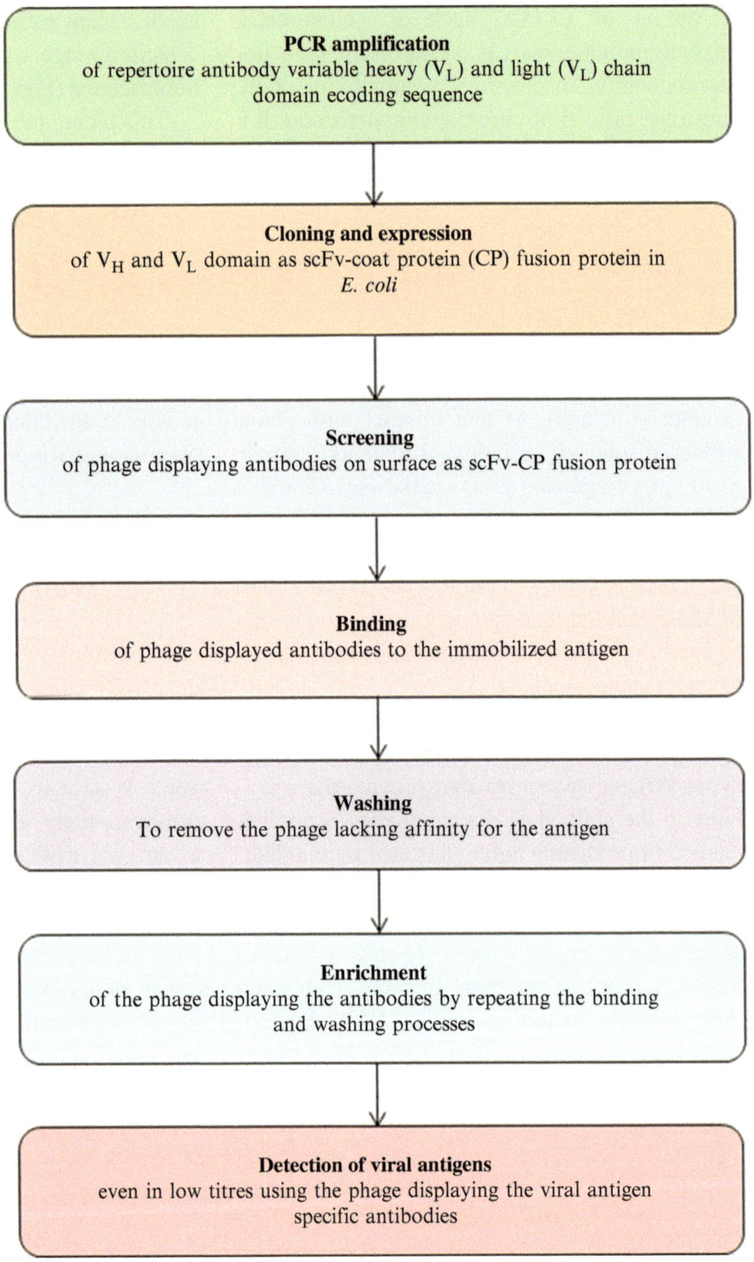

Fig. 7.3 Workflow of the phage display for the plant virus diagnosis

7.2.5 Quartz Crystal Microbalance (QCM) Immunosensors

The principle of QCM is that the adsorption of any substance on a quartz crystal surface modulates its resonance oscillation frequency (Cooper and Singleton 2007; Vashisht and Vashisht 2011). This technique utilizes virus-specific antibodies coated on quartz crystal disc (0.5 in. in diameter) for plant virus diagnostics. Purified virus or the crude sap from infected plant is allowed to bind on the immobilized antibodies. A voltage is applied across the disc that warps the disc slightly under a piezoelectric effect. Interaction of specific viral antigen with coated antibody leads to the increase in mass at surface of QCM which reduces the frequency of resonance oscillation. Change in frequency of resonance oscillation is directly proportional to the amount of virus interacted with immobilized antibody on the quartz crystal disc (Fig. 7.4).

Therefore, it is a rapid, economic and sensitive method suitable for qualitative and quantitative detection. The first report describing QCM method detected as little as 1 ng of *Cymbidium mosaic virus* and *Odontoglossum ringspot virus* particles in crude extracts (Eun et al. 2002). *Tobacco mosaic virus* has also been detected successfully using QCM from sap of infected plants (Dickert et al. 2004).

7.2.6 Lateral Flow Immune Assay (LFIA)

LFIA is a rapid immunochromatographic technique applied to plant virus detection mainly for viruses of vegetable and floriculture crops (Danks and Barker 2000; Salomone and Roggero 2002; Salomone et al. 2002, 2004).

The major attractions associated with LFIA are easy to perform even by inexperienced person by the use of very simple device and allow infield detection of viruses generating quick results.

Thus, it has been used reliably to develop kits for plant virus diagnostics.

Lateral flow enables the mass express diagnosis of viral infections for virus-free plant culture industry or individual use requires inexpensive, rapid and simple technologies that make possible analysis without special skill and equipment even under field conditions. The principle of LFIA is based on immunoreaction between virus in question with antibodies against viral proteins and chromogenic colloidal substances applied on lateral flow devices that are marketed as test strips. When lateral flow device/strip is spotted with the infected/symptomatic plant tissue extract/sap, the liquid sample flows through it and initiates antigen-antibody interaction leading to the chemical reaction that is visible as chromogenic bar/line for test and control samples. Byzova et al. (2009)

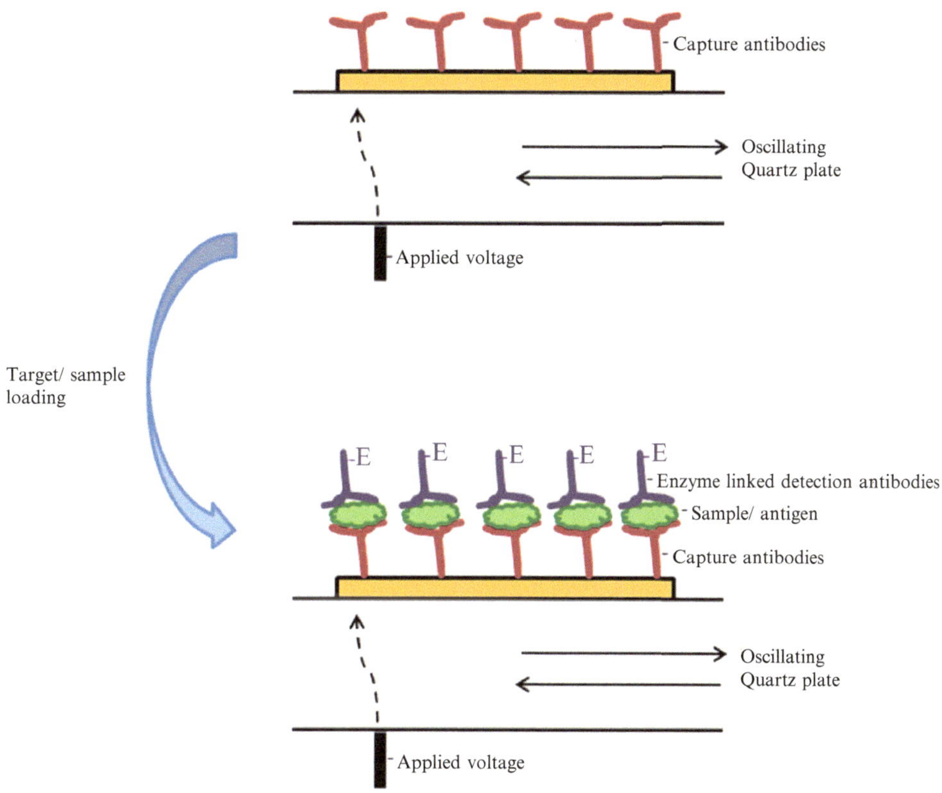

Fig. 7.4 QCM workflow showing binding of the target antigen/virus on immobilized antibodies causes a frequency change in the oscillation of the quartz crystal (Adapted from Grimme and Cropek 2013)

developed a test strip to detect five plant viruses simultaneously which are different in shape and size, namely, spherical (*Carnation mottle virus*, *Bean mild mosaic virus*), rod shaped (*Tobacco mosaic virus*: TMV) and filamentous (PVY, PVX) viruses. In India, CPRI has developed LFIA devices for the detection of five potato viruses, viz. PVX, PVA, PVS, PVM and PVY, either individually or in combination of two viruses, viz. PVA and PVX, PVA and PVS and PVY and PVM. Few recent reports on plant viruses using LFIA include the detection of PVX (Drygin et al. 2011; Safenkova et al. 2012), TMV and CMV (Lamptey et al. 2013).

7.2.7 Immunocapture Transmission Electron Microscopy (ICTEM)

ICTEM is a technique that combines the serological specificity and electron microscopy (Derrick 1973). It has been used for a wide variety of polyhedral and rod-shaped plant viruses for their detection by combining serological specificity for morphological visualization of virus particles (Lima and Purcifull 1980; Banttari et al. 1991; Milne 1991; Naidu and Hughes 2001). ICTEM has been used for the detection of low titres of viruses in potato nucleus stocks/mericlones, etc. (Khurana 1990). The ultrathin sections in electron microscopy add high sensitivity and specificity to morphological determination of virus (Khurana and Garg 1993). The process is based on the selective entrapment of virus particles from samples almost/completely devoid of host material, on grids coated with virus-specific antibodies (Lima and Purcifull 1980).

The electron microscopy uses copper grids with thin-layer Parlodion that are coated with 1000–6000-fold diluted virus-specific antibodies. The excess of antibody is washed off and treated with diseased plant tissue extract for 3–4 h at 25 °C. The grid is then washed thrice, stained with 1.0 % uranyl acetate in 50 % ethanol and dried followed by final scanning in electron microscope (Fig. 7.5).

The main advantage with the process is the requirement of small quantities of antibodies and viruses; moreover, labelling of antibodies is not required. It is a highly sensitive method for the detection and diagnosis of virions in leaf extracts in comparison with leaf dip method (Lima and Purcifull 1980). The additional advantages of using polyclonal antiserum has no interference with the virus particle visualization and immobilization of virus particle on grids that can be used even later for electron microscope analysis. The use of an electron microscope for the analysis of sample makes it suitable for small-scale samples for confirmation of the virus. Immunoelectron microscopy was found to detect twice as many banana streak-causing viruses from Nigeria and Ghana than detected through ELISA or IC-PCR (Agindotan et al. 2006).

7.3 Restriction Fragment Length Polymorphism (RFLP)

Combinatorial use of RFLP and PCR has enabled the identification of differences among the amplified virus genomes/genes by the presence or absence of restriction site(s) of a particular restriction enzyme within the amplified sequence. Initially PCR amplification is performed to yield amplified DNA followed by digestion with restriction enzyme(s) and analyses of the size of the fragments obtained by gel electrophoresis. Thus, RFLP enables to differentiate the virus isolates even without cloning and sequencing. This method accurately and effectively demonstrates the polymorphisms among the virus genomes/genes. The subgroup 2 of CMV in Western Australian lupin crops were detected and diagnosed using RFLP (Wylie et al. 1993).

The RFLP has been used to distinguish the related viruses and viroids including diversity among *African streak mastreviruses* (Willment et al. 2001) and *Citrus tristeza virus* isolates using coat protein nucleotide sequences (Corazza et al. 2012), for the screening of potato seeds for *Potato virus* M (Xu et al. 2010) and for the detection and discrimination of *Pospiviroids* (Luigi et al. 2014).

RFLP followed by PCR has been used to distinguish the 18 isolates of PVY (Glais et al. 1996), for large-scale screening of potato plants and tubers for *Alfalfa mosaic virus* (Xu and Nie 2006), to distinguish different strains of *Soybean*

Fig. 7.5 Workflow of ICTEM

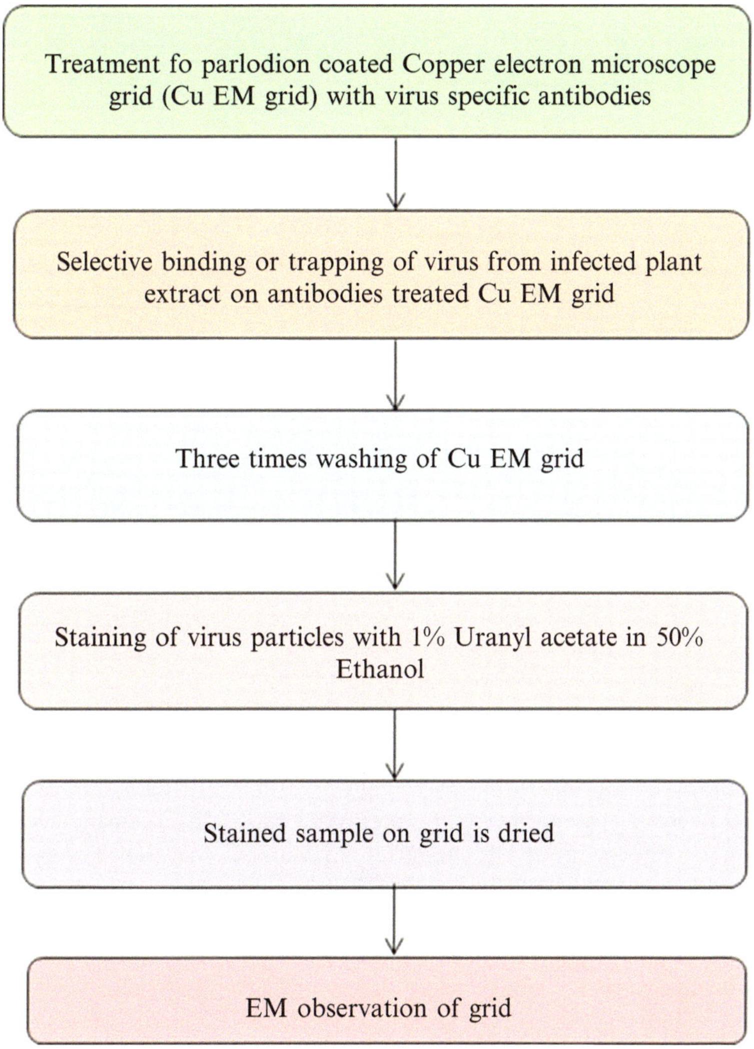

mosaic virus (Kim et al. 2004), for the detection and discrimination of *Melon necrotic spot virus* (MNSV) (Kubo et al. 2005) and for the identification and distinction of two strains of *Grape fan leaf virus* (GFLV) in Tunisia based on coat protein region (Fattouch et al. 2005) and viroids infecting pome fruits (Hadidi and Yang 1990).

7.4 Thermostable Amplification-Based Methods

PCRs and reverse transcriptase polymerase chain reactions (RT-PCRs) are the most popular thermostable amplification-based techniques for the detection and diagnosis of DNA and RNA plant viruses, respectively (Fig. 7.6).

These procedures are highly sensitive in comparison with serological methods and suitable for routine detection of plant viruses and hence have been widely applied for molecular detection. These methods are efficient in specific detection of the virus through sequence-specific primers, so availability of sequence information is a prerequisite. Also, with the growing number of sequence information available in the database, access to the viral genome sequences is now easy and can be utilized for designing the viral species or strain-specific primers (Chen and Adams 2001; Gibbs and Mackenzie 1997; Langeveld et al.

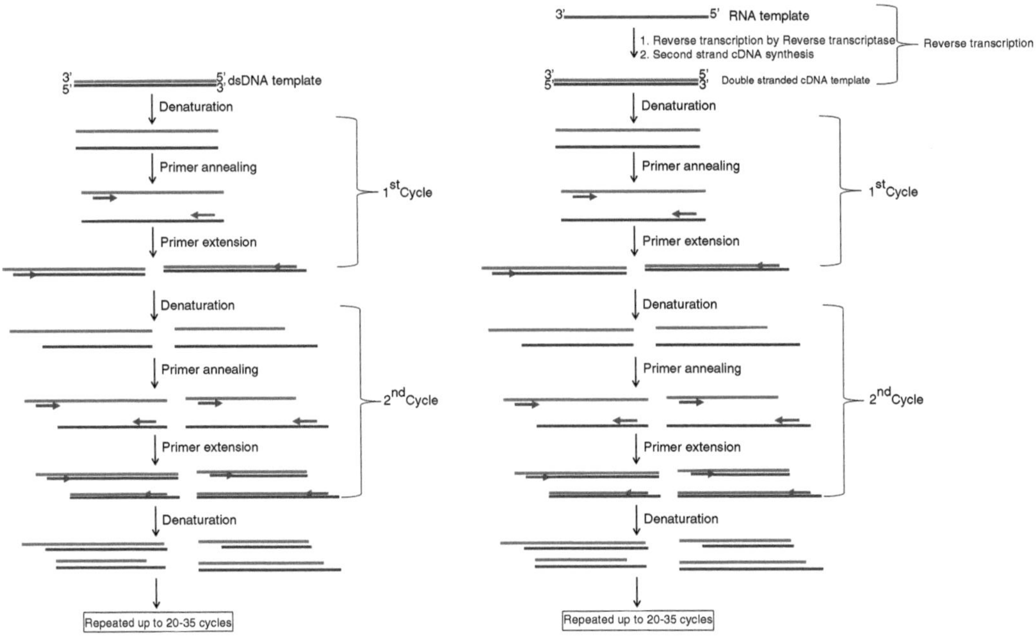

Fig. 7.6 Diagrammatic representation of (**a**) PCR and (**b**) RT-PCR

1991; Zheng et al. 2010). Being a highly sensitive technique, it essentially requires positive or negative controls to avoid false negatives or positives. PCRs are simpler and effective than ELISA as it is easy to design and synthesize new primers than the production of novel antibodies. Moreover PCR products if required can be cloned into suitable cloning vector and sequenced. Certainly these methods have stringent polymerization conditions and are performed at higher temperatures which utilize thermostable polymerase like *Taq* DNA polymerase from *Thermus aquaticus*. A variety of alternatives to *Taq* DNA polymerases are now available; the advantages related to them include *Thermococcus litoralis* (*Til*) polymerase with high fidelity than *Taq* polymerase, *Thermus thermophilus* (*Tth*) polymerase having DNA-/RNA-dependent DNA polymerization activity and *Sulfolobus acidocaldarius* (*Sac*) DNA polymerase have being identified to be highly thermostable than *Taq*.

The prerequisite for the optimization and development of any PCR-based diagnostic assay is to identify the DNA sequences for designing of primers enabling efficient and specific detection of the target pathogen. Virus genomes are relatively small; therefore a number of viral genomes have been completely sequenced and submitted to nucleotide databases. Coat protein gene is a highly preferred target sequence for virus diagnostics. However, other regions of virus genome have also been used for diagnostics providing significant sequence data from related viruses on the same region available on the database. RNA virus assay involves additional step to convert genomic RNA to cDNA before PCR for detection. Nucleic acid-based diagnosis uses two approaches: the first approach is based on targeting of known gene sequences in target and nontarget viruses followed by the identification of unique regions to design primers, while the second approach is based on the use of randomly selected DNA fragments. PCR-based diagnostic methods have certain limitations. They require prior sequence information of target viruses to design primers and a thermocycler. Intelligent primer designing is critical for the successful detection of single or multiple strains of a genus. Table 7.2 summarizes the list of plant viruses and viroids identified by PCR.

Table 7.2 Short list of common viruses and viroids detected by PCR

Organism/agent	Hosts	Target sequence/sample treatment	Comments
Viroids			
Hop stunt	Hops	Viroid sequence; total nucleic acid extracted from infected host tissue	Amplification and sequencing of viroid cDNA
Apple scar skin Dapple apple Pear rusty skin	Pome fruits	Viroid sequence; total nucleic acid extracted from infected host tissue	Amplification of viroid cDNA
Citrus exocortis Cachexia Citrus viroid IIa	Citrus	Viroid sequence; total nucleic acid extracted from infected host tissue	Amplification of viroid cDNA
Viruses			
Cauliflower mosaic virus (CaMV)	Cabbage	CaMV sequence; DNA extracted from single aphids fed on infected plant tissue	Detection of viral DNA in a single aphid; *Caulimovirus*
Bean golden mosaic virus (BGMV)	Bean	BGMV sequence; DNA extracted from infected bean	Detection and sequencing of viral DNAs; *Geminivirus*
Cucumber mosaic virus (CMV)	Cucumber	CMV sequence; total nucleic acid from infected cucumber	Sequence amplified with *Taq* polymerase; *Cucumovirus*
Tomato ringspot virus (ToRSV)	Cucumber	ToRSV sequence; total nucleic acid from infected cucumber	Sequence amplified with *Taq* polymerase; *Nepovirus*
Tobacco mosaic virus (TMV)	Tobacco	Transport protein gene sequence; RNA extracted from infected leaf tissue	Sequence amplified with *Tth* polymerase; *Tobamovirus*
Luteoviruses	Cereals, sugar beet, potato	Conserved luteovirus sequences; total plant nucleic acid extracts	Detection of viral cDNA of barley yellow dwarf, beet western yellows and potato leaf roll luteoviruses Differentiation by RFLP analysis
Bean yellow mosaic virus (BYMV)	Gladiolus	BYMV-specific sequences; total plant nucleic acid extracts	Detection of viral cDNA in leaves and corns; *Potyvirus*
Pea seed-borne mosaic virus (PSbMV)	Pea	PSbMV-specific sequences; total plant nucleic acid extracts	Detection of viral cDNA; *Potyvirus*
Plum pox virus (PPV)	Pome fruit	PPV-specific sequences; total plant nucleic acid extracts	Detection of viral cDNA using biotin label in PCR reaction; *Potyvirus*
Potyviruses	Various	Conserved potyvirus sequences; extracts from purified virus or total plant nucleic acid extracts	Detection of viral cDNA, used degenerated primers
Cherry leafroll virus (CLRV)	Walnut	CLRV-specific sequences; crude tissue extract	Detection of viral cDNA; sample not treated with phenol or protease; *Nepovirus*

Modified after Henson and Roy (1993)

7.4.1 PCR and RT-PCR

The major breakthrough in detection, discovery and characterization of viruses was with the discovery and use of PCR (Mullis et al. 1986; Mullis and Faloona 1987). It has been successfully used to characterize viruses, utilizing sequence information available for the viruses of related genera or families (Chen and Adams 2001; Yadav and Khan 2008). It is a highly sensitive and specific diagnostic procedure for the detection of virus that relies on the presence of unique sequences in the virus genome (Bartlett 2003; Waterhouse and Chu 1995). In viruses containing RNA as the genome, a first-strand cDNA is synthesized with the help of reverse transcriptase (RT) enzyme, which in turn can be used a template for in vitro amplification using heat-stable polymerase enzyme. The PCR and RT-PCR have been widely employed successfully for the discovery and characterization of plant viruses. For the broad detection of virus infections in the plant samples, group-specific degenerate primers are utilized (Martin et al. 2000; Zheng et al. 2010). The degenerate primers are designed to target the conserved sites in the known sequences of viral genera. The conserved blocks or sites in the sequences are found by comparing the available sequences of related genera or families through the bioinformatics. Degenerate primers targeting the conserved sites in the genome of potyviruses, geminiviruses, closteroviruses, badnaviruses, etc., have been widely used for the initial screening and characterization of new variants or species infecting diverse crop species (Anderson et al. 2004; Baranwal et al. 2010; Yang et al. 2003). The amplification of specific portion of the viral genome obtained in PCR or RT-PCR needs to be sequenced to confirm its exact identity. The most commonly used sequencing strategy has been the capillary electrophoresis-based Sanger's dideoxy chemistry.

Multiple variations of the basic technique have been designed to improve the sensitivity and specificity with facilitated automation for the detection of plant viruses (Waterhouse and Chu 1995).

7.4.2 Multiplex PCR/RT-PCR

Multiplex PCR procedures, invented by Chamberlain et al. (1988), had revolutionized the PCR-based molecular detection of plant viruses. Such procedures were developed for convenient and sensitive detection of several DNA/RNA viruses, in one reaction/tube simultaneously (Rosenfield and Jaykus 1999; Nassuth et al. 2000; James 1999). The technique is very useful where plant viruses exist as multiple or mixed infection in the same host. Multiplex PCR involves the use of many sets of oligonucleotides targeting the genomic regions of multiple viruses (within the same genus, strains of the same species or viruses from taxonomically different groups) in one PCR amplification reaction mixture. Primers are designed in such a way that amplification can be carried out at the same annealing temperature and do not have complementarity with other primers. The PCR products of different sizes from various targets ensure easy analysis of amplicons by agarose gel electrophoresis. The multiplex PCR thus has advantages of saving reagents against individual PCR, convenient, rapid and cost-effective. This technique has been applied for the detection of a varied range of plant viruses (Deb and Anderson 2008; Nie and Singh 2000; Roy et al. 2005).

Thus, multiple species or strains of multiple viruses from different plant samples are detected in a single tube by adding the specific primers. Multiplex PCR is of great utility in plant virology as different RNA viruses are known to infect a single host and thus need sensitive detection methods of such viruses to raise virus-free propagules.

While performing the multiplex PCR, one must keep in mind that the amplified targets should be of different lengths/sizes also different primer pairs used should neither share any complementarity nor cross-reactivity/specificity.

Ito et al. (2002) have reported the multiplex RT-PCR assay for screening of diseased citrus for the detection of six citrus viroids and *Apple stem grooving virus* (ASGV) or *Citrus tatter leaf virus* (CTLV). Also it has been used for simultaneous detection of five different viruses from diseased tobacco plants (Dai et al. 2012). It has also been successfully used for the one-step-one-tube detection of four viroid genera, *viz. Apscaviroid*, *Pelamoviroid*, *Pospiviroid* and *Hostuviroid* (Ragozzino et al. 2004), as well as for ascertaining the presence of eight viruses in suspected stone fruit trees (Sanchez-Navarro et al. 2005) and their effective amplification through the multiplex PCR has been reported. Recently, Kwon et al. (2014) using dual priming primer sets developed two multiplex PCR systems for the simultaneous detection of seven different viruses in cucurbits. The first system efficiently enabled to detect *Papaya ringspot virus*, *Watermelon mosaic virus* and *Zucchini yellow mosaic virus* whereas the other one was to detect *Cucumber green mottle mosaic virus*, *Cucumber fruit mottle mosaic virus*, *Kyuri green mottle mosaic virus* and *Zucchini green mottle mosaic virus*. The use of multiple primer sets designed against various target sequences in the same reaction saves time and reduces running cost of the optimized multiplex PCR assays (Bertolini et al. 2001; Walsh et al. 2001).

7.4.3 Immunocapture PCR (IC-PCR)

This variant of PCR combines advantages of serology to capture virus particles by antibodies and PCR to target a molecule using specific primers (Fig. 7.7). The process begins with the adsorption of the virus on antibody-coated microtitre plate or microfuge tube which is then removed by heating in the presence of Triton X-100, a non-ionic surfactant, followed by amplification of nucleic acid using RT-PCR. Capturing of viruses by antibody helps to purify the virions from the plant sample even when the titre of virus is low or when nucleic acids extracted have certain inhibitor compounds that hamper PCR.

This procedure is of significance for the diagnosis of plant viruses with very low concentration utilizing the sensitivity and accuracy of both immunological and molecular approaches. The virion particles trapped in tubes using the specific antisera are used for PCR amplification as template, providing a sensitive tool for the detection of plant viruses having low titre in diseased host or even the viral genomes integrated with genome of the host (LeProvost et al. 2006).

IC-PCR has been used for its high sensitivity and specific detection of PPV in plum tree saps (Wetzel et al. 1992) and *Sugarcane streak mosaic virus* from sugarcane sap (Hema et al. 2003). Harper et al. (1999a) reported the efficient detection of the episomal *Banana streak virus*, known to have its genome present naturally within the genome of banana through IC-PCR, which otherwise results in higher false positives upon conventional PCR. The IC-PCR has been

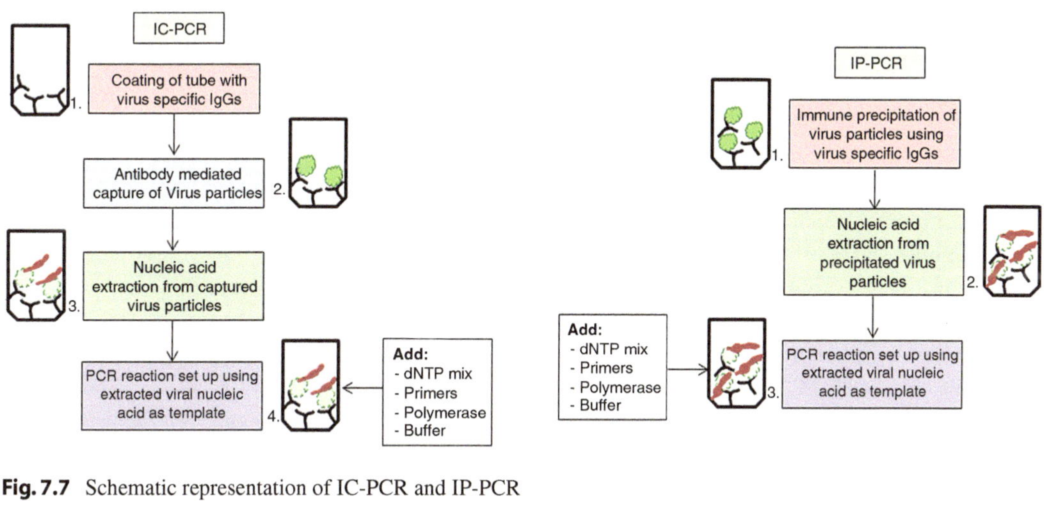

Fig. 7.7 Schematic representation of IC-PCR and IP-PCR

successfully used to validate the presence of many viruses, viz. *Pepper mild mottle virus* (PMMV), BYMV, CLRV, CMV, CTV, GFLV, PLRV and TSWV along with CMV satellite RNA (Narayanasamy 2011).

Thus, IC-PCR involving virus immobilization by antibody has been used for the identification and molecular characterization of plant viruses of various families and genera to assign appropriate taxonomic position. The major advantage of this highly specific technique includes no encounter with problems associated with RNA extraction through the combination of serological efficacy and nucleic acid amplification of many viruses, viz. PVY, tobamoviruses, etc. (Jacobi et al. 1998; Mulholland 2009; Gawande et al. 2011).

7.4.4 Immuno-precipitation PCR (IP-PCR)

IP-PCR is the modification of PCR technology involving the precipitation of virus particles via antibodies (Fig. 7.7). Like IC-PCR, it has been used for the identification and molecular characterization of plant viruses of various families and genera enabling to taxonomic positioning (Lima et al. 2011). The major advantage of this technique is its specificity and no requirement of RNA extraction. So it possesses the advantage of serological specificity and PCR amplification, making it a quick and specific detection method (Fig. 7.8).

The process begins with the precipitation of the virus particles by polyclonal antibodies followed by RNA extraction, and first-strand cDNA is synthesized and amplified by PCR. IP-PCR reduces the co-contamination of plant RNAs. This technique is based on partial virus purification by immuno-precipitation; thus it is especially significant for efficient detection of viruses present in low or variable titres in plant species containing PCR amplification inhibitors. It has remarkably contributed in the detection of various plant viruses. IP-PCR has been used to detect *Papaya lethal yellowing virus* (PLYV) in 10,000-fold diluted tissue extracts of the diseased plants without amplifying host cDNA from healthy plant extracts (Lima et al. 2012). IP-PCR because of its sensitivity and specificity has been used efficiently to detect the five virus species of different families, viz. Bromoviridae, Comoviridae, Potyviridae and Sobemoviridae (Lima et al. 2011).

7.4.5 Nested PCR

It is a modification of PCR that reduces nonspecific amplification of products due to the additional primer binding sites and is completed by performing two PCR reactions using two sets of primer pairs. The first reaction is usually carried out with external primers pair to increase the concentration of target sequence which in turn acts as a template in the second PCR reaction. The external primers are mostly degenerate primers with low specificity. The second step PCR reaction is set up using an aliquot of the amplification product from the first round and the internal primers which necessarily anneal at a target sequence internal/within the first-round amplicon (Fig. 7.9). Hence, it has been reported for the successful detection of species belonging to *Vitivirus* and *Foveavirus* in grapevines using nested PCR (Dovas and Katis 2003).

Therefore, this approach is profound particularly for the infections where the viral load/titre is very low; also it is capable of increasing the target DNA molecule as well as results in dilution of inhibitors of DNA polymerase that are hard to remove from plant extract with nucleic acid purification (Fig. 7.9).

7.4.6 Multiplex Nested PCR

It is a single-tube reaction optimized for the sensitive and parallel detection of few viruses (such as CMV, CLRV, SLRSV, ArMV) and bacterium from diseased olive plant samples using 20 sequence-specific primers compatible with the target sequence (Bertolini et al. 2003). This method combines in a tube the sensitivity and accuracy of nested PCR with the advantage of simultaneous detection of multiple targets of

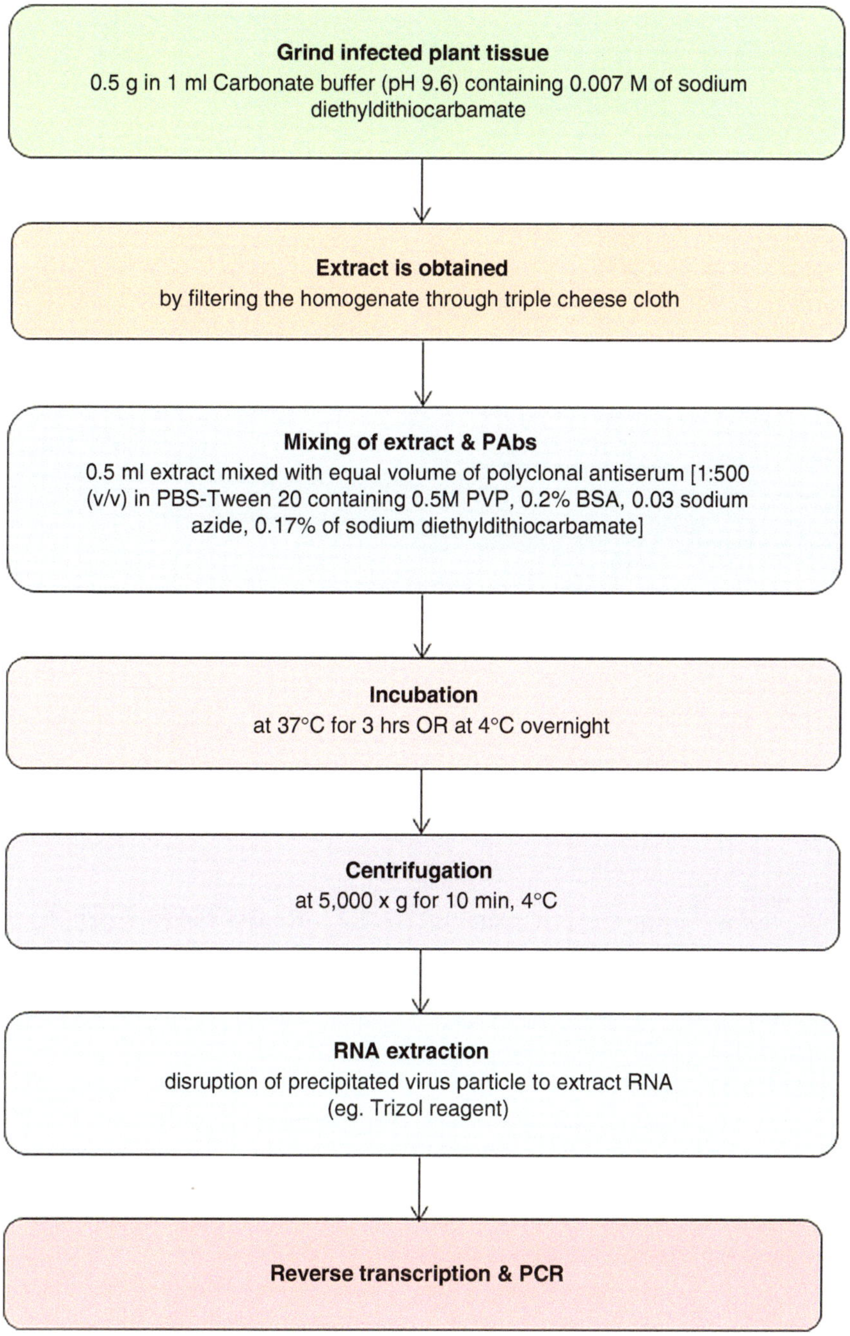

Fig. 7.8 Schematic representation of IP-PCR method

multiplex RT-PCR, thus making it a powerful tool to detect multiple viruses, bacteria and phytoplasmas (Biswas et al. 2013) simultaneously from angiosperms in one tube.

This is a cost-effective and time-saving method which works in a single tube but requires critically designed accurate internal and external primers. These external and internal primers used in multiplex PCR should not interfere with the

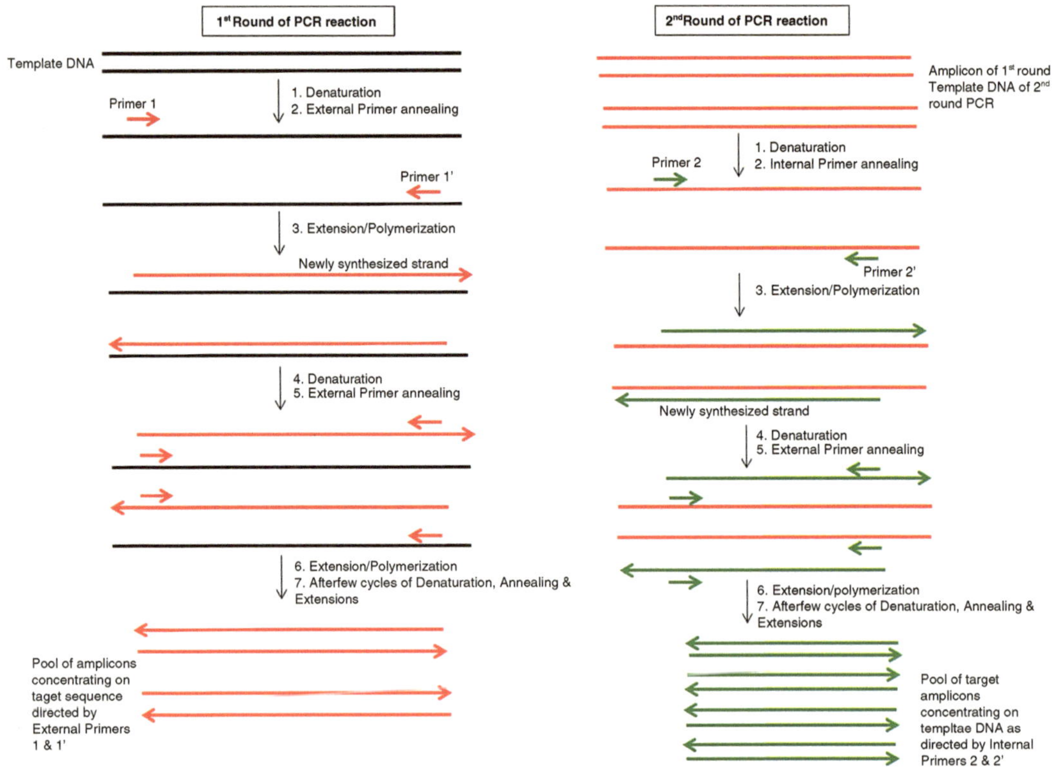

Fig. 7.9 Schematic representation of nested PCR

amplification process. Once the first round of amplification is completed with external sets of primers, the second round of the PCR is performed using internal sets of primers and amplicon of the first round of PCR as template to have nested multiplex amplification. Relatively higher concentrations of internal primers in the second round of PCR performed with an aliquot of the first round amplicon as template have nearly negligible concentrations of external primers, hence enabling the minimum interference in progress of nested multiplex PCR. The nested multiplex impart about 100 times more sensitivity and accuracy than routine multiplex-PCR/-RT-PCR used for the diagnosis of plant viruses. This approach enables the discrimination of different amplicons of even the same size representing various targets, when integrated with colorimetric detection.

7.4.7 Real-Time PCR (qPCR)

Quantitative PCR (qPCR) is an improved and highly sensitive protocol over standard PCR which is used to detect as well as quantify specific targeted DNA or RNA or cDNA molecule.

The principle underlying qPCR is that the time taken by the reaction to enter exponential phase of amplification is directly proportional to the quantity target DNA in question. It avoids the need for agarose gel electrophoresis after completion of reaction as amplified products produced at each cycle are detected and measured by a built-in fluorimeter. It involves the use of fluorescent dyes like SYBR Green that binds to any dsDNA, thus a simple and cheap way to quantify amplification. The major drawback with the SYBR Green is its incapability to discriminate between specific and non-specific dsDNA;

therefore non-specific amplifications and primer-dimers also produce signals. This can be overcome by another approach based on the use of sequence-specific fluorescent DNA probes like TaqMan DNA probes (Livak et al. 1995). The use of TaqMan DNA probes (Fig. 7.10b) reduces signals due to non-specific primer binding at target DNA or primer-dimer formation. It helps to detect only the specific amplifications in the reaction mixture even at early stage without requiring completion of reaction for observing and confirming amplification of precise region (Wittwer et al. 1997).

This approach uses two primers flanking target sequence and a fluorescent-tagged probe annealing within the target sequence being amplified. These probes contain fluorescent dyes, viz. a reporter dye and a quencher dye at 5′-end and 3′-end, respectively, reducing net fluorescence produced. While annealing of PCR cycle when fluorescent probes bind within the target sequence being amplified, the reporter dye is removed by 5′-exonuclease action of DNA polymerase. Hence, quencher and reporter dyes are separated, resulting in increased fluorescent signals proportional to quantity of product amplified. Additionally, few other fluorescent-tagged probes and primers that utilize reporter and quencher dyes include Molecular Beacons (Fig. 7.10a) (Mackay et al. 2002; Tyagi and Kramer 1996) and Scorpions (Thelwell et al. 2000).

The cycle threshold (C_t) is determined for all test samples, showing cycle number where statistically notable raised fluorescence was recorded. C_t value is inversely proportional to the target DNA amplified. The standard curve is generated for the C_t values of amplified reference DNA to compare with the concentration of unknown amplified sample DNA that allows its quantification. The qPCR machines have inbuilt software that enable quick analyses of data generated.

Real-time PCR is increasing its applications in plant virus diagnostics with many reported assays to detect viruses and viroids (Schaad et al.

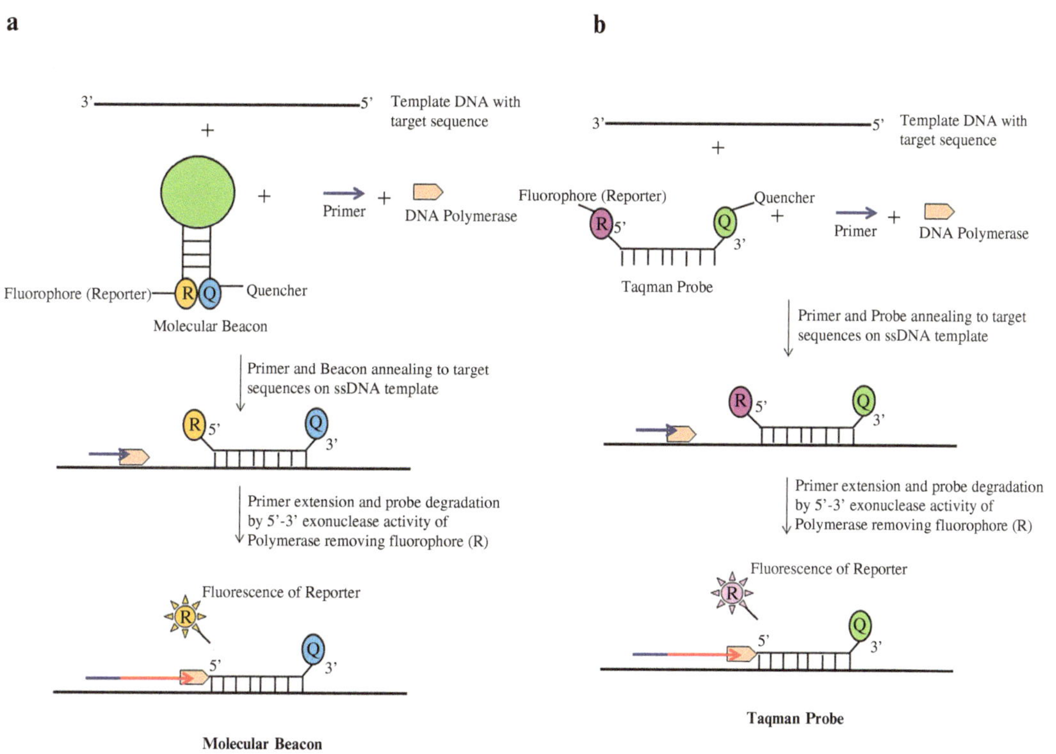

Fig. 7.10 Diagrammatic representation of (**a**) molecular beacons and (**b**) TaqMan probes in qPCR

2002; Boonham et al. 2004; Mumford et al. 2000; Schaad and Frederick 2002). TaqMan assay has been used to detect *Potato spindle tuber viroid* with ~1000 times more sensitivity than chemiluminescent assays (Boonham et al. 2004). This technique has been utilized for the successful screening of the virulent vectors and thrips for the presence of TSWV (Boonham et al. 2002) and has been efficiently contributed in the detection, quantitation and discrimination of *Wheat yellow mosaic virus* (WYMV) in wheat plants (Liu et al. 2013). The qPCR is finding its wide employability for quantitative detection of virus load and transcript level in diagnostics and gene expression studies, respectively. Other than diagnostics, it has also been used to analyse the consequences of viral infection on host gene expression by quantifying mRNAs and miRNAs in host plants.

7.4.8 Multiplex Real-Time PCR

This technique allows the simultaneous detection of multiple target sequences by the use of labelled probes designed for one target region in addition to the primer pair. The important feature of the labelling of probe is that the excitation/emission wavelength should not overlap with other fluorophores used in the same multiplex reaction as reporter dye. The major advantage of multiplex real-time PCR is that it can precisely detect and quantify multiple putative viruses in a sample. It is extensively used to determine the viral load(s) and single-nucleotide polymorphism (SNP) in plant viruses. It has been reported for the successful detection and differentiation of the immunologically cross-reactive *Chrysanthemum stem necrosis virus* and the closely related TSWV. Multiplex PCRs have been developed for the identification of multiple organisms and for identifying resistance- and virulence-specific sequences simultaneously. Simultaneous detection of multiple infections in plants is mainly possible by the use of degenerate primers having broad-spectrum specificity for the group or family of viruses. It has been used for the detection of *Tobacco rattle virus* and *Potato mop top virus* (Mumford et al. 2000). The major limitation of conventional multiplex PCR is the resolution of amplification products which can be dealt with unique fluorescent tagging of different probes (Mumford et al. 2000). This procedure has been used for the detection of two orchid viruses (Eun et al. 2000) and six viruses from olive trees (Bertolini et al. 2001). It has the ability to discriminate between severe and mild strains of CTV using unique oligonucleotide probes each for severe and mild isolates (Ruiz-Ruiz et al. 2009). Saponaria et al. (2013) have reported the use of multiplex qPCR assays for high-throughput and simultaneous detection of few important invasive citrus pathogens, namely, *Candidatus* Liberibacter asiaticus (*C*Las), CTV genotypes (VT and T3), *Hop stunt viroid* (HSVd) and *Citrus exocortis viroid* (CEVd).

7.5 Isothermal Amplification Methods

Isothermal amplification methods depend on the nonthermal separation or melting of the double strands of DNA. Such methods are similar to the conventional PCRs which also aim to amplify the target DNA using sequence-specific primers to carry out extensions with the DNA polymerase or in limited cases by RNA polymerase. Thus isothermal amplifications have primer binding to the target DNA without following thermal cycling or repeated denaturation and annealing steps.

Different procedures for nonthermal denaturation and annealing of template sequences and primer include transcription, strand displacement of circular DNA, nicking or partial degradation of primer extension products to allow extension or further priming and formation of secondary structure in single-stranded primer binding sites. These methods are being carried out at single temperature instead of thermal cycling and hence are cost-effective and require less power-consuming equipment. Being cheaper and simpler to perform, they can be easily used for field detection of viruses (Boonham et al. 2014). Table 7.3 summarizes the comparison of different isothermal methods.

Table 7.3 Comparison of several methods of isothermal amplification

	HDA	RPA	NASBA	LAMP	RCA
Reaction temperature	65 °C	37–42 °C	41 °C	60–65 °C	30 °C
Enzymatic denaturation	Yes; helicase	Yes; recombinase	No	No	No
Thermal denaturation	No	No	Initial thermal denaturation required	No	No
Primers used	Sequence specific	Sequence specific	A pair of target specific primer with T7 promoter sequence	3–4 pairs; including external and internal loop primers	Exo-resistant random hexamers
DNA polymerase used/source	Exo⁻ Klenow Fragment of DNA polymerase I; Bst DNA polymerase	Bsu DNA polymerase I (large fragment)/*Bacillus subtilis*	By simultaneous action of Avian myeloblastosis virus reverse transcriptase (AMV-RT), T7 DNA-dependent RNA polymerase (DdRp), RNaseH	Bst DNA pol./*Bacillus stearothermophilus*	*phi29* polymerase/ Bacteriophage φ
Reaction time	30–60 min	30 min or less	90 min	16–60 min	1 h
True isothermal process	Yes	Yes	No	Yes	Yes

7.5.1 Helicase-Dependent Amplification (HDA)

HDA is the isothermal amplification method that relies on the enzymatic duplex strand denaturation via helicase (Fig. 7.11).

The helicase-dependent separation of the DNA strands allows binding of primers followed by amplification by DNA polymerase at ~65 °C. The HDA reaction completes in 30–90 min. This method is convenient for diagnostics as it is better used for the amplification of short target DNA amplicons of ca. 70–120 bp (Vincent et al. 2004). The whole process of HDA is carried out at a single temperature. Reverse transcriptase-helicase-dependent amplification (RT-HDA) has been used for accurate, sensitive and early detection of *High plains virus* (HPV) damaging wheat crop at early developmental stages (Arif et al. 2014).

7.5.2 Recombinase Polymerase Amplification (RPA)

RPA methodology is based on enzymatic or non-thermal denaturation by the use of recombinase that forms primer-recombinase complex for the initiation of amplification (Fig. 7.12).

This method amplifies the target DNA by the use of recombinase, DNA polymerase and SSB proteins. It works with the complex of nucleoprotein and primers where recombinase scans dsDNA to facilitate primer binding to target sequence and displaces non-template strand which is stabilized by SSB proteins. The recombinase then dissociates from 3′-end of the primer and makes it accessible to DNA polymerase. This process leads to exponential amplification of the target sequence by cyclic replication. Usually this does not require initial denaturation step. But for better results, manual tube agitation for at least for 5 min is

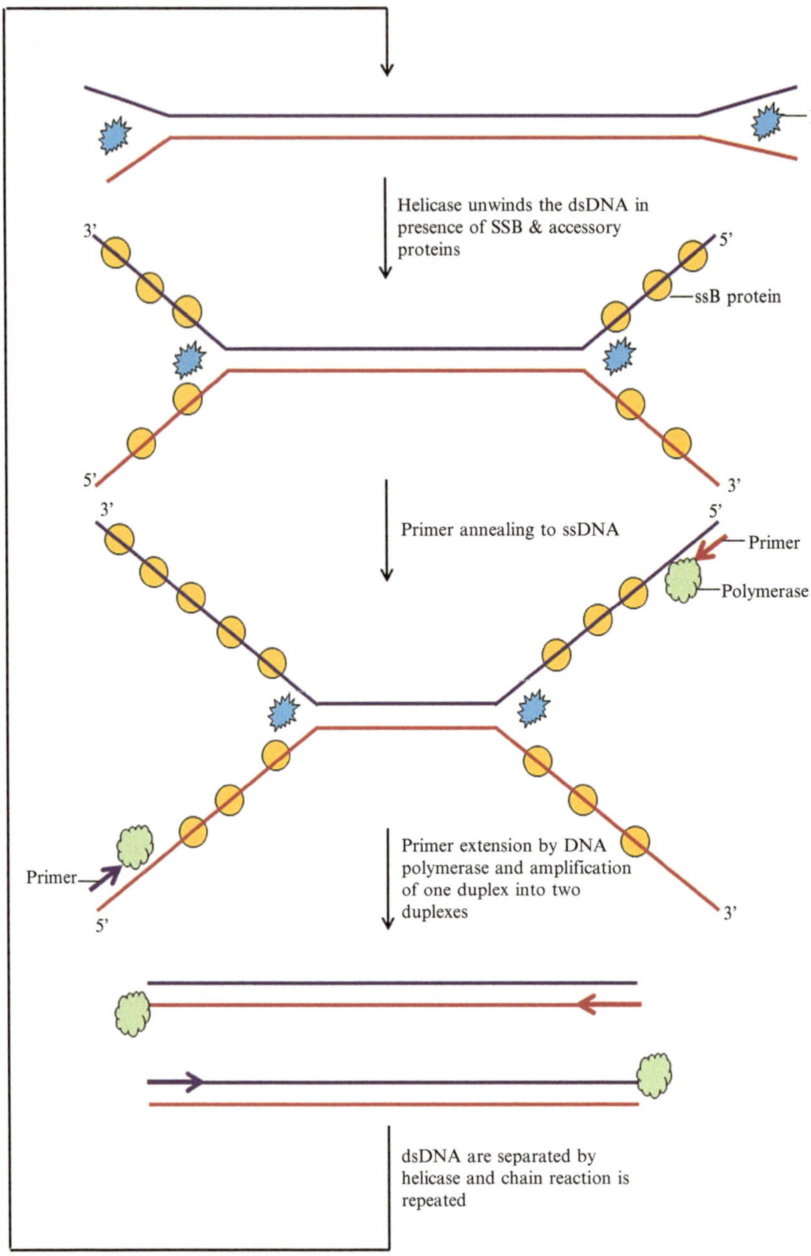

Fig. 7.11 Schematic representation of HDA (Adapted from Vincent et al. (2004))

preferred after amplification. The reaction temperature for the RPA usually ranges from 37°C to 42 °C and thus eliminates the need of high power-consuming instruments for thermal denaturation and incubation. At the same time, such low reaction temperatures may result in non-specific amplification in comparison with the isothermal amplification methods using relatively high reaction temperature. The major advantage associated with RPA is short reaction time of 30 min or less. Zhang et al. (2014) used RPA for the rapid detection of PPV in *Prunus* plants.

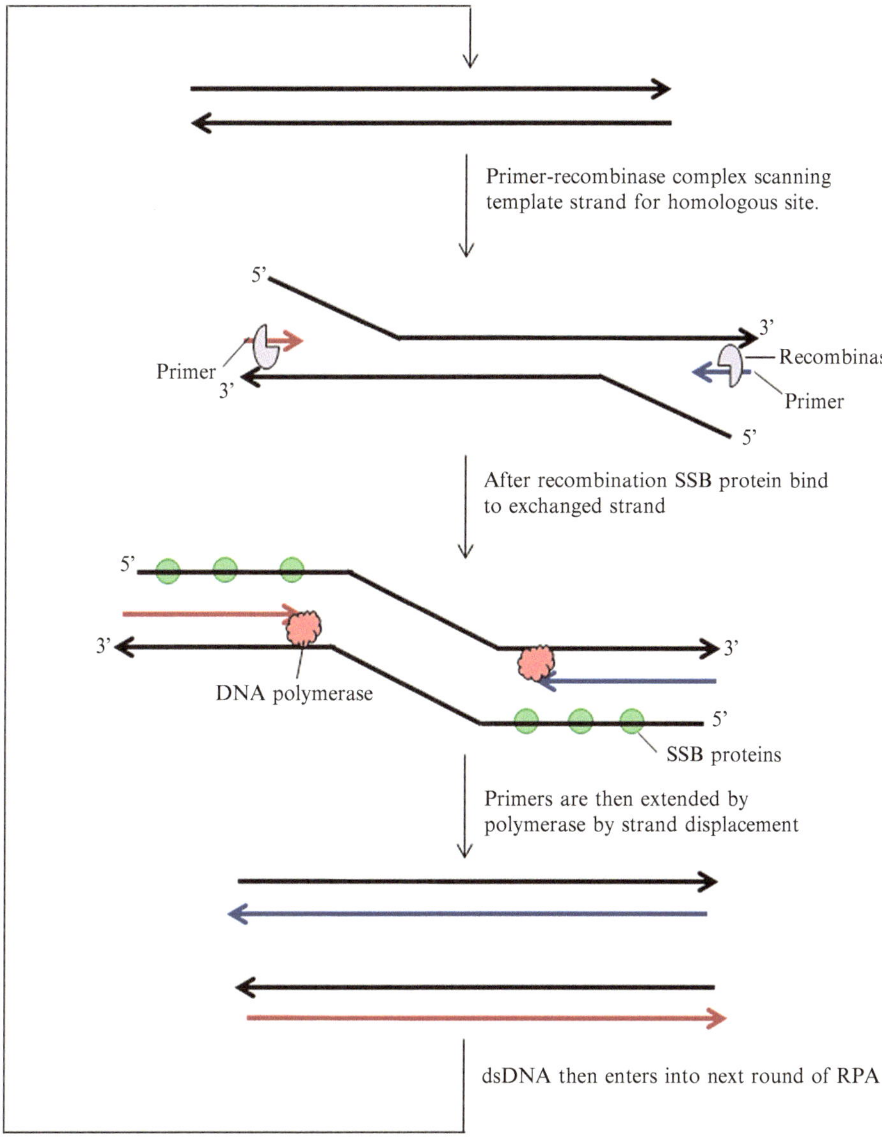

Fig. 7.12 Schematic representation of RPA (After Piepenburg et al. (2006))

7.5.3 Nucleic Acid Sequence-Based Amplification (NASBA)

In this method, RNA is isothermally amplified on the principles of transcription (Compton 1991). Primers are modified in a way that they can incorporate T7 RNA polymerase promoter sequence in the dsDNA intermediate. It makes the promoter functional and transcription reaction of the ssRNA product is performed at 41 °C (Fig. 7.13).

Olmos et al. (2007) used NASBA to detect *Plum pox virus*.

The modification of this approach using molecular beacons called AmpliDet is reported to detect various phytopathogens including ASPV (Klerks et al. 2001), PVY and *Arabis mosaic virus* (ArMV) (Szemes and Schoen 2003) and *Strawberry vein banding virus* (Vaskov et al. 2004). The AmpliDet is based on real-time measurement of fluorescence representing the

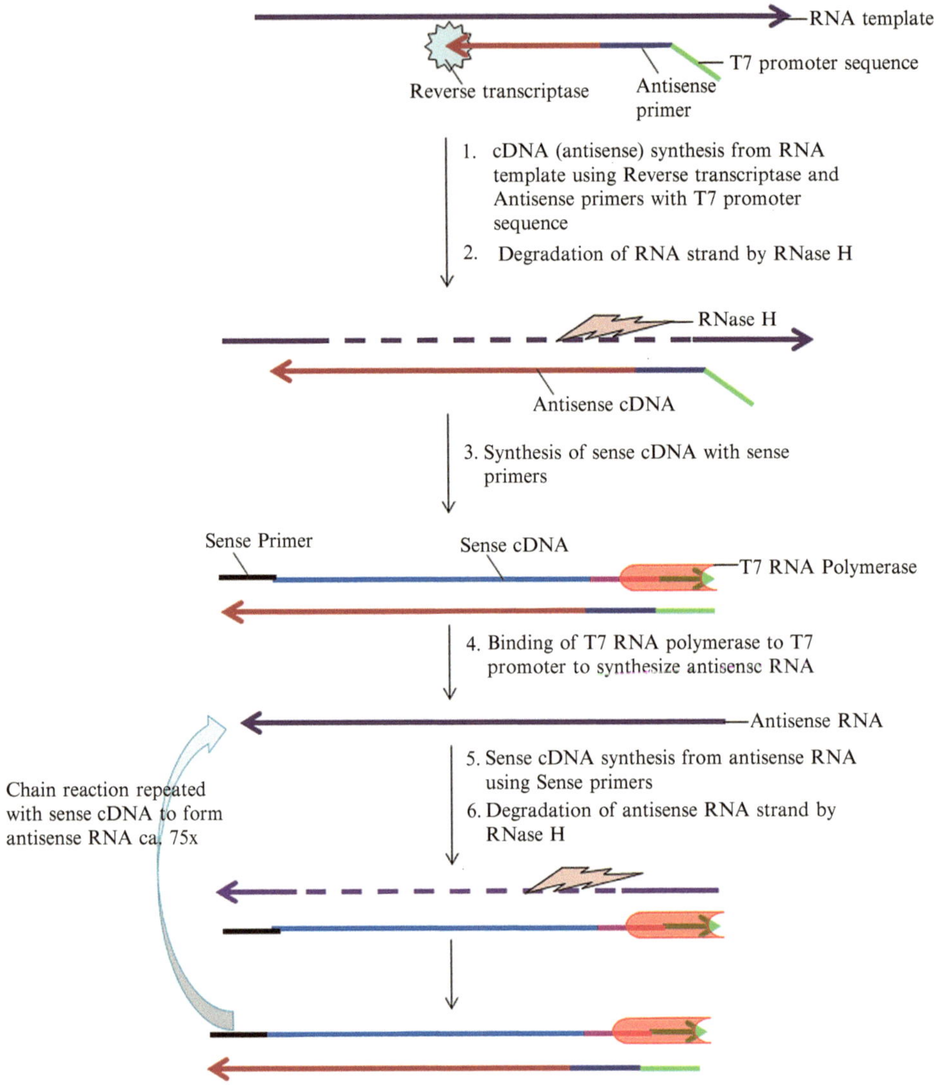

Fig. 7.13 Schematic representation of NASBA (Adapted from Compton (1991)

hybridization of the single-stranded target amplicon with the probes or beacon but requires specialized machine for monitoring. This method requires the initial denaturation of the template DNA for annealing of the primer followed by the addition of the non-thermostable polymerase and thus is not truly isothermal. The use of probe or molecular beacon makes it a very sensitive method for detection, but it has long reaction time of at least 90 min.

7.5.4 Loop-Mediated Isothermal Amplification (LAMP)

PCR and RT-PCR-based approaches have certain limitations, viz. the requirement of thermal cycler, low specificity and amplification efficiency in some instances. Therefore, LAMP is being developed that can amplify very low copies of target DNA in a very short time. It is rapid, highly sensitive and accurate. In a LAMP reaction, four primer pairs complementary to six specific regions in template DNA are used along with *Bst* DNA polymerase possessing high strand-displacement activity (Notomi et al 2000). The reaction is carried out at 65 °C without requiring repeated denaturation and is completed in 16–60 min. LAMP uses three pair of primers, i.e. internal, external and loop primers, generating single-stranded loop regions enabling the annealing of primers without template denaturation at ~60–65 °C. Therefore, LAMP can be used for accurate detection of viral targets in a short time without expensive thermocyclers. The LAMP primers can be designed using software PrimerExplorer V4 (https://primerexplorer.jp/elamp4.0.0/). LAMP Assay Versatile Analysis (LAVA) is a program for signature designing for LAMP primers. It identifies six LAMP primer signature regions and eight LAMP primers including loop primers (http://lava-dna.googlecode.com/).

Initially, LAMP or reverse transcriptase LAMP (RT-LAMP) has been also used in the diagnosis of infectious diseases. LAMP has been used to detect plant viruses and virus-like pathogens such as *Japanese yam mosaic virus* (Fukuta et al. 2003a), *Tomato yellow leaf curl virus* (Fukuta et al. 2003b), *Plum pox virus* (Varga and James 2006), *Banana bunchy top virus* (Peng et al 2012) and many more. RT-LAMP has been used for the diagnosis of *Zucchini yellow mosaic virus* (Kuan et al. 2014), *Potato leafroll virus* (Ahmadi et al. 2013) and *Wheat yellow mosaic virus* (Zhang et al. 2011). Nine rice viruses in Asia have been detected using RT-LAMP (Le et al. 2010). It has been used to detect and distinguish SRBSDV and RBSDV from host plants and insect vectors using RT-LAMP (Zhou et al. 2012). There is a great scope to use this technique for highly sensitive and specific detection of broad range of viral pathogens.

7.5.5 Rolling Circle Amplification (RCA)

The other major breakthrough which revolutionized the discovery and characterization of single-stranded circular DNA viruses especially begomoviruses is RCA (Haible et al. 2006).

It was used for the first time in cloning of a single-stranded circular DNA genome segment of a *Begomovirus* (Inoue-Nagata et al. 2004). RCA-based amplification protocol involves sequence-independent amplification at isothermal temperature using the *phi*29 polymerase. The technique has specific advantage of amplifying the novel sequence variants as it does not require the use of specific primers, but instead exo-resistant random hexamer primers are used (Johne et al. 2009; Jones 2009). RCA has been used to detect episomal badnavirus in banana (Baranwal et al. 2013). RCA has proved to be highly sensitive in comparison with PCR and serological methods for the detection and differentiation of integrated and episomal viral sequences of *Banana streak virus* (James et al. 2011; Wambulwa et al. 2012). Being isothermal, amplification method does not require expensive device such as thermocycler and thus the cost per reaction is low. The RCA-amplified fragment can be characterized by RFLP analysis and direct sequencing (Haible et al. 2006).

7.6 Nucleic Acid Sequence Hybridization Techniques

Nucleic acid hybridization facilitates the detection of plant viruses with single- or double-stranded genome using DNA/RNA probes. Polyprobe systems are designed for concurrent detection of several plant viruses.

7.6.1 In Situ Hybridization

This procedure is used for the detection of specific viral sequences or proteins. The combination of microscopy and hybridization has found significant application in plant virus diagnostics and in virus-host interactions. The experimental analysis gives an insight about the localization of target viral nucleic acid within a cell or tissue. Thus, it has been successfully used for the identification of replication-associated viral sequences

(Cillo et al. 2002). Also it has been used to track viral movement and detect any viral sequences integrated into the plant chromosome as in case of *Banana streak virus* infecting *Musa sp.* (Harper et al. 1999b).

7.6.2 Macroarray (Dot Blot/ Slot Blot)

DNA strands have very high affinity based on the sequence complementarity which gives high degree of specificity to interaction. Such specific interactions are the backbone of the assays based on nucleic acid hybridization where the two nucleic acid strands can be DNA-DNA/DNA-RNA/RNA-RNA. It uses a single-stranded labelled DNA or RNA to detect the hybridization. The label/tag can be radioisotope or nonradioactive tag so that target-probe hybrid molecules are detected depending on tag or reporter molecule attached to probe. Dot/spot and slot blot are the major techniques based on the hybridization of nucleic acids for plant virus detection and diagnosis (Maule et al. 1983; Garger et al. 1983; Owens and Diener 1984; Rosner et al. 1986; Baulcombe and Fernandez-Northcote 1988). Nucleic acid hybridization is a tool for the detection of single-/double-stranded viral nucleic acids using DNA or RNA probes (Fig. 7.14). The cDNA clones prepared from any segment of viral genome can be used as probes for the detection of viruses (Eweida et al. 1990) which maintain continuous supply of pure DNA and yield uniform results.

DNA probes can be labelled by PCR, random primer labelling or nick translation, while RNA probes are prepared by in vitro transcription. The cRNA probes with radioactive isotopes or nonradioactive label are preferred over cDNA probes, during the detection of RNA viruses due to the high stability of RNA-RNA duplex over DNA-RNA duplex on hybridization. But the major limitations of RNA probes are the risk of its degradation by RNAase contamination during hybridization and high cost. Thus, DNA probes become a more common choice for assays developed for plant virus detection. Macroarray has been used to detect 50 and 250 pg of *Cymbidium mosaic virus* and *Odontoglossum ringspot virus*, respectively, in the sap of diseased tissues used to immobilize viral nucleic acid on nylon membrane (Hu and Wong 1998).

Radiolabelled probes produce radioactive signals detected by autoradiography but do not have longevity due short half-life of ^{32}P and emit hazardous radioactive rays, thus requiring extensive and expensive procedures for safety maintenance. Hence, nonradioactive label probes such as biotin/streptavidin and digoxigenin (DIG) are popular (Dietzgen et al. 1994; Singh and Singh 1995; Wesley et al. 1996). The disadvantage of biotin/streptavidin probe is intense background produced by endogenous biotin in sap due to nonspecific binding with streptavidin. The digoxigenin (DIG)-labelled probes have been used widely for plant virus diagnostics, where hybridization is followed by exposure to anti-DIG antibodies conjugated with enzyme (ALP, HRP) to produce insoluble, chromogenic product from suitable substrate.

7.6.3 Microarray

DNA microarray chips are prepared by photolithography by etching a very large number of specific DNA/RNA sequences as the capture probes. Each nucleic acid fragment is specific for a DNA or RNA sequence in single assay. Glass, nylon or other polymers are used for the biochip preparation. A single microarray chip contains up to 70,000 short and unique DNA fragments. These arrayed probes can be short synthetic oligonucleotide of 30–50 bp or PCR products. The arrayed chips are allowed to hybridize with fluorescently tagged DNA/RNA in test sample using several fluorophores that reveal the presence/absence of target sequence in sample when read with laser technology.

Nanochips developed by Nanogen (San Diego, USA) contain streptavidin immobilized on the solid support for increased affinity of biotin-labelled DNA. The limitation of microarray is that concurrent detection of multiple targets such as bacteria and virus is ambiguous.

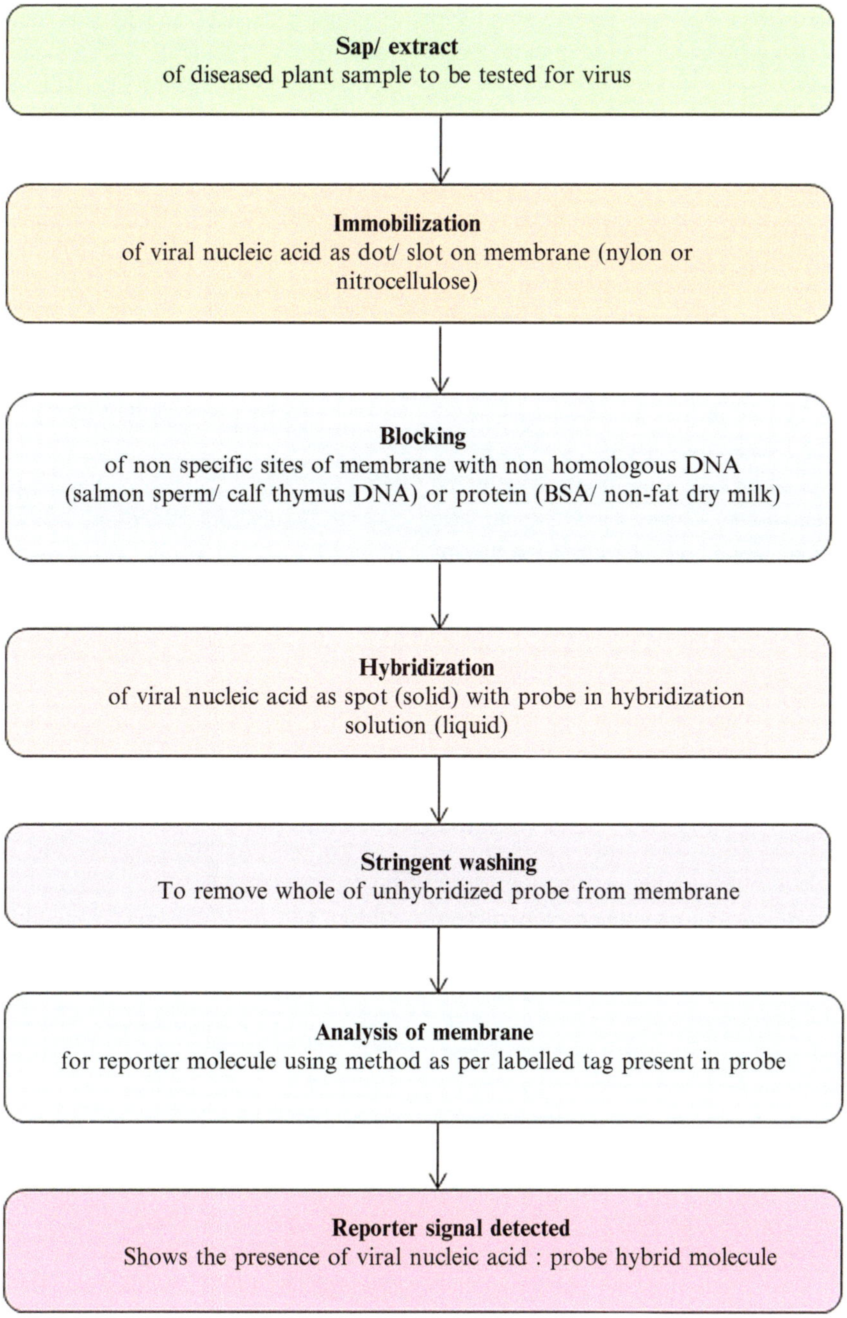

Fig. 7.14 Flow chart depicting steps of dot/slot blot hybridization assay

Different procedures used in various laboratories for microarray-based detection of plant pathogens demand the PCR-amplified fragments which reduces the sensitivity. Thus, utility of microarray is substantially reduced as a diagnostic tool as compared to its widespread use for functional genomics. Although this method is efficient in generating quick information about several pathogens, experimentation cost is very high and demands high skill to interpret the huge data generated.

Microarray is an emerging method offering alternative to conventional serodiagnostic methods (Boonham et al. 2007). DNA microarray can detect large number of virus-specific serotypes using specific probes in single assay with high precision and facilitate a comprehensive and unbiased analyses of viral prevalence. This new approach uses single-stranded short synthetic oligomers encrypted on microchips. It has been used for the detection of various potato viruses such as PLRV, PVA, PVY, PVX, PVM and PVS (Boonham et al. 2003) from single and mixed infected potato plants. This method discriminates sequences that show <80 % identities but detects variants that differ in sequence with >90 % identities. Thus, microarray is useful for discriminating viruses at the species level, especially RNA viruses having high degree of genome variability. DNA microarray chip is developed to screen diseased tomato plants for the presence of ten different economically significant viruses known to infect and damage the tomato cultivation by using "CombiMatrix" that uses 40-mer oligoprobes for detection. This platform was efficient and specific and could concurrently detect, differentiate and genotype multiple RNA viruses, their strains and viral satellite infecting tomato, viz. CMV, CMV-satellite RNA, *Tomato mosaic virus*, *Tomato infectious chlorosis virus*, *Tomato spotted wilt virus*, *Pepino mosaic virus*, *Tomato chlorosis virus*, PVY and TMV (Tiberini et al. 2010). A single-nucleotide polymorphism (SNP) utilizing single-base primer extension generated genotype of *Wheat streak mosaic virus* on DNA microarray chip. Among the commonly used techniques used for testing the plant materials including seeds and vegetative propagules, microarray gives the highest capabilities for specific and parallel screening. It provides a platform for that has much higher sensitivity than ELISA for testing samples for individual or mixture of viruses. DNA microarrays have potentially impartial ability to detect plant viruses using 30-mer and 50–70-mer oligoprobes. The assay uses random primers for the amplification of target sequences and its subsequent in vitro transcription to generate complementary RNA to be used for hybridization after labelling (Grover et al. 2010). The DNA virus chip provides a reliable useful tool for quarantine and certification purposes. It can also be used as unbiased tool for virus discovery on the basis of sequence similarities of new viral genome with the available microarray probes.

7.7 Next-Generation Sequencing (NGS)

NGS is a very recent, high-throughput comprehensive sequencing technology which has greatly reduced DNA sequencing cost than routine automated chain termination procedure. It generates millions or billions of DNA sequences which range from 25 to 400 nt. The sequences obtained are much smaller than those of conventional Sanger's sequencing reads which usually range from 300 to 750 nt. This technology has now been improved to generate sequences >750 nt in length.

Most recently, NGS platforms have been used in the discovery and characterization of novel viruses and their variants. It involves the rapid sequencing of hundreds of gigabases of sequences in single sequencing reaction using the highly sophisticated technology (Prabha et al. 2013). NGS platforms provide a high-throughput technology where the results can be obtained with high resolutions. NGS has been used as the most powerful tool for the discovery and characterization without the prior knowledge of macromolecular sequences (Barzon et al. 2011). The different NGS platforms used in the plant virus research are 454 FLX, Illumina, HeliScope, Ion Torrent, etc. (Schadt et al. 2010; Barzon et al. 2011; Prabha et al. 2013). This strategy includes either the extraction of total genome from the infected plant sample and its subsequent massive sequencing or the isolation of dsRNA (representing the viral RNA) followed by sequencing. The other strategies involved the isolation of siRNAs from plant samples having virus infection and its sequencing following NGS platforms; the reads thus obtained are then mapped to the particular viral genome (Kreuze et al. 2009). The use of NGS and metagenomics has been successfully demonstrated in the discovery of novel viruses from the infected samples of sweet potato, tomato, grapevine, pepper, cucumber, cotton, etc. (Prabha et al. 2013).

NGS has been used as a powerful tool for studies on plant pathogens, especially for plant virus identification. NGS is sequence-independent and culture-independent approach, used for concurrent detection of RNA/DNA viruses and viroids having very low titre in plant samples. It is a revolutionary technology for plant virus diagnostics aiming for easy identification of novel unidentified viruses as compared to classical amplification and immunological diagnostic procedures which uniquely target a species/strain, whereas other alternatives like electron microscopy and sap inoculation of indicator species support identification only at genus level. Thus, NGS has successfully proved its utility for survey and identification of viruses in numerous plant species (Barba et al. 2014; YaJuan et al. 2014). Also, it is an alternate method for the identification of viruses using sequence generated in non-specific manner by similarity search against GenBank.

NGS and conventional techniques have been used for the identification of a new disease of maize in Kenya. ELISA and TEM could not detect the presence of a virus, while sap inoculation on few cereal species showed that the sample was infected with unidentified sap transmissible virus(es). Symptomatic tissues were used for RNA purification and subsequent sequencing by Roche 454 GS-FLX+ generated reads which on database search recognized the resultant sequence as *Maize chlorotic mottle virus* and *Sugarcane mosaic virus* that are collectively reported to cause maize lethal necrosis disease (Adams et al. 2013). Such NGS data even help to characterize viral strains and develop specific diagnostic assays for qPCR. NGS data has enabled the identification of viral strains in infected materials from various fields/regions in Kenya (Adams et al. 2013). Thus, NGS is an advantageous tool that allows quick detection and identification/characterization of known and novel viruses/viral strains, respectively.

7.8 Conclusion

Plant virus diagnostics is gaining importance with the increased spread of viruses and threats of epidemics by known and/or novel/unidentified viruses. Proper diagnostic assays therefore help in the timely identification or detection of viruses to help enforcement of quarantine. The diagnostic methods discussed above broadly include various serological and molecular approaches. Among molecular approaches, PCR has gained a lot of popularity. Nowadays, qPCR is preferred because of its qualitative and quantitative detection free from post-PCR contamination. Developments of isothermal amplification-based techniques including HDA, RPA, NASBA, etc., are gaining acceptance for their suitability for poorly resourced laboratories and being cost-effective, quick methodologies supporting on-site testing of plant samples. Arrays are also a good choice of those aiming for accurate and reliable detection of broad range of viruses that migrated to new areas or infected new hosts. NGS is a revolutionary technique generating huge sequence data that helps to detect and identify new viruses/viral strains.

However, the challenge remains in the development of technologies that are cost-effective and easy to perform supporting infield detection/screening of suspected plants material and are affordable to underdeveloped and developing countries.

References

Adams IP, Miano DW, Kinyua ZM, Wangai A, Kimani E, Phiri N, Reeder R, Harju V, Glover R, Hany U, Souza-Richards R, Deb Nath P, Nixon T, Fox A, Barnes A, Smith J, Skelton A, Thwaites R, Mumford R, Boonham N (2013) Use of next-generation sequencing for the identification and characterization of *Maize chlorotic mottle virus* and *Sugarcane mosaic virus* causing maize lethal necrosis in Kenya. Plant Pathol 62(4):741–749

Agindotan B, Winter S, Lesemann D, Uwaifo A, Mignouna J, Hughes J, Thottappilly G (2006) Diversity of banana streak-inducing viruses in Nigeria and Ghana: twice as many sources detected by immunoelectron microscopy (IEM) than by TAS-ELISA or IC-PCR. Afr J Biotechnol 5(12):1194–1203

Ahmadi S, Almasi MA, Fatehi F, Struik PC, Maradi A (2013) Visual detection of potato leafroll virus by one-step reverse transcription loop-mediated isothermal amplification of DNA with hydroxynaphthol blue dye. J Phytopathol 161:120–124

Anderson PK, Cunningham AA, Patel NG, Morales FJ, Epstein PR, Daszak P (2004) Emerging infectious diseases of plants: pathogen pollution, climate change and agrotechnology drivers. Trends Ecol Evol 10:535–544

Arif A, Aguilar-Moreno GS, Wayadande A, Fletcher J, Ochoa-Corona FM (2014) Primer modification improves rapid and sensitive in vitro and field-deployable assay for detection of high plains virus variants. Appl Environ Microbiol 80(1):320–327

Bai FW, Zhang HW, Yan J, Qu ZC, Xu J, Wen JG, Ye MM, Shen DL (2002) Selection of phage-display peptides that bind specifically to the outer coat protein of *rice black streaked dwarf virus*. Acta Virol 46(2):85–90

Balamurlikrishnan M, Doraisamy S, Ganapathy T, Vishwanathan R (2002) Serological specificity and titre of sugarcane mosaic virus polyclonal antisera raised under varying immunization procedures and bleeding time. J Plant Dis Protection 109:646–654

Banttari EE, Khurana SMP (1998) The potato viruses and their management. In: Khurana SMP (ed) Pathological problems of economic crop plants and their management. Scientific Publishers, Jodhpur, India, pp 489–509

Banttari EE, Clapper DL, Hu SP, Daws KM, Khurana SMP (1991) Rapid magnetic microsphere enzyme immunoassay for potato virus X and potato leaf roll virus. Phytopathology 81:1039–1042

Baranwal VK, Arya M, Singh J (2010) First report of two distinct badnaviruses associated with *Bougainvillea spectabilis* in India. J Gen Plant Pathol 76:236–239

Baranwal VK, Sharma SK, Khurana D, Varma R (2013) Sequence analysis of shorter than genome length episomal banana streak OL virus like sequences isolated from banana in India. Virus Genes. doi:10.1007/s11262-013-0984-1

Barba M, Czosnek H, Hadidi A (2014) Historical perspective, development and applications of next-generation sequencing in plant virology. Viruses 6(1):106–136

Bartlett J (2003) PCR protocols. Humana Press, Totowa

Barzon L, Lavezzo E, Militello V, Toppo S, Palu G (2011) Applications of next-generation sequencing technologies to diagnostic virology. Int J Mol Sci 12: 7861–7884

Baulcombe DC, Fernandez-Northcote EN (1988) Detection of strains of potato virus X and of a broad spectrum of potato virus Y isolates by nucleic acid spot hybridization (NASH). Plant Dis 72:307–309

Bertolini E, Olmos A, Martinez MC, Gorris MT, Cambra M (2001) Single-step multiplex RT-PCR for simultaneous and colorimetric detection of six RNA viruses in olive trees. J Virol Methods 96:33–41

Bertolini E, Olmos A, Lopez MM, Cambra M (2003) Multiplex nested reverse-transcription polymerase chain reaction in a single tube for sensitive and simultaneous detection of four RNA viruses and Pseudomonas savastanoi pv. savastanoi in olive trees. Phytopathology 93:286–292

Bhat A, Varma A, Jain RK, Khurana SMP (1997) Differentiation of potato virus Y strains by N-terminal serology and HPLC peptide profiling. Indian Phytopathol 50:89–96

Biswas C, Dey P, Satpathy S (2013) A multiplex nested PCR assay for simultaneous detection of Corchorus golden mosaic virus and a phytoplasma in white jute (Corchorus capsularis L.). Lett Appl Microbiol 56(5):373–378. doi:10.1111/lam.12058

Boonham N, Barker I (1998) Strain specific recombinant antibodies to potato virus Y potyvirus. J Virol Methods 74:193–199

Boonham N, Smith P, Walsh K, Tame J, Morris J, Spence N, Bennison J, Barker I (2002) The detection of tomato spotted wilt virus (TSWV) in individual thrips using real-time fluorescent RT PCR (TaqMan). J Virol Methods 101:37–48

Boonham N, Walsh K, Smith P, Madagan K, Graham I, Barker I (2003) Detection of potato viruses using microarray technology: towards a generic method for plant viral disease diagnosis. J Virol Methods 108:181–187

Boonham N, Perez LG, Mendez MS, Peralta EL, Blockley A, Walsh K, Barker I, Mumford RA (2004) Development of a real-time RT-PCR assay for the detection of potato spindle tuber viroid. J Virol Methods 116:139–146

Boonham N, Tomlinson J, Mumford RA (2007) Microarrays for rapid identification of plant viruses. Annu Rev Phytopathol 45:307–328

Boonham N, Kreuzeb J, Winterc S, Vlugtd R, Bergervoetd J, Tomlinsona J, Mumforda R (2014) Methods in virus diagnostics: from ELISA to next generation sequencing. Virus Res. doi:10.1016/j.virusres.2013.12.007

Bowers JH, Bailey BA, Hebbar PK, Sanogo S, Lumsden RD (2001) The impact of plant diseases on world chocolate production. Online. Plant Health Prog. doi:10.1094/PHP-2001-0709-01-RV

Bragard C, Duncan GH, Wesley SV, Naidu RA, Mayo MA (2000) Virus-like particles assemble in plants and bacteria expressing the coat protein gene of Indian peanut clump virus. J Gen Virol 81(1):267–272

Byzova NA, Safenkova IV, Chirkov SN, Zherdev AV, Blintsov AN, Dzantiev BB, Atabekov IG (2009) Development of immunochromatographic test systems for express detection of plant viruses. Appl Biochem Microbiol 45(2):204–209

Chamberlain JS, Gibbs RA, Ranier JE, Nguyen PN, Caskey CT (1988) Deletion screening of the Duchenne muscular dystrophy locus via multiplex DNA amplification. Nucleic Acids Res 16:11141–11156

Chen J, Adams MJ (2001) A universal PCR primer to detect members of the Potyviridae and its use to examine the taxonomic status of several members of the family. Arch Virol 146:757–766

Cillo F, Roberts IM, Palukaitis P (2002) In situ localization and tissue distribution of the replication-associated proteins of cucumber mosaic virus in tobacco and cucumber. J Virol 76:10654–10664

Clackson T, Hoogenboom HR, Griffiths AD, Winter G (1991) Making antibodies using phage display libraries. Nature 352:624–628

Clark MF, Adams AN (1977) Characteristics of the microplate method of enzyme linked immunosorbent assay for the detection of plant viruses. J Gen Virol 34:475–483

Clark MF, Lister RM, Bar-Joseph M (1988) ELISA techniques. In: Weissbaum A, Weissbaum H (eds) Methods for plant molecular biology. Academic Press, San Diego, California, p 527

Compton J (1991) Nucleic acid sequence-based amplification. Nature 350(6313):91–92

Cooper JI, Edwards ML (1986) Variations and limitations of enzyme-amplified immunoassays. In: Jones RAC, Torrance L (eds) Developments and applications in virus testing. Association of Applied Biologists, Wellesbourne, pp 139–154

Cooper MA, Singleton VT (2007) A survey of the 2001 to 2005 quartz crystal microbalance biosensor literature: applications of acoustic physics to the analysis of biomolecular interactions. J Mol Recognit 20:154–184

Corazza MJ, Zanutto CA, Zanineli-Ré ML, Muller GW, Nunes WMC (2012) Comparison of *citrus tristeza virus* (CTV) isolates by RFLP analysis of the coat protein nucleotide sequences and by the severity of the symptoms. Trop Plant Pathol 37(3):179–184

Crosslin JM, Hamm PB, Shiel PJ, Hane DC, Brown CR, Berger PH (2005) Serological and molecular detection of tobacco veinal necrosis isolates of potato virus Y (PVYN) from potatoes grown in the western United States. Am J Potato Res 82:263–269

Dai J, Chen J, Huang T, Zheng X, Wu Y (2012) A multiplex reverse transcription PCR assay for simultaneous detection of five tobacco viruses in tobacco plants. J Virol Methods 183(1):57–62. doi:10.1016/j.jviromet.2012.03.029

Danks C, Barker I (2000) On-site detection of plant pathogens using lateral-flow devices. EPPO Bull 30:421–426

Deb M, Anderson JM (2008) Development of a multiplexed PCR detection method for barley and cereal yellow dwarf viruses, wheat spindle streak virus, wheat streak mosaic virus and soil-borne wheat mosaic virus. J Virol Methods 148:17–24

Derrick KS (1973) Quantitative assay for plant viruses using serologically specific electron microscopy. Virology 56:652–653

Desbiez C, Wipf-Scheibel C, Lecoq H (2002) Biological and serological variability, evolution and molecular epidemiology of Zucchini yellow mosaic virus (ZYMV, Potyvirus) with special reference to Caribbean islands. Virus Res 85:5–16

Dickert FL, Hayden O, Bindeus R, Mann KJ, Blaas D, Waigmann E (2004) Bioimprinted QCM sensors for virus detection screening of plant sap. Anal Bioanal Chem 378(8):1929–1934

Dietzgen RG, Xu Z, Teycheney PY (1994) Digoxigenin-labeled cRNA probes for the detection of two potyviruses infecting peanut (Arachis hypogaea). Plant Dis 78:708–711

Dijkstra J, DeJager CP (1998) Practical plant virology, protocols and exercises. Springer, Berlin/Heidelberg/New York

Dovas CI, Katis NI (2003) A spot nested RT–PCR method for the simultaneous detection of members of the Vitivirus and Foveavirus genera in grapevine. J Virol Methods 107:99–106

Drygin YF, Blintsov AN, Grigorenko VG, Andreeva IP, Osipov AP, Varitzev YA, Uskov AI, Kravchenko DV, Atabekov JG (2011) Highly sensitive field test lateral flow immunodiagnostics of PVX infection. Appl Microbiol Biotechnol. doi:10.1007/s00253-011-3522-x

Eun AJ, Seoh M, Wong S (2000) Simultaneous quantitation of two orchid viruses by the TaqMan real-time RT-PCR. J Virol Methods 87:151–160

Eun AJ, Huang L, Chew FT, Li SFY, Wong SM (2002) Detection of two orchid viruses using quartz crystal microbalance (QCM) immunosensors. J Virol Methods 99:71–79

Eweida M, Xu H, Singh BP, Abouhaidar MG (1990) Comparison between ELISA and biotin-labelled probes from cloned cDNA of potato virus X for the detection of virus in crude tuber extracts. Plant Pathol 30:623–628

Fajardo TVM, Barros DR, Nickel O, Kuhn GB, Zerbini FM (2007) Expression of *Grapevine leafroll-associated virus 3* coat protein gene in *Escherichia coli* and production of polyclonal antibodies. Fitopatol Bras 32:496–500

Fattouch S, Acheche H, M'hirsi S, Mellouli L, Bejar S, Marrakchi M, Marzouki N (2005) RT-PCR-RFLP for genetic diversity analysis of Tunisian Grapevine fan leaf virus isolates in their natural host plants. J Virol Methods 127:126–132

Fukuta S, Iida T, Mizukami Y, Ishida A, Ueda J, Kanbe M, Ishimoto Y (2003a) Detection of Japanese yam mosaic virus by RT-LAMP. Arch Virol 148(9):1713–1720

Fukuta S, Kato S, Yoshida K, Mizukami Y, Ishida A, Ueda J, Kanbe M, Ishimoto Y (2003b) Detection of tomato yellow leaf curl virus by loop-mediated isothermal amplification reaction. J Virol Methods 112(1–2):35–40

Garger SJ, Turpin T, Carrington JC, Morris TJ, Jordan RL, Dodds JA, Grill LK (1983) Rapid detection of plant RNA viruses by dot blot hybridization. Plant Mol Biol Report 1:21–25

Gawande SJ, Shukla A, Chimote VP, Kaushal N, Kaundal P, Garg ID, Chimote KP (2011) Development of PCR-based techniques for the detection of immobilised potato virus Y virions. J Plant Pathol 93(1):127–132

Gibbs A, Mackenzie A (1997) A primer pair for amplifying part of the genome of all 'potyvirids' by RT-PCR. J Virol Methods 63:9–16

Glais L, Kerlan C, Tribodet M, Tordo VMJ, Robaglia C, Astier-Manifacier S (1996) Molecular characterization of potato virus Y^N isolates by PCR-RFLP. Eur J Plant Pathol 102(7):655–662

Gonsalves D (1979) Detection of tomato ring spot virus in grapevines: a comparison of Chenopodium quinoa and enzyme linked immunosorbent assay (ELISA). Plant Dis Rep 63:962–965

Griep RA, van Twisk C, Schots A (1999) Selection of beet necrotic yellow vein virus specific single-chain Fv antibodies from a semi-synthetic combinatorial antibody library. Eur J Plant Pathol 105:147–156

Griep RA, Prins M, van Twisk C, Keller HJHG, Kerschbaumer RJ, Kormelink R, Goldbach RW, Schots A (2000) Application of phage display in selecting *tomato spotted wilt virus* specific single-chain antibodies (scFvs) for sensitive diagnosis in ELISA. Virology 90(2):183–190

Grimme JM, Cropek DM (2013) Biomimetic sensors for rapid testing of water resources. doi:10.5772/52438

Grover V, Pierce ML, Hoyt P, Zhang F, Melcher U (2010) Oligonucleotide-based microarray for detection of plant viruses employing sequence-independent amplification of targets. J Virol Methods 163:57–67

Hadidi A, Yang X (1990) Detection of pome fruit viroids by enzymatic cDNA amplification. J Virol Methods 30:261–270

Haible D, Kober S, Jeske H (2006) Rolling circle amplification revolutionizes diagnosis and genomics of geminiviruses. J Virol Methods 135:9–16

Harper G, Dahal G, Thottappilly G, Hull R (1999a) Detection of episomal banana streak badnavirus by IC–PCR. J Virol Methods 79:1–8

Harper G, Osuji J, Heslop-Harrison JS, Hull R (1999b) Integration of banana streak badnavirus into the Musa genome: molecular and cytogenetic evidence. Virology 255:207–213

Hema M, Kirthi N, Sreenivasulu P, Savithri HS (2003) Development of recombinant coat protein antibody based IC–PT–PCR for detection and discrimination of sugarcane streak mosaic virus isolates from southern India. Arch Virol 148:1185–1193

Heng CK, Noor SM, Yee TS, Othman RY (2007) Biopanning for *banana streak virus* binding peptide by phage display peptide library. J Biol Sci 7(8):1382–1387

Henson JM, Roy F (1993) The polymerase chain reaction and plant disease diagnosis. Annu Rev Plant Pathol 31:81–109

Hoogenboom HR, Griffiths AD, Johnson KS, Chiswell DJ, Hudson P, Winter G (1991) Multi-subunit proteins on the surface of filamentous phage: methodologies for displaying antibody (Fab) heavy and light chains. Nucleic Acids Res 15:4133–4137

Hu WW, Wong SM (1998) The use of DIG-labelled cRNA probes for the detection of Cymbidium mosaic potexvirus (CymMV) and Odontoglossum ringspot tobamovirus (ORSV) in orchids. J Virol Methods 70:193–199

Hull R (2002) Matthews' plant virology, 4th edn. Academic, San Diego

Inoue-Nagata AK, Albuquerque LC, Rocha WB, Nagata T (2004) A simple method for cloning the complete begomovirus genome using the bacteriophage ϕ29 DNA polymerase. J Virol Methods 116:209–211

Ito T, Ieki H, Ozaki K (2002) Simultaneous detection of six citrus viroids and apple stem grooving virus from citrus plants by multiplex reverse transcription polymerase chain reaction. J Virol Methods 106:235–239

Jacobi V, Bachand GD, Hamelin RC, Castello JD (1998) Development of a multiplex immunocapture RT-PCR assay for detection and differentiation of tomato and tobacco mosaic tobamoviruses. J Virol Methods 74:167–178

James D (1999) A simple and reliable protocol for the detection of apple stem grooving virus by RT-PCR and in a multiplex PCR assay. J Virol Methods 83:1–9

James AP, Geijskes RJ, Dale JL, Harding RM (2011) Development of a novel rolling-circle amplification technique to detect *banana streak virus* that also discriminates between integrated and episomal virus sequences. Plant Dis 95(1):57–62

Johne R, Muller H, Rector A, Van Ranst M, Stevens H (2009) Rolling-circle amplification of viral DNA genomes using phi29 polymerase. Trends Microbiol 17(5):205–211

Jones RA (2009) Plant virus emergence and evolution: origins, new encounter scenarios, factors driving emergence, effects of changing world conditions, and prospects for control. Virus Res 141:113–130

Joseph J, Savithri HS (1999) Determination of 3 ′-terminal nucleotide sequence of pepper vein banding virus RNA and expression of its coat protein in *Escherichia coli*. Arch Virol 144:1679–1687

Karande AA, Savithri HS, Khurana SMP (1998) Monoclonal antibodies production and application for the detection and diagnosis of potato viruses. In: Khurana SMP et al (eds) Comprehensive potato biotechnology. MPH, New Delhi, pp 163–186

Karyeija RF, Kreuze JF, Gibson RW, Volkonen JPT (2000) Two serotypes of sweetpotato feathery mottle virus in Uganda and their interaction with resistant sweetpotato cultivars. Phytopathology 90:1250–1255

Katul L (1992) Characterization by serology and molecular biology of bean leaf roll virus and Faba Bean Necrotic yellows virus. PhD thesis, Gottingen, Germany: University of Gottingen, 115 pp

Khan JA, Lohuid H, Goldbach RW, Dijkstra J (1990) Distinction of strains of *bean common mosaic virus* and *blackeye cowpea mosaic virus* using antibodies to N- and C- or N-terminal peptide domains of coat proteins. Ann Appl Biol 117:583–593

Khurana SMP (1990) Modern approaches for detection and management of the potato viruses and viroid. In: Grewal JS et al (ed) Current facets in potato research. IPA, CPRI, Simla, India, pp 98–108

Khurana SMP, Garg ID (1993) New techniques for detection of viruses and viroids. In: Chadha KL, Grewal JS (eds) Advances in horticulture, vol 7. MPH, New Delhi, pp 529–566

Kim YH, Kim OKS, Roh JH, Moon JY, Sohn SI, Lee SC, Lee JY (2004) Identification of *soybean mosaic virus* strains by RT-PCR/RFLP analysis of cylindrical inclusion coding region. Plant Dis 88(6):641–644

Klerks MM, Leone G, Lindner JL, Schoen CD, van den Heuvel JFJM (2001) Rapid and sensitive detection of apple stem pitting virus in apple trees through RNA amplification and probing with fluorescent molecular beacons. Phytopathology 91:1085–1091

Kreuze JF, Perez A, Untiveros M, Quispe D, Fuentes S (2009) Complete viral genome sequence and discovery of novel viruses by deep sequencing of small RNAs: a generic method for diagnosis, discovery and sequencing of viruses. Virology 388:1–7

Kuan CP, Deng TC, Huang HC, Chi HH, Lu YL (2014) Use of reverse transcription loop-mediated isothermal amplification for the detection of *Zucchini yellow mosaic virus*. J Phytopathol 162:238–244

Kubo C, Nakazono-Nagaoka E, Hagiwara K, Kajihara H, Takeuchi S, Matsuo K, Ichiki TU, Omura T (2005) New severe strains of *Melon necrotic spot virus*: symptomatology and sequencing. Plant Pathol 54:615–620

Kumari SG, Makkouk KM, Attar N (2006) An improved antiserum for sensitive serologic detection of Chickpea chlorotic dwarf virus. J Phytopathol 154:129–133

Kumari SG, Makkouk KM, Loh MH, Negassi K, Tsegay S, Kidane R, Kibret A, Tesfatsion Y (2008) Viral diseases affecting chickpea crops in Eritrea. Phytopathol Mediterr 47:42–49

Kushwaha R, Payne CM, Downie AB (2013) Uses of phage display in agriculture: a review of food-related protein-protein interactions discovered by biopanning over diverse baits. Comput Math Methods Med 2013:12

Kwon JY, Hong JS, Kim MJ, Choi SH, Min BE, Song EG, Kim HH, Ryu KH (2014) Simultaneous multiplex PCR detection of seven cucurbit-infecting viruses. J Virol Methods 206:133–139. doi:10.1016/j.jviromet.2014.06.009

Lamptey JNL, Osei MK, Mochiah MB, Osei K, Berchie JN, Bolfrey-Arku G, Gilbertson RL (2013) Serological detection of tobacco mosaic virus and cucumber mosaic virus infecting tomato (Solanum Lycopersicum) using a lateral flow immunoassay technique. J Agric Stud 1(2):102–113

Langeveld SA, Dore JM, Memelink J, Derks AF, Van der Vlugt CI, Asjes CJ, Bol JF (1991) Identification of potyviruses using the polymerase chain reaction with degenerated primers. J Gen Virol 72:1531–1541

Le DT, Netsu O, Uehara-Ichiki T, Shimizu T, Choi IR, Omura T, Sasaya T (2010) Molecular detection of nine rice viruses by a reverse-transcription loop-mediated isothermal amplification assay. J Virol Methods 170:90–93

LeProvost G, Iskra-Caruana ML, Acina I, Teycheney P-Y (2006) Improved detection of episomal banana streak viruses by multiplex immunocapture PCR. J Virol Methods 137:7–13

Lima JAA, Purcifull DE (1980) Immunochemical and microscopical techniques for detecting blackeye cowpea mosaic and soybean mosaic viruses in hypocotyls of germinated seeds. Phytopathology 70(2):142–147

Lima JAA, Nascimento AKQ, Radaelli P, Silva AKF, Silva FR (2011) Immune precipitation polymerase chain reaction for identification of plant viruses. In: Proceedings of XXII national meeting of virology, vol 16(1), pp 56, Atibaia, SP

Lima JAA, Nascimento AKQ, Radaelli P, Purcifull DE (2012) Serology applied to plant virology. In: Moslih Al-Moslih. (Org.) Serological diagnosis of certain human, animal and plant diseases. InTech, 1, Rijeka. pp 71–94

Liu W, Zhao X, Zhang P, Mar TT, Liu Y, Zhang Z, Han C, Wang X (2013) A one step real-time RT-PCR assay for the quantitation of wheat yellow mosaic virus (WYMV). Virol J 10:173

Livak KJ, Flood SJ, Marmaro J, Giusti W, Deetz K (1995) Oligonucleotides with fluorescent dyes at opposite ends provide a quenched probe system useful for detecting PCR product and nucleic acid hybridization. PCR Methods Applic 4:357–362

Luigi M, Costantini E, Luison D, Mangiaracina P, Tomassoli L, Faggioli F (2014) A diagnostic method for the simultaneous detection and identification of Pospiviroids. J Plant Pathol 96(1):151–158

Mackay IM, Arden KE, Nitsche A (2002) Real-time PCR in virology. Nucleic Acids Res 30:1292–1305

Makkouk KM, Kumari SG (1996) Detection of ten viruses by the tissue-blot immunoassay (TBIA). Arab J Plant Prot 14:3–9

Martin RR, James D, Le'vesque CA (2000) Impacts of molecular diagnostic technologies on plant disease management. Annu Rev Phytopathol 38:207–239

Maule AJ, Hull R, Donson J (1983) The application of spot hybridization to the detection of DNA and RNA viruses in plant tissues. J Virol Methods 6:215–224

McCafferty J, Griffiths AD, Winter G, Chiswell DJ (1990) Phage antibodies: filamentous phage displaying antibody variable domains. Nature 348(6301):552–554

Milne RG (1991) Immunoelectron microscopy for virus identification. In: Mendgen K, Lesemann DE (eds) Electron microscopy of plant pathogens. Springer, Berlin, pp 87–120

Mulholland V (2009) Immunocapture-PCR for plant virus detection. Methods Mol Biol 508:183–192

Mullis K, Faloona F (1987) Specific synthesis of DNA *in vitro* via a polymerase-catalyzed chain reaction. Methods Enzymol 155:335–350

Mullis K, Faloona F, Scharf S, Saiki R, Horn G, Erlich H (1986) Specific enzymatic amplification of DNA in vitro: the polymerase chain reaction. Cold Spring Harb Symp Quant Biol 51:263–273

Mumford RA, Walsh K, Barker I, Boonham N (2000) Detection of potato mop top virus and tobacco rattle virus using a multiplex real-time fluorescent reverse-transcription polymerase chain reaction assay. Phytopathology 90:448–453

Naidu RA, Hughes Jd'A (2001) Methods for the detection of plant virus diseases. In: Hughes Jd'A, Odu BO (eds) Plant virology in Sub-Saharan Africa. Proceedings of a conference organized by IITA. International Institute of Tropical Agriculture, Ibadan, pp 233–260

Narayanasamy P (2011) Detection of virus and viroid pathogens in plants. Microbial plant pathogens-detection and disease diagnosis. Springer Dordrecht Heidelberg London New York, pp 7–220

Nassuth A, Pollari E, Helmeczy K, Stewart S, Kofalvi SA (2000) Improved RNA extraction and one-tube RT-PCR assay for simultaneous detection of control plant RNA plus several viruses in plant extracts. J Virol Methods 90:37–49

Nie XH, Singh RP (2000) Detection of multiple potato viruses using an oligo(dT) as a common cDNA primer in multiplex RT-PCR. J Virol Methods 86:179–185

Notomi T, Okayama H, Masubuchi H, Yonekawa T, Watanabe K, Amino N, Hase T (2000) Loop-mediated isothermal amplification of DNA. Nucleic Acids Res 28, e63

Olmos A, Bertolini E, Cambra M (2007) Isothermal amplification coupled with rapid flow-through hybridization for sensitive diagnosis of plum pox virus. J Virol Methods 139(1):111–115

Ounouna H, Kerlen C, Lafaye P, Loukili MJ, ElGaaied A (2002) Production of monoclonal antibodies against synthetic peptides of the N-terminal region of potato virus Y coat protein and their use in PVY strain differentiation. Plant Pathol 51:487–494

Owens RA, Diener TO (1984) Spot hybridization for the detection of viroids and viruses. In: Maramorosch K, Koprowski H (eds) Methods in virology, vol VII. Academic, New York, pp 173–189

Pappu HR (1997) Managing tospoviruses through biotechnology: progress and prospects. Biotechnol Dev Monit 32:14–17

Peng J, Zhang J, Xiaa Z, Li Y, Huang J, Fan Z (2012) Rapid and sensitive detection of *banana bunchy top virus* by loop-mediated isothermal amplification. J Virol Methods 185:254–258

Piepenburg O, Williams CH, Stemple DL, Armes NA (2006) DNA detection using recombination proteins. PLoS Biol 4(7), e204

Prabha K, Baranwal VK, Jain RK (2013) Applications of next generation high throughput sequencing technologies in characterization, discovery and molecular interaction of plant viruses. Indian J Virol 24:157–165

Ragozzino E, Faggioli F, Barba M (2004) Development of a one tube-one step RT-PCR protocol for the detection of seven viroids in four genera: Apscaviroid, Hostuviroid, Pelamoviroid and Pospiviroid. J Virol Methods 121:25–29

Riedel D, Lesemann DE, Mai E (1998) Ultrastructural localization of nonstructural and coat proteins of 19 potyviruses using antisera to bacterially expressed proteins of *plum pox potyvirus*. Arch Virol 143:2133–2158

Rodoni BC, Dale JL, Harding RM (1999) Characterization and expression of the coat protein-coding region of *banana bract mosaic potyvirus*, development of diagnostic assays and detection of the virus in banana plants from five countries in Southeast Asia. Arch Virol 144:1725–1737

Rosenfield SI, Jaykus LAA (1999) Multiplex reverse transcription polymerase chain reaction method for the detection of foodborne viruses. J Food Prot 62:1210–1214

Rosner A, Lee RF, Bar-Joseph M (1986) Differential hybridization with cloned cDNA sequences for detecting a specific isolate of citrus tristeza virus. Phytopathology 76:820–824

Roy A, Fayad A, Barthe G, Brlansky RH (2005) A multiplex polymerase chain reaction method for reliable, sensitive and simultaneous detection of multiple viruses in citrus trees. J Virol Methods 129:47–55

Ruiz-Ruiz S, Moreno P, Guerri J, Ambrós S (2009) Discrimination between mild and severe *citrus tristeza virus* isolates with a rapid and highly specific real-time reverse transcription-polymerase chain reaction method using TaqMan LNA probes. Virology 99(3):307–315

Safenkova I, Zherdev A, Dzantiev B (2012) Factors influencing the detection limit of the lateral-flow sandwich immunoassay: a case study with potato virus X. Anal Bioanal Chem 403(6):1595–1605

Salomone A, Roggero P (2002) Host range, seed transmission and detection by ELISA and lateral flow of an Italian isolate of Pepino mosaic virus. J Plant Pathol 84:65–68

Salomone A, Bruzzone C, Minuto G, Minuto A, Roggero P (2002) A comparison of lateral flow and ELISA for the detection of tomato mosaic virus in tomato. J Plant Pathol 84:193

Salomone A, Mongelli M, Roggero P, Boscia D (2004) Reliability of detection of citrus tristeza virus by an immunochromatographic lateral flow assay in comparison with ELISA. J Plant Pathol 86(1):43–48

Sanchez-Navarro JA, Aparicio F, Herranz MC, Minafra A, Myrta A, Pallas V (2005) Simultaneous detection and identification of eight stone fruit viruses by one-step RT- PCR. Eur J Plant Pathol 111:77–84

Saponaria M, Loconsoleb G, Liaoc HH, Jiangd B, Savinob V, Yokomie RK (2013) Validation of high-throughput real time polymerase chain reaction assays for simultaneous detection of invasive citrus pathogens. J Virol Methods 193:478–486

Schaad NW, Frederick RD (2002) Real-time PCR and its application for rapid plant disease diagnostics. Can J Plant Pathol 24:250–258

Schaad NW, Opgenorth D, Gaush P (2002) Real-time polymerase chain reaction for one-hour on-site diagnosis of Pierce's disease of grape in early season symptomatic vines. Phytopathology 92:721–728

Schadt EE, Turner S, Kasarskis A (2010) A window into third-generation sequencing. Hum Mol Genet 19:227–240

Sherwood JL, German TL, Moyer JW, Ullman DE (2003) Tomato spotted wilt. Plant Health Instruct. doi:10.1094/PHI-I-2003-0613-02

Singh M, Singh RP (1995) Digoxigenin-labelled cDNA probes for the detection of potato virus Y in dormant potato tubers. J Virol Methods 52:133–143

Smith HG, Stevens M, Hallsworth PB (1991) The use of monoclonal antibodies to detect beet mild yellowing virus and beet western yellows virus in aphids. Ann Appl Biol 119:295–302

Sun W, Jiao K, Zhang S, Zhang C, Zhang Z (2001) Electrochemical detection for horseradish peroxidase-based enzyme immunoassay using p-aminophenol as substrate and its application to detection of plant virus. Anal Chim Acta 434:43–50

Susi P, Ziegler A, Torrance L (1998) Selection of single-chain variable fragment antibodies to black currant reversion associated virus from a synthetic phage display library. Phytopathology 88:230–233

Szemes M, Schoen CD (2003) Design of molecular beacons for AmpliDet RNA assay characterization of binding stability and probe specificity. Anal Biochem 315:189–201

Thelwell N, Millington S, Solinas A, Booth J, Brown T (2000) Mode of action and application of Scorpion primers to mutation detection. Nucleic Acids Res 28:3752–3761

Tiberini A, Tomassoli L, Barba M, Hadidi A (2010) Oligonucleotide microarray-based detection and identification of 10 major tomato viruses. J Virol Methods 168(1–2):133–140

Torrance L (1995) Use of monoclonal antibodies in plant pathology. Eur J Plant Pathol 101:351–363

Tyagi S, Kramer FR (1996) Molecular Beacons: probes that fluoresce upon hybridization. Nat Biotechnol 14:303–308

Uhde K, Kerschbaumer RJ, Koenig R, Hirschl S, Lemaire O, Boonham N, Roake W, Himmler G (2000) Improved detection of Beet necrotic yellow vein virus in a DAS ELISA by means of antibody single chain fragments (scFv) which were selected to protease-stable epitopes from phage display libraries. Arch Virol 145:179–185

Van den Heuvel JFJM, Peters D (1989) Improved detection of potato leafroll virus in plant material and in aphids. Phytopathology 79:963–967

Van Regenmortel MHV, Dubs MC (1993) Serological procedures. In: Matthews REF (ed) Diagnosis of plant virus diseases. CRC, Boca Raton, Florida pp 159–214

Varga A, James D (2006) Use of reverse transcription loop-mediated isothermal amplification for the detection of *plum pox virus*. J Virol Methods 138:184–190

Vashisht SK, Vashisht P (2011) Recent advances in quartz crystal microbalance-based sensors. J Sensors. doi:10.1155/2011/571405

Vaskov D, Spak J, Klerks MM, Schoen CD, Thompson JR, Jelkmann W (2004) Real-time NASBA for detection of strawberry vein banding virus. Eur J Plant Pathol 110:213–221

Villmor DEV, Bajet NB, Lockhart BE (2003) Purification and production of antiserum against abaca mosaic potyvirus. Philipp Agric Sci 86:368–376

Vincent M, Xu Y, Kong H (2004) Helicase-dependent isothermal DNA amplification. EMBO Rep 5:795–800

Walsh K, North J, Barker I, Boonham N (2001) Detection of different strains of potato virus Y and their mixed infections using competitive fluorescent RT–PCR. J Virol Methods 91:167–173

Wambulwa MC, Wachira FN, Karanja LS, Muturi SM (2012) Rolling circle amplification is more sensitive than PCR and serology-based methods in detection of banana streak virus in Musa germplasm. Am J Plant Sci 3:1581–1587

Waterhouse PM, Chu PWG (1995) Nucleic acid based approaches to plant virus and viroid diagnostics. In: Skerritt JH, Appels R (eds) New diagnostics in crop sciences, biotechnology in agriculture, vol 13. CAB International, Wallingford, pp 195–114

Webster CG, Wylie SJ, Jones MGK (2004) Diagnosis of plant viral pathogens. Curr Sci 86(12):1604–1607

Wesley SV, Miller JS, Devi PS, Delfosse P, Naidu RA, Mayo MA, Reddy DVR, Jana MK (1996) Sensitive broad-spectrum detection of Indian peanut clump virus by nonradioactive nucleic acid probes. Phytopathology 86:1234–1237

Wetzel T, Condresse T, Macquaire G, Ravelonandro M, Dunez J (1992) A highly sensitive immunocapture polymerase chain reaction method for plum pox potyvirus detection. J Virol Methods 39:27–37

Willment JA, Martin DP, Rybicki EP (2001) Analysis of the diversity of African streak mastreviruses using PCR-generated RFLP's and partial sequence data. J Virol Methods 93:75–87

Wittwer CT, Herrmann MG, Moss AA, Rasmussen AP (1997) Continuous fluorescence monitoring of rapid cycle DNA amplification. Biotechniques 22:130–138

Wylie SJ, Wilson CR, Jones RAC, Jones MGK (1993) A polymerase chain reaction assay for cucumber mosaic virus in lupin seeds. Aust J Agric Res 44:41–51

Xu H, Nie J (2006) Identification, characterization, and molecular detection of Alfalfa mosaic virus in potato. Virology 96(11):1237–1242

Xu H, D'Aubin J, Nie J (2010) Genomic variability in *potato virus M* and the development of RT-PCR and RFLP procedures for the detection of this virus in seed potatoes. Virol J 7:25

Yadav N, Khan JA (2008) Identification of a potyvirus associated with mosaic disease of *Narcissus* sp. in India. Plant Pathol 57:394

Yadav N, Khan JA (2009) Over expression of *Narcissus* potyvirus coat protein in *E. coli*. Indian J Virol 21:44

YaJuan Q, Yi X, Qi Z, XuePing Z (2014) Application of next-generation sequencing technology for plant virus identification. Scientia Sinica Vitae 44(4):351–363

Yang IC, Hafner GJ, Revill PA, Dale JL, Harding RM (2003) Sequence diversity of South Pacific isolates of *Taro bacilliform virus* and the development of a PCR-based diagnostic test. Arch Virol 148:1957–1968

Zhang ZY, Liu XJ, Li DW, Yu JL, Han CG (2011) Rapid detection of wheat yellow mosaic virus by reverse transcription loop-mediated isothermal amplification. Virol J 8:550

Zhang S, Ravelonandro M, Russell P, McOwen N, Briard P, Bohannon S, Vrient A (2014) Rapid diagnostic detection of plum pox virus in Prunus plants by isothermal AmplifyRP(®) using reverse transcription-recombinase polymerase amplification. J Virol Methods 207:114–120

Zheng L, Rodoni BC, Gibbs MJ, Gibbs AJ (2010) A novel pair of universal primers for the detection of potyviruses. Plant Pathol 59:211–220

Zhou T, Du L, Fan Y, Zhou Y (2012) Reverse transcription loop mediated isothermal amplification of RNA for sensitive and rapid detection of southern rice black-streaked dwarf virus. J Virol Methods 180:91–95

Ziegler A, Torrance L (2002) Applications of recombinant antibodies in plant pathology. Mol Plant Pathol 3: 401–407

Application of Microbial Enzymes in Dissolving Pulp Production

8

Prabhjot Kaur, Nishi K. Bhardwaj, and Jitender Sharma

Abstract

The uprising demand in dissolving pulp, a special chemical pulp, during the last decade has fascinated the researchers to develop a modern biotechnology which could either improve its existing processes or facilitate novel processes of production in eco-friendly manner. These include the use of different microbial enzymes or the microorganisms themselves in various bioprocesses such as biopulping and/or biobleaching of sulphite pulp or bioconversion of kraft pulp to dissolving pulp. The hydrolytic enzymes specifically xylanases and cellulases have been used as the process tools rendering the benefit of eco-friendly and economic bioprocess. Special emphasis is paid to convert kraft pulp, originating from both wood and nonwood, into dissolving pulp by using xylanases and cellulases to selectively reduce hemicelluloses and improve pulp reactivity, respectively. The viscose process being a major consumer of dissolving pulp has drawn more attention. Extensive research work has been conducted to achieve high pulp reactivity as well as accessibility towards solvent and reagent for reducing the carbon disulphide (CS_2) consumption in viscose process. Here, the various characteristic properties of dissolving pulp and its end use with various processes, including

P. Kaur
Kurukshetra University,
Kurukshetra 136 119, Haryana, India

Avantha Centre for Industrial Research and Development, Paper Mill Campus,
Yamuna Nagar 135 001, Haryana, India
e-mail: prabh_jot99@rediffmail.com

N.K. Bhardwaj (✉)
Avantha Centre for Industrial Research and Development, Paper Mill Campus,
Yamuna Nagar 135 001, Haryana, India
e-mail: nishi.bhardwaj@gmail.com; bhardwaj@avanthresearch.org

J. Sharma
Department of Biotechnology, Kurukshetra University,
Kurukshetra 136 119, Haryana, India
e-mail: jksharmakuk@rediffmail.com

P. Shukla (ed.), *Frontier Discoveries and Innovations in Interdisciplinary Microbiology*,
DOI 10.1007/978-81-322-2610-9_8

existing and novel, for its production are reviewed. Microbial enzymes, namely, xylanases and cellulases, for their immense potential as process tools are briefly discussed along with their mode of action.

Keywords
Dissolving pulp • Xylanases • Cellulases • Kraft pulp • Reactivity

8.1 Introduction

Nowadays, the increasing demand for textile fibres due to ever-growing population is largely covered by petroleum-based synthetic fibres. However, due to continuous depletion of fossil resources and other environmental concerns, there is tremendously growing interest in the utilisation of renewable resources for production of various products as a sustainable approach. For decades, the forest industry is mainly focused on the production of pulp for paper and board; however, a small niche is contributed towards the production of dissolving pulp. Dissolving pulp is a special chemical pulp and a key material to manufacture rayon, cellophane, cellulose acetate and other cellulose derivatives. Rayon is a main product of regenerated cellulose and a natural man-made fibre used in textile industry. More than three fourth of dissolving pulp is consumed in viscose process for the production of cellulosic fibres including viscose rayon staple, acetate staple, filament yarn, etc. (Bajpai 2012). The uprising demand in dissolving pulp during the last decade has fascinated the researchers to develop a modern biotechnology which could either improve its existing processes or facilitate novel processes of production in eco-friendly manner. To attain this purpose, the use of microbial enzymes in the production of dissolving pulp has become an active area of research.

Enzymes are the special catalysts that find a vast number of applications in various industries apart from playing a vital role in metabolic reactions. Except a small group of catalytic RNA molecules, all enzymes are proteins in their primary composition. They display a wide variety at supramolecular level due to different folding patterns. The folding involves both covalent and non-covalent interactions within the polypeptide chain and is directed by its amino acid sequence. It is because of this structural distinctiveness that they exhibit the unique specificity for their substrate and are able to speed up the reactions at a good pace without being consumed. Moreover, they are eco-friendly and infallible in their performance at their optimum conditions. They are considered as the main tools for the application of basic biotechnological techniques (genetic engineering and cell fusion), the target of therapeutics drugs and the indispensible intermediate in all bioprocesses (fermentation and cell culture) (Vitolo 2009).

Development of industrial processes for the production of rayon for textiles and cellophane film for packaging led to the development of the dissolving pulp industry in 1911. Until then, the pulp industry was concerned only with the manufacture of fibres for the paper industry; chemical derivatives of cellulose were produced from purified rags or cotton linters. Development of dissolving pulp from wood made available a relatively cheap and pure source of cellulose as a raw material for the expanding chemical industry (McGinnis and Shafizadeh 1979).

8.1.1 Dissolving Pulp

Dissolving pulp is a special chemical pulp consisting of 90–98 % cellulose with very little amounts of hemicelluloses, extractives, minerals and residual lignin. It exhibits uniform molecular weight distribution and high brightness. Dissolving pulp is a low yield pulp in comparison to other chemical pulps. In chemical pulping, raw material (generally wood) is cooked in the presence of acid or alkali at controlled conditions of temperature, pressure and time to selectively remove lignin and other impurities, thereby separating the fibres (in the form of pulp). The digested raw material turns into a brown-coloured pulp due to reprecipitation of lignin and is further bleached to get the desired pulp properties. In general the resulting pulp containing both cellulose and hemicellulose is utilised for papermaking. Hemicellulose is retained in paper grade pulp for better yield and strength, while in dissolving grade pulp, its higher content causes detrimental effects in end product quality. Therefore, the high content of cellulose is desirable. Moreover, cellulose molecule properties are important, while fibre properties are not directly of interest. As shown in Fig. 8.1, dissolving pulp is manufactured either by the kraft process using a pre-hydrolysis step to remove hemicelluloses or by an acid sulphite process from both hardwoods and softwoods. Currently there are mainly two types of processes to produce dissolving pulp at mill scale. The acid sulphite process, being a major type, is covering nearly 65 % of production, whereas pre-hydrolysis kraft process contributes approximately 25 % of the total production (Sixta 2006; Bajpai 2012).

The acid sulphite process removes the major polymeric constituents other than cellulose such as hemicelluloses and lignin in single step, which later encumbers in the recovery process of hemicelluloses from the spent liquor. Still the recovery rate of the inorganic cooking chemicals is higher along with the advantage of totally chlorine-free (TCF) bleaching. The usage of raw material in the process is restricted to a narrow range of wood species because of its limited ability to dissolve extractives. Besides, more pollution is generated due to high BOD (biochemical oxygen demand) of sulphite mill effluents and high sulphur dioxide losses (Gustafsson et al. 2011). Also the pulp so produced shows broad molecular weight distribution.

The other process utilises alkaline cooking conditions after a pre-hydrolysis stage in which the hemicelluloses are withdrawn from the wood chips. Pre-hydrolysis is technically a pre-extraction stage in which the low DP (degree of polymerisation) carbohydrates specifically hemicelluloses undergo auto-hydrolysis caused by the release of acetic acid. In addition to the benefits of high pulp quality and high cooking capacity, the PHK (pre-hydrolysis kraft) cooking also has

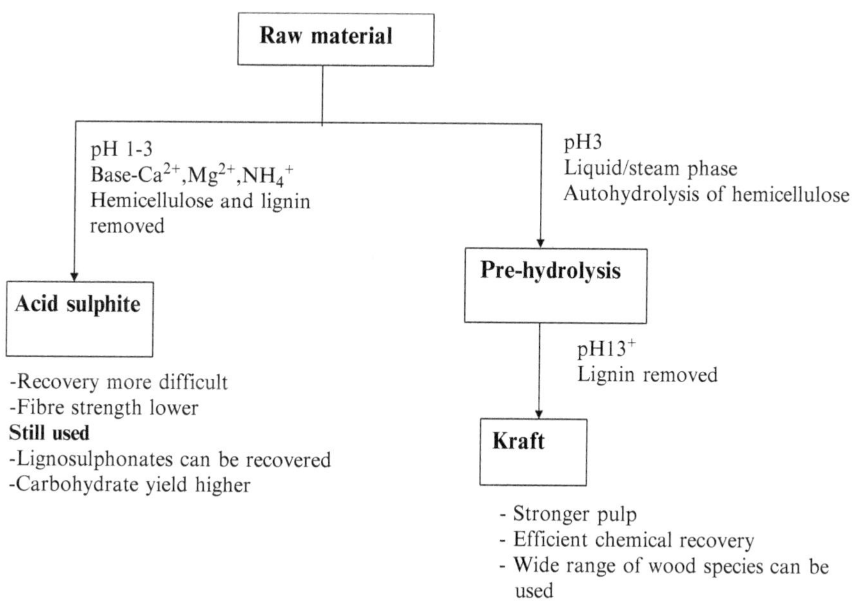

Fig. 8.1 Pulping processes for the production of dissolving pulp

a disadvantage of the formation of pitch-like compounds which hampers the drainage process.

Dissolving pulps have higher production costs in comparison to common kraft pulps as a consequence of costs of wood, capital and chemicals required for a given pulp yield. The lower yield of dissolving pulp than kraft pulp is attributed to retention of hemicelluloses in the pulping process of the latter. Low dissolving pulp yield requires more equipment to be employed to increase production rate. Moreover, this speciality pulp requires inventories and storage spaces prior to its delivery to specific customers with particular requirements (Hillman 2006).

Besides the high costs of dissolving pulp production, this industry is also confronted by the stringent environmental laws that have triggered interest in using enzymes for (1) advancement of paper grade pulps into dissolving grade pulp by selectively reducing the hemicellulose content and subsequently activating the pulps, (2) reactivity improvement of dissolving pulp for better end product quality and (3) investigation of other biotechniques, e.g. biopulping, biobleaching etc., in order to facilitate or improve the performance of conventional processes for dissolving grade pulp production.

With the advent of enzyme technology in dissolving pulp industry, a number of bioprocesses have been developed, the implementation of which can result in:

1. Savings in chemicals for pulping and bleaching
2. Improvement in the quality of dissolving pulp
3. Improvement in the effluent quality
4. Improvement in resource utilisation with less impact on the environment

Although the enzyme technology for dissolving pulp production is new, it is gaining momentum all over the world, with a major aim of developing a new bioprocess for the production of dissolving pulp, the implementation of which would lead to positive changes in the dissolving pulp properties as well as mill effluent quality in an environmentally friendly and cost-effective way.

8.1.1.1 Dissolving Pulp and Cellulose

Dissolving pulp is predominantly cellulosic in its chemical composition. Depending upon the process of production, it exhibits wide changeability in performance. Although cellulose is the only desired constituent, it would be of no use if the form of native cellulose is modified. Here, the molecular properties of cellulose chiefly determine the end use and quality of product. The plants originally produce cellulose I (native cellulose) during wood formation, and both celluloses I and II have been found in pulp. Other polymorphs of crystalline cellulose are celluloses III and IV (Christoffersson-Elg 2005). The polymorphism is most typical of crystals of organic compounds whose molecule contains groups capable of hydrogen bonding (Bernstein 2002). The basic chemical structure of cellulose exhibits β-1, 4-linked D-glucopyranose rings arranged so that glucosidic oxygens point in opposite directions giving rise to cellobiose as repeating unit (Fig. 8.2). This rotation of the adjacent monomer units in the cellulose chain results in intramolecular hydrogen bonding between the ring oxygen O(5) atom of one glycosyl unit and the hydrogen atom of C-3 hydroxyl group of the preceding ring and similarly between the hydroxyl group on C(2) and the primary hydroxyl group on C(6).

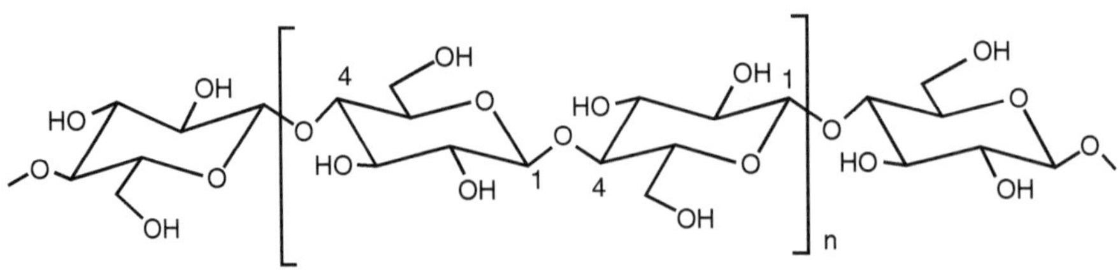

Fig. 8.2 Primary structure of cellulose molecule showing β-1, 4-linked D-glucopyranose residues

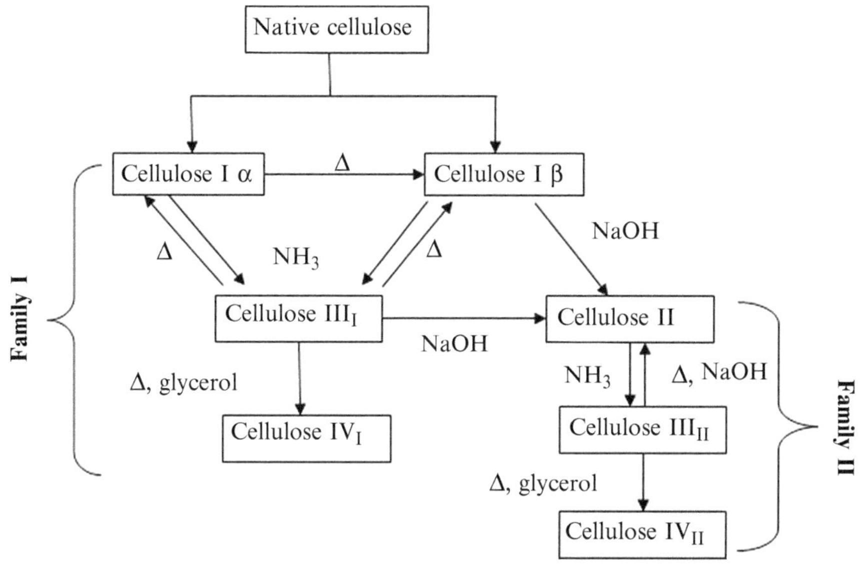

Fig. 8.3 Cellulose polymorphism

In cellulose I, the intermolecular hydrogen bonding between hydroxyl groups on C(6) and C(3′) in parallel chains results in the formation of cellulose sheets, whereas in cellulose II, the chains of cellulose are oriented antiparallel to each other and with different hydrogen bond patterns. Moreover, it carries an extra hydrogen bond per anhydroglucose unit (Henriksson and Lennholm 2009).

As the monomers bear three hydroxyl groups, their hydrogen bonding ability plays a major role in directing crystalline packing and in governing important physical properties (Dufresne 2012). More crystallinity and less amorphous regions in cellulose render poor accessibility to the solvents or reactants for any modification to occur. An efficient dissolution of cellulose in viscose process or its derivatisation hence require the availability of hydroxyl groups and the further access of the reactants to avail these hydroxyl groups for respective reactions to take place. In viscose process, regenerated cellulose filaments are obtained in an acid bath by spinning cellulose dissolved in CS_2 and alkali. Cellulose II is formed when pulp is immersed in strong alkali (18 % NaOH) solution during mercerisation and also when cellulose is regenerated in the acidic bath in the viscose manufacturing process. Cellulose III_{II} and cellulose IV_{II} are formed upon the subsequent treatment of cellulose II with liquid ammonia and with glycerol at high temperature (Henriksson and Lennholm 2009). As shown in Fig. 8.3, this constitutes family II which is converted from family I in an irreversible reaction since the former is thermodynamically more stable (Isogai et al. 1989).

8.1.1.2 Derivatives and End Products

The end products of dissolving pulp either require its dissolution in a suitable solvent or undergo various derivatisation reactions such as esterification, nitration, etherification, etc. Various end-use products derived from dissolving pulp and the corresponding manufacturing processes have been listed in Table 8.1. In the preparation of cellulose derivatives, e.g. cellulose acetate, basic reaction introduces an acetyl functional group $COCH_3$ on cellulose upon the addition of acetic anhydride in the presence of a small amount of catalyst (perchloric acid or sulphuric acid). Esterification reaction involves the addition of ester functional group COO on cellulose by condensation of a carboxylic acid group COOH and alcohol group OH (Dufresne 2012).

The dissolution and subsequent regeneration of cellulose in viscose process results in regener-

Table 8.1 Dissolving pulp end-use product and demand

Manufacturing process	Derivatives	End-use product	Demand by end use (million tons)
Xanthation	*Viscose rayon*		3.381
	Tyre cord	Tyre and belting reinforcement	
	Nonwovens	Filters, surgical packs, drapes, gown	
	Regular staple	Apparel	
	Continuous filament	Apparel	
	High-wet-modulus staple	Apparel, furnishing, parachute cord	
	Cellophane	Packaging	
	Industrial yarn	Heat ablative, brake	
	Miscellaneous	Sponges, sausage casing	
Acetylation	*Acetate*		0.735
	Tow	Filter cigarettes	
	Filament	Apparel, furnishing	
	Plastic mouldings	Film, sheet, extruded articles	
Ethering	*Cellulose ethers*		0.588
	Carboxymethyl cellulose	Detergents, cosmetics, food, textile and paper sizes, well drilling muds	
	Hydroxyethyl cellulose	Latex paints, oil well drilling muds, emulsion polymerisation	
	Methyl cellulose	Food, paints, pharmaceuticals	
	Ethyl cellulose	Coating and inks	
	Hydroxylpropyl cellulose	Foods and pharmaceuticals	
	Carboxymethyl hydroxyethyl cellulose	Liquid detergents	
Nitration	*Nitrates*		0.147
	Lacquers		
	Explosives		
	Celluloid		
	Film		
Others			0.049
Zinc chloride	Vulcan fibre	Lamination, insulation, tight machining	
Latex	Artificial leather	Fabrics, clothing	
Resin	Laminating and impregnation papers		
			Total = 4.9[a] million tons in 2010

[a]Including 1.1 million tons cotton linter pulp. Based on data from Hinck et al. (1985) and Floe (2011)

ated cellulose (mainly rayon). The most common solvent to dissolve cellulose is CS_2, which is very toxic, volatile and malodorous. Depending upon the conversion process used, there is a wide range of end products of dissolving pulp in various industries such as textile industry, pharmaceutical industry and food industry. In textile production, it has a unique advantage of complete recyclability. Cellulose nitrates containing higher N-content have been used extensively for military purposes. In addition, it can be used as a raw material of numerous food additives such as E-additives 460–469 except E462. Moreover a high-quality dissolving pulp finds its application in the manufacture of non-discolouring frames, cellophane and plastics as cellulose acetate

Table 8.2 Some important quality parameters of dissolving pulp

Chemical properties	Description
α-cellulose	Represents undamaged long-chain cellulose content; should be >90 %
Alkali solubility	Measures solubility of pulp (S10, S18, S21.5) in different concentrations of NaOH, represents the amount of hemicellulose and degraded cellulose present in the pulp
S10 (%)	Optimum swelling of pulp; both hemicellulose and degraded cellulose (up to <150 DP) are dissolved
S18 (%)	Hemicellulose (<50 DP) dissolved. Sometimes degraded cellulose of very low DP (usually after cellulase treatment) also dissolves. **Viscose process yield = 0.88(100-S18)** (Hinck et al. 1985)
S21.5 (%)	(100-S21.5) is proportional to viscose process yield
Reactivity (%)	Most critical parameter; methods include viscose filter value (Treiber 1987) and more common Fock method (Fock 1959); Fock reactivity should be >65%
Pentosan (%)	Represents major hemicellulose fraction of hardwoods; should be 0.5–10 %
Viscosity (ml/g)	Corresponds to the length of cellulose chain; gives a relative indication of the degradation (decrease in cellulose molecular weight)
Molecular weight distribution	Determined by gel permeation chromatography (GPC); should be uniform
Degree of polymerisation	Related to the molecular weight (M) by the formula **DP = M/162**, where 162 is the molecular weight of the anhydroglucose unit (Christoffersson-Elg 2005) and to intrinsic viscosity [η] for cellulose dissolved in cupriethylenediamine by $\mathbf{[\eta] = 1.33*DP^{0.905}}$ Mark–Houwink equation (Gustafsson et al. 2011)
Ash (%)	Should be <0.06 %
Extractives (%)	Should be <0.5 %
Residual lignin	Should be <0.2 %
Brightness (% ISO)	Should be >90 %
Yield (%)	30–35 %

(Mäntyranta 2013). Recently, dissolving pulps seem to be the preferred substrate for the manufacture of nanofibrillated cellulose (NFC), a future precursor of advanced materials (Sixta et al. 2013).

8.1.2 Dissolving Pulp Properties

Dissolving pulp is a high-quality pulp intended for the manufacture of various industrial products. Although the end-use product defines the quantitative constitution and the purity required, nevertheless there are certain fundamental requirements (Table 8.2) that must be fulfilled prior to use of dissolving pulp as a raw material.

8.1.2.1 Pulp Reactivity

The reactivity of pulp is certainly an important quality parameter of dissolving pulp since it signifies the processability or suitability of dissolving pulp to be used as raw material for the viscose and other derivatisation processes (Sixta 2006). The reactivity of a pulp directly affects the subsequent derivatisation reactions, such as acetylation, xanthation or nitration, and is mainly measured by the availability of C2 and C6 hydroxyl groups of the monomeric units of cellulose to the reactants (Sixta 2006). These hydroxyl groups are therefore mainly responsible for the reactions of cellulose. The hydroxyl group at the 6th position of the carbon atoms in anhydroglucose unit acts as a primary alcohol, whereas at the 2nd and 3rd positions, hydroxyl groups behave as secondary alcohols. From the esterification studies, it has been established that the relative reactivity is ten times more for the hydroxyl group at the 6th position than the other OH groups (Hebeish and Guthrie 1981; Dufresne 2012). Meanwhile, hydroxyl group at the 2nd position has been found to react two times faster than the hydroxyl group at the 3rd position (Dufresne 2012). Also, more amorphous cellulose in a pulp may exhibit higher reactivity, but for a quality product, homogenous substitution along the cellulose molecule is desired Zhu et al. (2006). Furthermore, the secondary structure of cellulose is also important as it gives rise to different hydrogen bond patterns and thus reactivity. Since an increased number of hydrogen bonds are present in cellulose II, its reactivity towards different derivatisation reactions is poor. Moreover the ratio of the two allomorphs, Iα and Iβ, present in cellulose I also influences the pulp reactivity as Iα is metastable and thus more reactive than Iβ, while the latter prevails in higher plants (Dufresne 2012).

It also seems that fibre wall supramolecular structure of cellulose hinders the efficient removal of hemicelluloses resulting in some residual xylan in the pulp. Besides, the presence of residual hemicelluloses is thought to impede the subsequent conversion processes by competing with cellulose in the reactions involved. Therefore, a detailed description of the fibre surface is required in order to gain the information on residual hemicelluloses and fibrillation as well as fibril structure. This further helps in understanding the associated phenomenon of hornification, the accessible inner surface and the pore volume to define the term reactivity (Krässig 1993).

8.1.2.2 Alkali Solubility

Low molecular weight carbohydrates (hemicellulose and degraded cellulose) can be extracted from pulps with sodium hydroxide solutions, referred to as alkali solubility. Solubility of a pulp in alkali thus provides information on the degradation of cellulose and on loss or retention of hemicelluloses during pulping and bleaching processes. The solubility of pulp in 10 % NaOH at 25 °C and 1.5 % consistency (S10) represents optimum swelling of pulp that results in dissolution of not only the degraded cellulose fraction (up to <150 DP) but the hemicellulose fraction as well. S18 or solubility in 18 % NaOH exhibits somewhat reduced swelling power, solubilising essentially the shorter chemical residual hemicellulose fraction (<50 DP). However, this is not always true; especially for pulps with low DP, both the low molecular weight cellulose and the hemicelluloses are dissolved (Christoffersson-Elg 2005). α-cellulose content is also a significant value for dissolving pulp and a measure of the long-chain cellulose content. For viscose pulps, 0.88(100-S18) as well as resistance of pulps to 21.5 % sodium hydroxide solution R21.5 (100-S21.5) is proportional to viscose process yield (Hinck et al. 1985).

8.1.2.3 Pentosan Content

Pentosan content in pulp is the indication of the amount of hemicellulose in general, lost or retained during pulping and bleaching processes. Pentosan content in softwoods is about 7–10 % and in hardwoods about 19–25 %. This is so because hardwood hemicellulose consists mainly of pentosans, whereas softwood hemicellulose consists of both pentosans and hexosans. Since hemicellulose contributes to the strength of paper pulps, high pentosan content is desirable, whereas in dissolving pulps, particularly acetate pulps, pentosan content should be kept low.

8.1.2.4 Viscosity/Degree of Polymerisation/Molecular Weight

Viscosity corresponds to the length of the cellulose chains and can be measured in different solvents. The solution viscosity of a pulp gives a measurement of the degree of polymerisation of the cellulose. Such an analysis therefore is useful in determining the relative indication of the degradation (decrease in cellulose molecular weight) caused by pulping and/or bleaching process. Molecular weight distribution and polydispersity, measured by gel permeation chromatography (GPC), viscosity measurements or other methods, are important parameters for characterising the pulp. The degree of polymerisation (DP) is related to the molecular weight (M) by the formula DP=M/162, where 162 is the molecular weight of the anhydroglucose unit (Christoffersson-Elg 2005), and to intrinsic viscosity $[\eta]$ for cellulose dissolved in cupriethylenediamine by the following Mark–Houwink equation (Gustafsson et al. 2011):

$$[\eta] = 1.33 \times DP^{0.905}$$

The molecular weight can be expressed in various ways, for example, the number average molecular weight (DP_n) or the weight average molecular weight (DP_w), depending on the method used. The polydispersity index, the ratio DP_n/DP_w, corresponds to the width of the molecular weight distribution.

8.1.2.5 Ash Content and Extractives

The inorganic impurities of the pulps, collectively described as the ash content, mainly consist of silicates, sulphates and carbonates, as well as metal ions combined to the acidic groups of the pulp. A common subdivision is acid-soluble and acid-insoluble ashes, the latter being mainly the silicates. These impurities affect the processing properties of the dissolving pulps. Some impurities, mainly iron, manganese and copper, catalytically accelerate the air oxidation and impair the brightness, brightness stability of pulps, rayon colour and aging of the alkali cellulose. The presence of carbonates and carbonate forming ions, such as calcium and magnesium, causes trouble in the pressing of the alkali cellulose, the filtration of viscose and the spinning of the rayon. Polyvalent metal ions cause viscosity anomalies and haze in the cellulose acetate prepared from impure acetate pulp. Extractives are those constituents that are soluble in neutral organic solvents, which can be both either lipophilic or hydrophilic. The compounds which are extracted by acetone are called acetone extractives (%). The content of extractives should be as low as possible, although they have positive effects on the viscose process. However, it is better to have low constant levels than uncontrolled varying levels in order to maintain constant viscose process parameters. For dissolving pulps to have good quality, they must not exceed 0.5 %.

8.1.2.6 Kappa Number and Brightness

Residual lignin in pulp is not easy to measure directly due to its complex and variable chemical structure. But the lignin content correlates well with the consumption of certain oxidants like potassium permanganate, and the amount consumed is expressed as kappa number (Gustafsson et al. 2011). For dissolving grade pulp, residual lignin should be below 1 %. Dissolving pulp exhibits high brightness and whiteness along with good dye ability.

8.1.3 Microbial Enzymes

Microbes have always been exploited as a major source of industrial enzymes. They have been used along the centuries in industrial processes as tannery, brewing, bakery, dairy, etc., but their potential use in the pulp and paper industry was realised in the late 1980s, and thereafter, it has shown a rapid growth. This growth has been triggered by various factors that include:

1. Better understanding and improved knowledge of the interactions between enzymes and the pulp constituents
2. An increased requirement of the industry to adopt eco-friendly approach
3. Improved production technology and process economics for the relevant enzymes

The enzymes can be obtained from any living form such as plants (e.g. papain, ficin, bromelain), animals (pepsin, trypsin and chimosin) and microorganisms (keratinases, pectinases and amylases, among others), but the tendency of using microbial enzymes on an industrial scale is prominent. Since they render plentiful supply, less sophisticated equipment/conditions for production and easy control on all the phases of production, they are economical than their counterparts (Vitolo 2009). Also the commercial processes are well standardised for the production. There exist more than 3000 different known enzymes of which commercially used enzymes are 150–170 only (Binod et al. 2013). Moreover, a diverse range of enzymes similar to those of plants and animals can be synthesised by genetically manipulating microbial strains through RDT (recombinant DNA technology).

Potential benefits of using microbial enzymes in pulp and paper industry have been realised for the past 30 years. The enzymes were first employed at mill scale during the 1980s, shortly after the discovery and validation of the xylanase-aided bleaching concept. Since then, the industrial enzymes with a wide pH and temperature ranges suitable for target processes have been developed rapidly. In addition, novel enzymatic and microbial applications have been investigated for improving the processing of wood fibres as well as for improving the sustainability of the pulping, bleaching and papermaking processes (Viikari 2002). The integration of hemicellulases and cellulases or microbial strains into processing of dissolving pulp production along with some chemical treatments for the attainment of better quality product in a cost-effective manner as well as reducing the environmental load could be of great benefit to the industry.

Enzymes are known to modify a number of pulp and paper properties. Of the different enzymes and their applications, cellulases have been used for deinking (Heitmann et al. 1992; Pathak et al. 2011; Prasad et al. 1993) and dewatering of recycled fibre (Verma et al. 2013), xylanases have been implemented for biobleaching and also to improve the mechanical properties of pulps and the refining process (Bhardwaj et al. 1996; Buchert et al. 1998; Cadena et al. 2010; Gil et al. 2009), lipases have been used for pitch control (Fischer et al. 1993) and laccases or the laccase-mediator system have been used in the bleaching process (Ibarra et al. 2006). Microorganisms have also been investigated for the biotechniques such as biopulping and biobleaching to produce dissolving pulp (Christov et al. 1998; Christov 1999; Ferraz et al. 1998 and Mosai et al. 1999).

The use of enzymes in pulp and paper industry is still being explored, either directly acting upon the pulp to selectively alter the properties for intended purpose or modification of materials used in the papermaking such as starch, nanocellulose, etc. One of the enzyme applications includes the reactivity improvement of dissolving pulp by using cellulases (Cao and Tan 2002, 2006; Engström et al. 2006; Henriksson et al. 2005; Kvarnlöf et al. 2005; Rahkamo et al. 1996, 1998). Furthermore, endoglucanases have been applied for surface treatments in the textile and laundry industries. Cellobiohydrolases are used to decrease the energy consumed during mechanical pulping, due to their good defibrillation properties (Teeri 1998).

On the other hand, the most widely used application of xylanases is in the pre-bleaching of kraft pulp to reduce the use of bleaching chemicals and the associated environmental problems along with an improved quality of product (Allison et al. 1993; Bajpai et al. 1994; Bim and Franco 2000; Senior and Hamilton 1991, 1992; Yang et al. 1992; Zhan et al. 2000). Studies have also shown their applicability in desizing and scouring of fabric (Csiszar et al. 2001; Dhiman et al. 2008a; Losonczi et al. 2005), increasing water retention, reducing beating times in virgin pulps, restoring bonding and increasing freeness in recycled fibres (Dhiman et al. 2008b). Xylan-degrading enzyme systems also have considerable potential in other biotechnological applications including the bioconversion of lignocelluloses and agricultural wastes into fermentative products, biobleaching in paper manufacture, liquefaction of coffee mucilage, alteration of the rheological and organoleptic properties of musts and wines, extraction of pigments and flavours, protoplast formation and production of modified hemicelluloses for industrial use (Coughlan 1992). Recently, xylanases have been used alone or in combination with alkaline extraction for dissolving grade pulp production as demonstrated by Bajpai and Bajpai (2001), Gehmayr et al. (2011), Jackson et al. (1998) and Köpcke (2010).

8.2 Mode of Action

8.2.1 Xylanases

Xylanases (EC3.2.1.8) are a group of hydrolytic enzymes known to cleave the 1, 4-β-D xylosidic bonds in xylan (Collins et al. 2005). Xylan is a major part of hemicelluloses in hardwood. Due to its high molecular mass, the penetration of xylan into the cell wall is not feasible, whereas its low molecular mass fragments can readily enter inside the cell and regulate the enzyme biosynthesis. These short fragments are actually the products of constitutively produced enzymes that are present in the small amounts in the cell. The liberated molecules released by the action of these enzymes are consisting of xylose, glucose and their heterodisaccharides and positional isomers together with fragmented xylan of different DP such as xylobiose, xylotriose and xylooligosaccharides (Motta et al. 2013). For the complete hemicellulose degradation, a complement of about 24 enzymes is required (Shrinivasan 1992; Battan et al. 2007). Hemicellulases are typically produced as multiple isozymes, i.e. enzymes virtually with the same catalytic action but display different kinetic parameters or different regulatory properties. On the basis of amino acid or nucleic acid sequence of their catalytic modules, they are classified into two groups, glycosidic hydrolases (GHs) or carbohydrate esterases (CEs), and further grouped into various families according to their primary sequence homology (Henrissat and Bairoch 1996; Rabinovich et al. 2002a, b; Battan et al. 2007). Moreover, the information provided by Carbohydrate-Active Enzyme (CAZy) (2014) database (http://www.cazy.org) on sequence annotations of enzymes related to carbohydrates and glycoconjugates metabolism has associated the xylanases with GH families 5, 7, 8, 9, 10, 11, 12, 16, 26, 30, 43, 44, 51 and 62. Among them, the families 5, 7, 8, 10, 11 and 43 possess a distinct catalytic domain exhibiting endo-1, 4-β-xylanase activity, whereas 16, 51 and 52 families seem to be consisting of two catalytic domains with bifunctional enzymes. Other families, namely, 9, 12, 26, 30 and 44, may exhibit side activity of xylanase (Motta et al. 2013).

Of several enzymes with different functions which participate in the hydrolysis of xylan to its monomers, the most significant are the endo-β-1, 4- xylanases (Sunna and Antranikian 1997). These enzymes degrade xylan to short-chain xylooligosaccharides of varying lengths. On the basis of their hydrophobic cluster analysis of the catalytic domain and sequence homology, these enzymes are chiefly grouped into families 10 and 11 in the numerical classification of glycosyl hydrolases (Battan et al. 2007). Although there have been extensive study done to understand the catalytic properties of the members of the families 10 and 11 and several models have been proposed to explain the mechanism of their action, little is known about the remaining families 5, 7, 8 and 43. Xylanases are hydrolytic enzymes and the bond cleavage takes place by either the retention or the inversion of the anomeric carbon of the reducing sugar monomer of the substrate. The families that undergo retention of the anomeric configuration are 5, 7, 10 and 11. The catalytic mechanism which is also a double-displacement mechanism requires the involvement of two glutamate residues resulting in the formation of a covalent glycosyl-enzyme intermediate for subsequent hydrolysis. In the first step, the carboxylate group of one glutamate residue acts as a general acid by donating a H^+ to O-4′ that links C-1 of the incipient reducing sugar and C-4′ of the departing glycosyl fragment, while the second carboxylate group carries out a nucleophilic attack leading to C_1–$O_{4'}$ bond cleavage and departure of aglycone. Thus, α-glycosyl-enzyme intermediate is formed. Meanwhile, a water molecule near the glycosidic oxygen side is dissociated by the first carboxylate group (now acts as a general base) and resulting hydroxyl ion reacts with the anomeric carbon leading to a second substitution. In this manner the β-configuration at the anomeric centre is retained. Also, xylanases that undergo retaining mechanism are believed to effect hydrolysis via boat transition state (Fig. 8.4a), rather than the half chair that appears to be used by glucosidases and cellulases. Considering the hypothesis that the bulky C5-hydroxymethyl substituent is placed in an unfavourable axial position in a boat conformation would explain the absolute xylan specificity of family 11, while the broad specificity of family 10 enzymes for cellulose/xylan is believed to be attributed by the half chair transition state (Rye and Withers 2000).

Members of GH families 8 and 43 operate through inversion mechanism in which the catalytic residues are perhaps glutamate and aspartate that lead to single-displacement reaction. As its name suggests, the reaction catalysed by these inverting enzymes occurs in single step in which one carboxylate protonates the substrate and the second triggers a nucleophilic water molecule which in turn attacks the anomeric carbon. As a

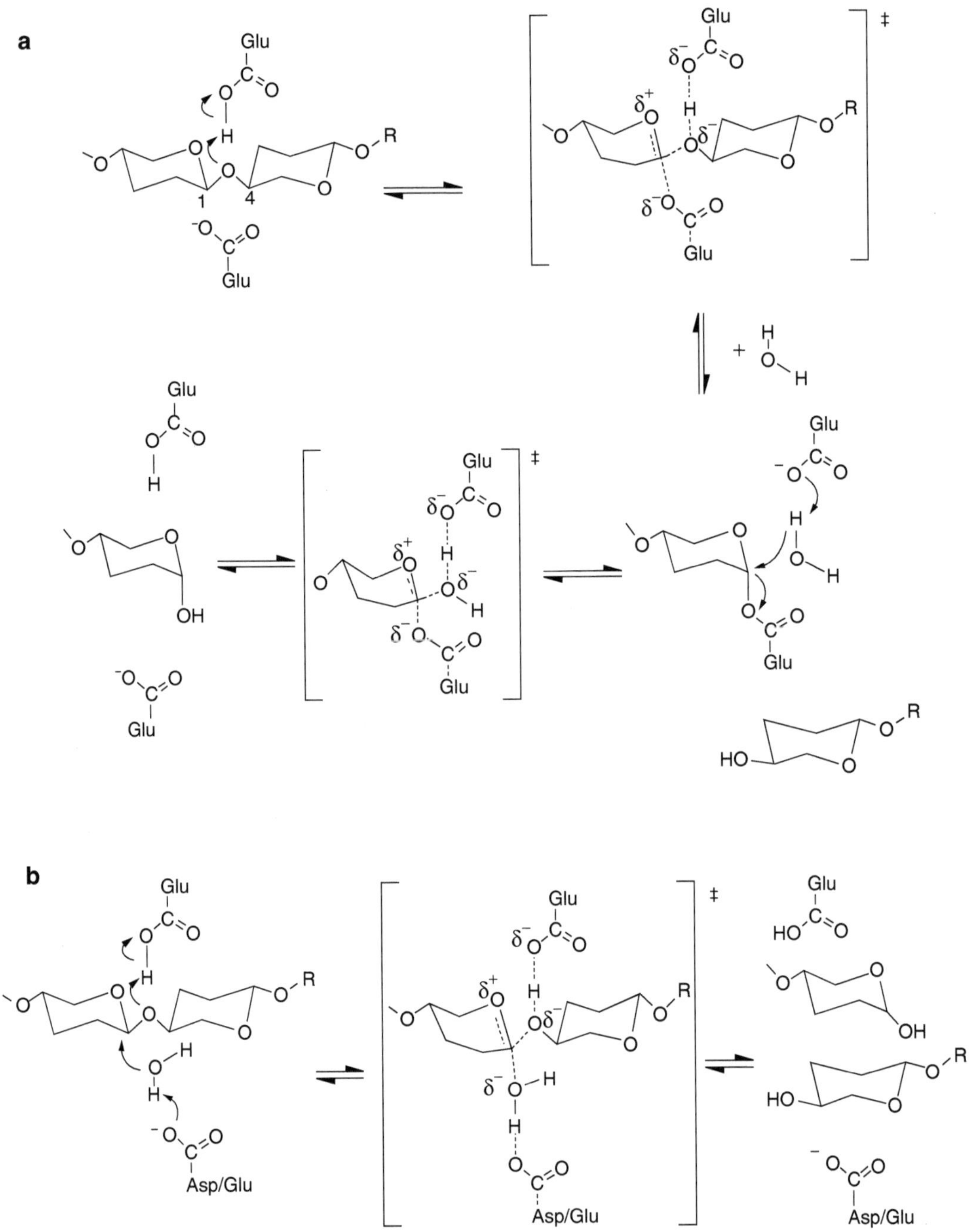

Fig. 8.4 Action mechanisms of xylanases (**a**) retention and (**b**) inversion (Based on Collins et al. 2005; figures were prepared with CS ChemDraw Ultra version 8.3)

consequence of these reactions, the aglycone is departed and the configuration at the anomeric centre is inverted simultaneously (Collins et al. 2005; Coughlan 1992; Paës et al. 2012). GH11 family is considered to be the family of true xylanases since its xylanases exploit D-xylose-containing compounds as their sole substrates. Moreover, their reaction products can be utilised

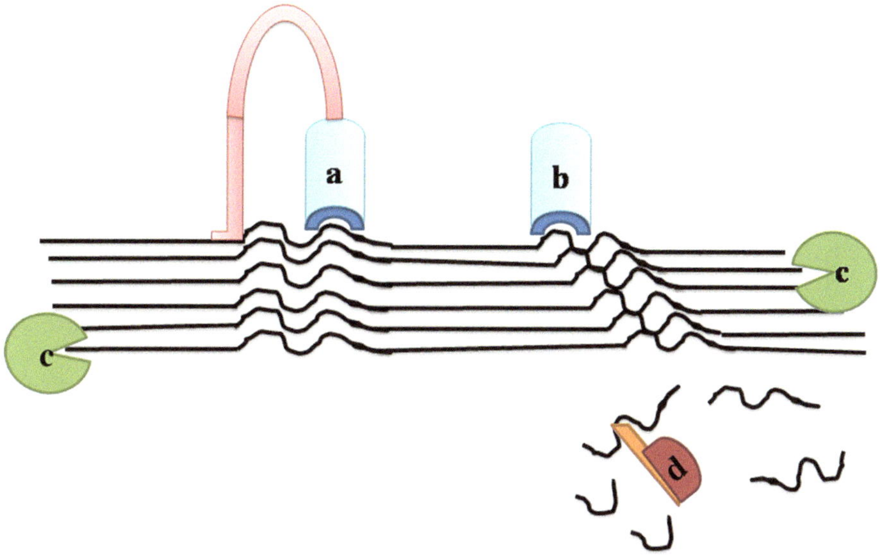

Fig. 8.5 Mode of action of different groups of cellulases. (**a**) Endoglucanase with cellulose-binding domain, (**b**) endoglucanase without cellulose-binding domain, (**c**) cellobiohydrolase, (**d**) glucosidase

as substrates by the enzymes of family 10. The sites where the xylose residues bind to xylanases are known as subsites and the bond cleavage occurs between the sugar residues at the −1 (non-reducing) and the +1 (reducing) ends of the polysaccharide substrate (Davies et al. 1997; Motta et al. 2013). In enzyme assays where arabinoxylan was used as a substrate, family 11 products were found to have arabinose residues substituted at the +2 subsite, while the family 10 products have arabinose residues substituted at the +1 subsite. Therefore, GH 10 enzymes are able to hydrolyse xylose linkages closer to the side-chain residues, whereas the GH 11 xylanases preferentially cleave the unsubstituted regions of the arabinoxylan backbone (Motta et al. 2013).

8.2.2 Cellulases

Recently, cellulases have been studied extensively for their various applications in pulp and paper industry. Especially monocomponent endoglucanases have been found to play a significant role in the enhancement of the cellulose reactivity and accessibility of dissolving pulps (Engström et al. 2006; Henriksson et al. 2005; Kvarnlöf et al. 2005; Köpcke 2010). The cellulase treatment results in the formation of additional openings/surface areas in the fibre structure via the possible action of 'etching'. As a result, the pore volume of pulp fibres increases, which lead to the increase in the accessibility to xanthation (Miao et al. 2014). Due to their synergistic action, cellulases have traditionally been considered detrimental to pulp and paper properties particularly with regard to yield and strength properties. However, the use of monocomponent cellulases or defined cellulase mixtures alone or together with hemicellulases has been shown to have potential in pulp and paper applications (Suurnäkki et al. 2000).

Cellulases are the hydrolytic enzymes that cleave the 1, 4-β-D-glucosidic linkages of the cellulose chain. The prerequisite for hydrolysis is the binding of cellulases to its substrate (Zhang and Lynd 2004). Further, all the three major types of cellulases must act together synergistically for the maximum degradation of cellulose. These groups are endoglucanases (EC. 3.2.1.4), glucosidases (EC. 3.2.1.21) and cellobiohydrolases or exo-1, 4-β-glucanases (EC. 3.2.1.91) as illustrated in Fig. 8.5. The reaction products of these enzymes may vary depending on whether the enzymes are used alone or in combination. Among all, the key component for the hydrolysis of cellulose is endo-1, 4-β-glucanases that generate new chain ends after randomly cleaving the polysaccharide chain, preferably amorphous regions. The cascade of hydrolysis then involves cellobiohydrolases to act upon the ends of polymeric chain to generate cellobiose, which in turn is substrate for β-glucosidases, ultimately leading to the end product glucose (Lynd et al. 2002. The degree of cellulose degradation depends upon the three primary parameters: the crystallinity, the specific surface area and the degree of polymerisation of the cellulose chain (Eremeeva et al. 2001; Mansfield et al. 1999).

In general, enzymes consist of structurally and functionally discrete units called domains that can fold independently of other regions of the same polypeptide chain or of a larger aggregate. Cellulases, in most cases, are composed of two domains. One is a catalytic domain, designed for the specific cleavage of 1, 4-β-D-glucosidic bonds of cellulose, while the other is cellulose-binding domain required to bring the catalytic domain in the proximity of substrate by binding to the cellulose chain. The endoglucanases are consisting of cleft-shaped catalytic domain, whereas the exoglucanases are comprised of tunnel-shaped catalytic domain. The two domains are connected to each other with the help of an inter-domain linker (Rabinovich et al. 2002a, b). The main factors that affect the hydrolytic efficiency of these enzymes involve their size and structure since a large enzyme complex would only be able to modify the surface of the substrate (Mansfield et al. 1999; Zhang and Lynd 2004).

8.3 Biopulping

Biopulping is the treatment of wood chips and other lignocellulosic materials with natural wood decay fungi or other isolated microorganisms since their enzymes such as xylanases, pectinases, cellulases, hemicellulases, ligninases and their combination prior to pulping process improve the pulp properties (Kirk and Jeffries 1996). Biopulping is preferred because:

- It reduces the chemical and energy utilisation.
- It reduces the pollutants.
- It increases the yield and strength properties of pulp.

The first experiment on fungal pretreatment of eucalyptus wood for biosulphite pulping was carried out using *Ceriporiopsis subvermispora* L-14807 SS-3 (Ferraz et al. 1998). The resulting pulp showed improved characteristics especially in terms of kappa number (reduced by 31 %) and pentosan content (reduced by 36 %) at the same level of pulp yield. This could be translated into increased pulp yield over the control at a given kappa number (Ferraz et al. 1998). Christov et al. 1998 investigated five different strains of *C. subvermispora* for biosulphite pulping of eucalyptus in conjunction with various bleaching sequences. Prior to the acid sulphite pulping, the wood chips were subjected to fungal treatment for 2 weeks. Table 8.3 presents the various properties such as alkali solubility, α-cellulose, viscosity, etc., of the pulp so produced. The resulting pulps showed reduction in both brightness and yield over the control. Although kappa number was reduced by

Table 8.3 Biosulphite pulping of eucalyptus wood using five strains of *C. subvermispora**

Strains	S10 (%,w/w)	S18 (%,w/w)	α-cellulose (%,w/w)	Kappa number	Brightness (%, ISO)	Viscosity (cP)	Yield (%,w/w)
Control	9.2	6.8	90.7	8.9	51.5	48.5	47.4
SS-1	10.2	6.9	90.0	8.0	47.7	46.1	40.8
SS-3	9.2	6.7	90.6	8.3	49.3	47.5	47.1
SS-4	9.8	6.8	90.5	8.7	49.4	47.8	45.2
SS-5	9.4	6.7	90.7	8.9	48.6	47.1	46.9
SS-10	9.6	6.9	90.3	8.7	46.8	47.5	45.6

*Data adopted from Christov et al. (1998)

10 %, it did not correlate with the corresponding brightness of the pulp samples.

Several other strains of white-rot fungi were also investigated by many researchers for their potential use in biosulphite pulping of *Eucalyptus grandis*, and the pulp characteristics in terms of kappa number, yield and brightness were evaluated (Christov et al. 1998; Christov 1999; Ferraz et al. 1998; Mosai et al. 1999). Mosai et al. (1999) evaluated the biopulping efficiency of ten different white-rot fungi strains on *Eucalyptus grandis* in terms of pulp yield, kappa number, brightness, alkali solubility (S10 and S18) and viscosity when subjected to different pretreatment time. Fungal pretreatment of eucalyptus wood chips with *C. subvermispora* SS-3 for 10 days without additional carbon source resulted in substantial decrease in kappa number (by 29 %) and increase in brightness (by 12 %) (Mosai et al. 1999). As shown in Table 8.4, the resulting pulps had lower pulp yield compared to control. Reduction in kappa number (24 %) and simultaneous increase in brightness (6 points) in comparison to control were only observed in *C. subvermispora* SS-3, whereas in other treated pulp, yield loss and apparent increase in lignin occurred (Christov 1999; Mosai et al. 1999). Scott et al. (1996) studied the effect of biosulphite pulping of pine when treated with *C. subvermispora* SS-3 for 2 weeks and observed the reduction in kappa number by 21 % over control. Also, the pretreatment of spruce with *C. subvermispora* CZ –3 for 2 weeks resulted in reduced kappa number by 22 % compared to control (Fischer et al. 1994).

Biopulping of eucalyptus wood with fungal strain SS-3 followed by bleaching resulted in a pulp of higher brightness and pulp yield comparable to control. However, the fungal strain SS-5 produced biopulp resulting in higher pulp yield with the same brightness as control (Table 8.5). Therefore, the pretreatment of eucalyptus wood with the *C. subvermispora* strains increased by 1 percentage point either the final pulp yield (SS-5) or pulp brightness (SS-3) relative to the controls (Christov et al. 1998; Christov 1999;

Table 8.4 Effect of biosulphite pulping of eucalyptus with selected strains of white-rot fungi[a]

Fungal strain	Yield (%)	Kappa number	Brightness (%)
Control	47.0	8.6	51.6
Phanerochaete sordida YK-624	46.9	10.7	45.6
Trametes versicolor 52 J	46.5	9.5	46.0
Laetiporus sulphureus SCC- 180	46.0	10.6	44.5
Phanerochaete chrysosporium BKMF-1767	45.8	10.0	49.0
Ceriporiopsis subvermispora SS-3	45.3	6.5	57.3
Ceriporiopsis subvermispora SS-5	44.4	8.6	49.9
Stereum hirsutum SCC-74	44.0	12.0	35.4
Trametes versicolor ATCC 20869	43.7	12.5	46.0
Sclerotinia sclerotiorum UFHRF	42.9	12.1	42.9

[a]Wood chips were pretreated with fungi for 10 days. Based on data from Christov (1999) and Mosai et al. (1999)

Table 8.5 Effects of biosulphite pulping of *Eucalyptus grandis* with two strains of *Ceriporiopsis subvermispora* on properties of unbleached and bleached dissolving pulp

	Unbleached dissolving pulp			Bleached dissolving pulp	
Fungal strain[a]	Kappa number	Yield (%)	Brightness (%)	Yield (%)	Brightness (%)
Control	8.6	47.0	51.6	40.8	93.1
Ceriporiopsis subvermispora SS-3	6.5	45.3	57.3	40.4	94.2
Ceriporiopsis subvermispora SS-5	8.6	44.4	49.9	41.8	93.1

[a]Wood chips were pretreated with fungi for 10 days. Based on data from Christov (1999)

Mosai et al. 1999). This could be translated into savings of chemicals or energy at a given pulp yield. The increased pulp yields thus indicate the better selectivity of the bleaching process resulting from the biopulping.

8.4 Biobleaching

Biobleaching of pulps is performed with either hemicellulolytic enzymes, in particular xylanases or lignin-degrading fungi or their enzymes, namely, ligninases and laccases. The action of xylanases on pulp results in the hydrolysis of hemicelluloses as a consequence of which the bleaching chemicals get an easier access to lignin, whereas the white-rot fungi and their ligninolytic enzymes depolymerise the lignin in pulp by a direct action on it. The main objective, however, in both cases is to increase delignification so as the amount of bleaching chemicals specifically chlorine and chlorine-containing compounds is reduced in the subsequent bleaching of pulp. Besides, biobleaching is considered as one of the preferred processes due to a number of benefits it offers over conventional chemical bleaching. One of the major advantages is the reduced adsorbable organic halide (AOX) levels in the discharged effluents resulting from the reduced requirement of chlorine during bleaching along with improved pulp quality, brightness gain and post colour (PC) number reduction. Most of the research on biobleaching for the production of dissolving pulp has been carried out using either white-rot fungi or xylanases. Ligninolytic enzymes are not yet studied for dissolving pulp production although they are extensively being explored for the production of paper grade pulp. Biobleaching with xylanases has been investigated for both kraft and sulphite pulps, whereas white-rot fungi are studied for sulphite pulp only.

Christov et al. (1996) reported that biopulping of eucalyptus wood chips in conjunction with bleaching/biobleaching of pulp to produce dissolving pulp results in a brightness gain over the control in most instances. The properties of biosulphite pulp produced by the pretreatment with SS-3 and SS-5 strains of *C. subvermispora* were evaluated with three different bleaching sequences: $OD_1E_OD_2H$, $OD_1E_OD_2P$ and X $OD_1E_OD_2P$ (O, oxygen; D_1 and D_2, chlorine dioxide; E_O, sodium hydroxide with reinforced oxygen; H, sodium hypochlorite; and P, hydrogen peroxide) (Christov et al. 1998). Biobleaching effect due to xylanase was only observed in fungus pretreated samples, not in control as shown in Table 8.6 (Ferraz et al. 1998). Brightness gain due to xylanase treatment was found maximum

Table 8.6 Effects of bleaching/biobleaching[a] of sulphite pulp produced by biopulping of wood chips with two different strains of *Ceriporiopsis subvermispora*

Fungal strain[b]	Bleaching sequence	S10 (%, w/w)	S18 (%, w/w)	α-cellulose (%, w/w)	Brightness (%, ISO)	Yield loss (%)
Control	$OD_1E_0D_2H$	7.9	4.7	92.8	93.1	14.1
	$OD_1E_0D_2P$	7.7	4.5	92.7	93.2	13.2
	$X\text{-}OD_1E_0D_2P$	8.1	4.5	92.5	93.2	13.4
SS-3	$OD_1E_0D_2H$	7.8	4.8	92.8	94.2	14.3
	$OD_1E_0D_2P$	8.3	5.4	92.6	92.9	14.0
	$X\text{-}OD_1E_0D_2P$	8.1	4.5	92.7	95.5	14.1
SS-5	$OD_1E_0D_2H$	7.9	4.7	92.9	93.1	11.6
	$OD_1E_0D_2P$	8.5	4.6	92.2	93.6	10.3
	$X\text{-}OD_1E_0D_2P$	8.4	4.5	92.3	94.0	10.4

[a]Xylanase pretreatment(X):4 IU/g pulp using Cartazyme HS-10; Charges at D_1 and D_2: 0.9 and 0.6 % as active Cl; P: 0.6 % H_2O_2; 0.4 % NaOH; 100 °C. Based on data from Ferraz et al. (1998)

[b]Fungal pretreatment of wood chips: 2 weeks

in SS-3 pretreated sample, whereas the yield loss resulting from bleaching of the biopulped samples was found minimum in SS-5 pretreated sample. The yield of unbleached pulp produced after biosulphite pulping with strain SS-5 was lower than that of control (Table 8.5); however, it was the highest after (bio)bleaching (Table 8.6). Therefore, the fungal pretreatment of wood chips not only increases the extent of lignin removal during pulping and bleaching, resulting in higher brightness of the dissolving pulp, but also improves the selectivity of the bleaching process, thus increasing the final pulp yield.

Biobleachability of industrial sulphite dissolving pulp was also studied by Christov et al. 1996 using white-rot fungi and xylanase from *A. pullulans*. It was reported that two strains of *C. subvermispora* (L-14807 SS-3 and CZ-3) effectively bleached the sulphite pulp. Both the strains were capable to remove lignin (as kappa number) with strain SS-3 by 85 % and strain CZ-3 by 88 %, in particular (Table 8.6). Also, the strain SS-3 significantly contributed to pentosan removal from pulp (13 %) than strain CZ-3 (5 %). The brightness of fungus-treated samples increased by 42 % (strain SS-3) and 47 % (strain CZ-3), respectively, as compared with the control samples. In addition, the α-cellulose content was reduced by 5 (SS-3) and 4 (CZ-3) points. In comparison the xylanase pretreatment lowered the pentosan content by 13 % and improved the α-cellulose content by nearly a point without significantly affecting the brightness and kappa number (Table 8.7).

Bajpai and Bajpai (2001) developed a bleaching process for dissolving grade pulp in which treatment with xylanase enzyme preceded the beaching sequence CEHED (C, chlorine; E, extraction with sodium hydroxide; H, hypochlorite; and D, chlorine dioxide). The unbleached pulp was produced involving a mild prehydrolysis step to retain relatively high pentosan content in the pulp that resulted in higher unbleached pulp yield at the same kappa number. Then after, the enzyme-treated pulp was bleached that led to the reduced requirement of bleaching chemicals. Further, improved brightness and bleach pulp yield were attained at desired levels of pentosan. The pulp reactivity and rayon yield were also found to be higher for the enzyme-treated pulp.

Enzyme treatment using commercial laccase from *Trametes villosa* with different natural and synthetic mediators has also been investigated on unbleached sulphite pulp for its biobleaching potential (Quintana et al. 2013). The studies reported the laccase–violuric acid (L–VA) system as the most effective enzyme mediator combination among others to obtain improved quality dissolving pulp. At first the laccase-mediator treatment reduced the ISO brightness and kappa number in all the samples due to delignification and formation of chromophore groups, but after P (P-hydrogen peroxide) stage, laccase–violuric acid treatment increased brightness (28 %) and reduced kappa number (87 %) while viscosity was reduced to 2.5 % of the control. Also, the effluent generated after laccase–violuric acid treatment was found to have least toxicity. An extended TCF biobleaching with L–VA stage was also assessed and could be translated in saving H_2O_2 consumption and reaction time.

Table 8.7 Treatment of sulphite pulp with *Ceriporiopsis subvermispora* (L-14807 SS-3 and CZ-3) and *A. pullulans* xylanase (X)[a]

Pulp treatment	Kappa number	α-cellulose (%, w/w)	Pentosan (%)	Brightness (%, ISO)
Control	6.7	89.9	3.9	56.0
SS-3	1.0	84.9	3.4	79.4
CZ-3	0.8	86.0	3.7	82.2
Xylanase	6.5	90.7	3.4	56.5

[a]Fungal treatment time: 2 weeks; X: 15 IU xylanase/g pulp: 3 h; 55 °C; pH 4.7; 9 % pulp consistency. Based on data from Ferraz et al. (1998)

8.5 Bioconversion of Paper Grade Pulp to Dissolving Pulp

The kraft process is known to be the world's major pulping method and is likely to remain so into the foreseeable future. During its evolution over a period of a century, it has become highly refined. Currently about 70 % of the world's annual pulp output of approximately 100 million tonnes is produced by the kraft process. Although some disadvantages are associated with it, still it is by far the most cost-effective, versatile and efficient wood delignification method available. Because of this fact and the large amount of capital already invested in kraft pulping, it is unlikely that the process will be replaced in the near future (Thakur et al. 2012). On the other hand, demands for dissolving pulp and fibres have boomed worldwide within the last decade, and thus, nowadays much effort is put into the production of dissolving pulps from kraft pulps besides the dissolving pulp production from the commonly applied technologies such as the acid sulphite process and the alkaline pre-hydrolysis kraft process (Sixta 2006).

With a view on the worldwide pulp production capacity of about 150×10^6 t per year, the production of dissolving pulps represents only a niche market of totally 2.5 % of the global chemical pulp production (Puls et al. 2006). This market situation and the high price difference between both pulp categories raises the question of why paper grade pulps cannot be converted into dissolving pulps in an economically feasible process by the selective reduction of hemicelluloses from kraft pulp and subsequently activating it. In this context, several studies have been conducted focusing on the development of economic and efficient process steps to upgrade paper grade pulp into dissolving pulp. Recently it has been found that treating the paper grade pulps with enzymes can give them a higher reactivity, enabling these pulps to be utilised as an economic alternative to dissolving grade pulps for viscose production (Gehmayr and Sixta 2011; Jackson et al. 1998; Köpcke et al. 2008; Wang et al. 2014). Also enzyme treatment and alkali extraction subjected to sisal pulps indicated that such pulps could also be used as dissolving pulps (Ibarra et al. 2010).

Cellulases are the most extensively used enzymes for the improvement of pulp reactivity and accessibility (Engström et al. 2006; Gehmayr and Sixta 2011; Gehmayr et al. 2011; Henriksson et al. 2005; Ibarra et al. 2010; Köpcke et al. 2008; Köpcke 2010; Kvarnlöf et al. 2007; Miao et al. 2014; Östberg 2012; Rahkamo et al. 1996, 1998; Wang et al. 2014; Wollboldt et al. 2010) of both dissolving grade pulp and paper grade pulp. Miao et al. (2014) reported that cellulase treatment at the dose as low as 2 u/g (based on the dry weight of pulp) subjected to hardwood kraft-based dissolving pulp resulted in increased Fock reactivity from 47.67 % to 79.9 %.

Upgradation of kraft paper pulps to dissolving pulps can also be achieved by a post-extraction of xylan by using an approximately 2 M sodium hydroxide solution at temperatures slightly above room temperature (cold caustic extraction or CCE; Jayme and Schenck (1949), Wallis and Wearne (1990)). Anyhow, pulp treatment at raised alkali concentration leads to the modification of cellulose I to cellulose II. Moreover, the residual xylan after CCE is characterised by rather high molecular weight and a resistance to dissolution in alkali (Gehmayr and Sixta 2011; Wollboldt et al. 2010). Both contribute to significant decrease in pulp reactivity, hence necessitating the gentle method of hemicelluloses extraction and reactivity enhancement. A number of treatments based on enzymatic treatments, hypochlorite treatments, alkali extraction and alkaline peroxide treatment either in conjunction or alone have been investigated in order to attain the commercial dissolving pulp quality in terms of cellulose reactivity, hemicelluloses content, pulp viscosity and molecular weight distribution. The treatment types, sequences and process parameters were further optimised for different raw materials used.

Jackson et al. (1998) reported that high-quality recovered paper sources rich in hardwood content after given an initial cold alkali extraction, a xylanase treatment and a final cold alkali extraction produced dissolving pulp with reduced hemicelluloses content, acceptable viscosity and alkali solubility. Henriksson et al. (2005) and Engström et al. (2006) studied the treatment effect of monocomponent endoglucanase and found that the reactivity of softwood sulphite pulp as determined by the Fock method (Fock 1959) was significantly increased with relatively low amounts of enzyme used, and the yield loss and decrease in viscosity were moderate.

Kvarnlöf et al. (2007) studied two different approaches, one chemical and one enzymatic, to enhance the pulp reactivity so as the carbon disulphide charge in the viscose process is reduced, thereby reducing the cost and indirectly the environmental impact. Different cellulases including either mixture or monocomponent endoglucanase were employed for enzymatic treatment of hardwood dissolving pulp and softwood dissolving and paper grade pulp. Kraft pulp has been shown an increase in reactivity of approximately 45 % units after being treated with enzyme. Finally, treating several cellulosic materials with enzymes showed that the reactivity of a paper grade pulp was increased from 40 % to about 80 %. Treating non-dissolving pulps with enzymes is therefore of great interest in the industrial manufacture of viscose. The work also demonstrated the reduced carbon disulphide consumption due to improved pulp reactivity in the viscose process while maintaining the good quality viscose dope. The best enzyme preparation among others was found to be the monocomponent endoglucanase, Carezyme, whose treatment was further optimised relating to viscose dope preparation. Moreover, the spent enzyme filtrate could be re-used several times without losing its efficiency under the industrially desired conditions of temperature, enzyme dose and reaction time, thus lowering the production cost. In the chemical method, pressurised oxygen was used after the mercerisation step resulting in the consumption of a carbon disulphide charge of 20 % instead of 30 % in the conventional viscose process. For enzyme-treated pulp to reach the lowered carbon disulphide charge, the dewatering and the sulphidation stage was modified accordingly. Also, a twofold increase in the reactivity of paper grade pulp was observed after the enzymatic treatment indicating that enzymes could be of great use in the upgradation of paper grade pulp to dissolving grade pulp.

Ibarra et al. (2010) and Köpcke (2010) evaluated different commercial monocomponent endoglucanases either with or without cellulose-binding domain (CBD) for their effect on the reactivity and accessibility of hardwood and softwood dissolving grade pulps for viscose production. It was observed that endoglucanase with a CBD notably enhanced the cellulose reactivity according to the Fock method by simultaneously decreasing the viscosity. This may be attributed to the proximity of catalytic domain with the cellulose chain for the required course of time after CBD has bound to the cellulose. Moreover, an inverting hydrolysis mechanism of endoglucanase with CBD had been the most effective among the different assayed enzymes to result in increased reactivity of both pulps (Ibarra et al. 2010). At the same time, more marked decrease in viscosity was observed for softwood dissolving pulp, whereas the effects of the different endoglucanase treatments were more pronounced on never-dried dissolving pulps.

Köpcke (2010) undertook research study to convert different wood and nonwood paper pulps to dissolving pulp. Commercial dried elemental chlorine-free (ECF) bleached kraft pulps from birch and eucalypt were studied along with several dried ECF bleached soda/anthraquinone pulps from flax, hemp, sisal, abaca and jute. Two alkali extraction steps followed by an endoglucanase treatment was found to be the most suitable sequence for birch, whereas for eucalyptus and sisal pulps, a combination of a xylanase treatment followed by an alkali extraction step prior to an endoglucanase treatment provided the best results. The xylanase treatment and alkali extraction steps reduced the hemicelluloses content, reaching values to those of a commercial dissolving grade pulp. The molecular weight distribution was found to be more uniform in converted dissolving pulps than that of a commercial dissolving pulp since only cellulose I was present in the former. Moreover the endoglucanase treatment resulted in improved cellulose reactivity of pulp and fulfilled the requirements of a commercial dissolving pulp for viscose production. However, the pulp viscosity exhibited values lower than those of commercial dissolving pulp but suitable for the production of viscose. Pretreatment of dissolving pulp with monocomponent xylanase and monocomponent cellulase has also been evaluated to improve the following dissolution of pulp in NaOH/ZnO (Ambjörnsson et al. 2014). The treated pulp showed increased solubility from 29 % to 81 % in NaOH/ZnO solution after subjecting to 16.7 AXU/g of xylanase dose followed by alkali extraction (with 7 % NaOH) and 10 ECU/g endoglucanase dose. Surprisingly the DP_η was as high as 1074.

Gehmayr and Sixta (2011) introduced Hem-Extra Process in which commercial oxygen delignified *E. globules* kraft pulp subjected to xylanase and CCE treatment prior to TCF bleaching followed by endoglucanase treatment resulted in the production of novel high-purity dissolving pulp. It was found that xylanase pretreatment and subsequent CCE at reduced alkalinity were sufficient to reach the desired residual xylan content of the pulp. Moreover, the enzymatic degradation by endoglucanase exhibited a direct correlation between the cellulose II content and the chain cleavage of the pulps. The final average degree of polymerisation was also adjusted by endoglucanase treatment. On the other hand, Fock reactivity and filter value both showed an increase after xylanase pretreatment because of the lower degree of fibre hornification, whereas the Fock reactivity was higher for enzyme-treated pulps when compared to an acidly hydrolysed pulp. It was also found that a substantial increase in yield at a given R18 content, ranging between 4 % and 8 % on oven-dried wood basis, could be achieved by the new Hem-Extra-pulps as compared to a conventional pre-hydrolysis kraft pulp.

Östberg (2012) investigated the effect of enzyme pretreatment, subjected to kraft and dissolving pulp, on carbon disulphide consumption for high-viscosity viscose preparation. It was found that the Fock reactivity of dissolving pulp when treated with Fibre Care R increased in all instances if compared to the reference pulp sample, but a higher enzyme charge or long treatment time could reduce the reactivity. On the other hand, the limiting viscosity number was reduced, necessitating relatively low enzyme charges combined with a relatively short treatment time to produce viscose grades intended for high-strength products. Moreover, xylanase pretreatment alone showed no change either in Fock reactivity or the limiting viscosity number of the dissolving pulp. No correlation between the gamma number of the viscose and the Fock reactivity of pulp was observed; rather the gamma number either was reduced or remained unchanged upon enzyme treatment except for the conventional softwood kraft pulp where it increased. It indicates that gamma number measures the xanthate groups in both the cellulose and hemicelluloses that undergo a certain degree of substitution in the viscose process, thus a poor indicator if the pulp also contains a high content of hemicelluloses.

Wang et al. (2014) introduced alkaline peroxide treatment after enzymatic treatment with EG-rich industrial cellulase to convert bleached softwood paper grade pulp to dissolving pulp for viscose production. With the enzyme dose of 300 IU/g bone dry pulp, 9 wt% NaOH and 1 wt% H_2O_2, the treated pulp exhibited 68.7 % of reactivity, 92.1 % of α-cellulose content and 506.9 mL/g of pulp viscosity. Furthermore, the reducing sugar in the spent liquor derived from the stage of enzymatic treatment can be recovered for the production of value-added products based on the concept of biorefinery.

8.6 Conclusions and Future Prospects

Enzyme technology has been contributing a vast number of new techniques and processes to pulp and paper industry and can provide new horizons in the future as well. One of its applications is producing dissolving pulp by using advanced techniques and employing microorganisms and their enzymes to lessen the environmental load and improve the product quality in a cost-effective manner. This includes the use of microorganisms/enzymes in biopulping and biobleaching for their selectivity that has been shown to result in either brightness gain or improved yield with reduced amount of pollutant generated. Also, the enzymes can be effectively used for reactivity improvement of dissolving pulp or techno-economic conversion of paper grade pulp to dissolve grade pulp. Various studies have shown that the upgradation of kraft pulp to a better quality dissolving pulp in terms of reactivity, brightness, yield, molecular weight with acceptable viscosity and hemicelluloses content could be achieved by enzymatic treatments and alkali extraction. Moreover, producing dissolving pulp from kraft pulp offers a pulp with higher market value along with the production of value-added bioproducts such as hemicelluloses. Hemicelluloses being one of the main biorefinery feedstocks can be further utilised to produce biopolymers, thermoplastic xylan derivatives, bioethanol/biofuels and immunomodulators or as dietary fibre depending upon the extent of degradation. Therefore, exploring more biotechnological applications and further implementing them in pulp and paper industry can be a solution for reducing environmental impact and producing better quality product simultaneously.

Acknowledgement We thank the Council of Scientific and Industrial Research (CSIR) of India for support by Grant-in-aid (grant no. 09/1089/2012) for Senior Research Fellows to Prabhjot Kaur.

References

Allison RW, Clark TA, Wrathall SH (1993) Pretreatment of radiate pine kraft pulp with a thermophilic enzyme. Part II effect on oxygen, ozone and chlorine dioxide bleaching. Appita 46(5):349–353

Ambjörnsson HA, Östberg L, Schenzel K, Larsson PT, Germgård U (2014) Enzyme pretreatment of dissolving pulp as a way to improve the following dissolution in NaOH/ZnO. Holzforschung 68(4):385–391

Bajpai P (2012) Production of dissolving grade pulp. Biotechnology for pulp and paper processing. Springer, New York

Bajpai P, Bajpai P (2001) Development of a process for the production of dissolving kraft pulp using xylanase enzyme. Appita J 54(4):381–384

Bajpai P, Bhardwaj NK, Bajpai PK, Jauhari MB (1994) The impact of xylanases in bleaching of eucalyptus kraft pulp. J Biotechnol 36(1):1–6

Battan B, Kuhar S, Sharma J, Kuhad RC (2007) Biodiversity of hemicelluloses degrading microorganisms and their enzymes. In: Kuhad RC, Singh A (eds) Lignocellulose biotechnology future prospects. IK International Publishing House Pvt. Ltd, New Delhi, pp 121–147

Bernstein J (2002) Polymorphism in molecular crystals. Oxford University Press, New York, p 14

Bhardwaj NK, Bajpai P, Bajpai P (1996) Use of enzymes in modification of fibers for improved beatability. J Biotechnol 51(1):21–26

Bim MA, Franco TT (2000) Extraction in aqueous two-phase systems of alkaline xylanase produced by *Bacillus pumilus* and its application in kraft pulp bleaching. J Chromatogr 743(1):349–356

Binod P, Palkhiwala P, Gaikaiwari R, Nampoothiri KM, Duggal A, Dey K, Pandey A (2013) Industrial enzymes-present status and future perspectives for India. J Sci Ind Res 72:271–286

Buchert J, Oksanen T, Viikari L (1998) Modification of fibre properties by enzyme treatments, improvement of recyclability and the recycling paper industry of the future. COST Action El paper recyclability, Las Palmas Gran Canaria, pp 306–311

Cadena EM, Chriac AI, Pastor FI, Diaz P, Vidal T, Torres AL (2010) Use of cellulases and recombinant cellulose binding domains for refining TCF kraft pulp. Biotechnol Prog 26(4):960–967

Cao Y, Tan H (2002) Effects of cellulase on the modification of cellulose. Carbohydr Res 337(14):1291–1296

Cao Y, Tan H (2006) Improvement of alkali solubility of cellulose with enzymatic treatment. Appl Microbiol Biotechnol 70(2):176–182

Carbohydrate-Active Enzyme (CAZy) Database. http://www.cazy.org. Accessed 27 Aug 2014

Christoffersson-Elg K (2005) Dissolving pulp – multivariate characterisation and analysis of reactivity and spectroscopic properties. PhD thesis, Umeå University, Umeå

Christov L (1999) Biotechnology in dissolving pulp manufacture. In: 1999 Pulping conference proceedings, Tappi

Christov LP, Akhtar M, Prior BA (1996) Biobleaching in dissolving pulp production. In: Srebotnik E, Messner K (eds) Biotechnology in the pulp and paper industry. Facultas- Universitätsverlag, Vienna, pp 625–628

Christov LP, Akhtar M, Prior BA (1998) The potential of biosulfite pulping in dissolving pulp production. Enzyme Microb Technol 23(1–2):70–74

Collins T, Gerday C, Feller G (2005) Xylanases, xylanase families and extremophilic xylanases. Microbiol Rev 29:3–23

Coughlan MP (1992) Towards an understanding of the mechanism of action of main chain hydrolyzing xylanases. In: Visser J, Beldman G, Someren MAK-V, Voragen AGJ (eds) Xylans and xylanases, vol 7, Progress in biotechnology. Elsevier Science Publishers BV, Amsterdam/New York, pp 111–139

Csiszar E, Urbanskzi K, Szakaes G (2001) Biotreatment of desized cotton fabric by commercial cellulase and xylanase enzymes. J Mol Catal B Enzym 11:1065–1072

Davies GL, Wilson K, Henrissat B (1997) Nomenclature for sugar-binding subsites in glycosyl hydrolases. Biochem J 321:557–559

Dhiman SS, Sharma J, Battan B (2008a) Pretreatment processing of fabrics alkalothermophilic xylanase from Bacillus stearothermophilus SDX. Enzyme Microb Technol 43:262–269

Dhiman SS, Sharma J, Battan B (2008b) Industrial applications and future prospects of microbial xylanases: a review. Bioresources 3(4):1377–1402

Dufresne A (2012) Nanocellulose from nature to high performance tailored materials. Walter de Gruyter GmbH, Berlin/Boston

Engström A-C, Ek M, Henriksson G (2006) Improved accessibility and reactivity of dissolving pulp for the viscose process: pretreatment with monocomponent endoglucanase. Biomacromolecules 7(6):2027–2031

Eremeeva T, Bikova T, Eismonte M, Viesturs U, Treimanis A (2001) Fractionation and molecular characteristics of cellulose during enzymatic hydrolysis. Cellulose 8(1):69–79

Ferraz A, Christov L, Akhtar M (1998) Fungal pretreatment for organosolv pulping and dissolving pulp production. In: Young RA, Akhtar M (eds) Environmentally friendly technologies for the pulp and paper industry. Wiley, New York

Fischer K, Puchinge L, Schloffer K, Kreiner W, Messner K (1993) Enzymatic pitch control of sulfite pulp on pilot scale. J Biotechnol 27(3):341–348

Fischer K, Akhtar M, Blanchette RA, Burnes TA, Messner K, Kirk TK (1994) Reduction of resin content in wood chips during experimental biological pulping processes. Holzforschung 48:285–290

Floe G (2011) Dissoving pulp: the great comeback "It's all about cotton!" TAPPI PEERS dissolving pulp forum. Pöyry management consulting 2011. Oregon Convention Center Portland, Ore., USA

Fock W (1959) A modified method for determining the reactivity of viscose-grade dissolving pulps. DasPapier 13:92–95

Gehmayr V, Sixta H (2011) Dissolving pulps from enzyme treated kraft pulps for viscose application. Lenzinger Berichte 89:52–160

Gehmayr V, Schild G, Sixta H (2011) A precise study on the feasibility of enzyme treatments of a kraft pulp viscose application. Cellulose 18:479–491

Gil N, Gil C, Amaral ME, Costa AP, Duarte AP (2009) Use of enzymes to improve the refining of a bleached *Eucalyptus globulus* kraft pulp. Biochem Eng J 46(2):89–95

Gustafsson J, Alén R, Engström J, Korpinen R, Kuusisto P, Leavitt A, Olsson K, Piira J, Samuelsson A, Sundquist J (2011) Pulping. In: Pedro F (ed) Chemical pulping part 1: fibre chemistry and technology second edition papermaking science and technology. Paperi ja Puu Oy, Finland

Hebeish A and Guthrie JT (1981) The chemistry and technology of cellulosic copolymers, Springer-Verlag, Berlin

Heitmann JA, Joyce TW, Prasad DY (1992) Enzyme deinking of newsprint waste. In: Srebotnik E, Messner K (eds) Biotechnology in the pulp and paper industry. Facultas- Universitätsverlag, Vienna, pp 175–180

Henriksson G, Lennholm H (2009) Cellulose and carbohydrate chemistry. In: Monica EK, Gellerstedt G, Henriksson G (eds) Pulp and pare chemistry and technology. Walter de gruyter GmbH & Co, Berlin, pp 72–99

Henriksson G, Christiernin M, Agnemo R (2005) Monocomponent endoglucanase treatment increases the reactivity of softwood sulphite dissolving pulp. J Ind Microbiol Biotechnol 32(5):211–214

Henrissat B, Bairoch A (1996) Updating the sequence-based classification of gylcosyl hydrolases. Biochem J 316:695–696

Hillman D (2006) Do dissolving pulps really dissolve? Paper Asia:12–18

Hinck JF, Casebier RL, Hamilton JK (1985) Dissolving pulp manufacture. In: Ingruber OV, Kocurek MJ, Wong A (eds) Pulp and paper manufacture. Joint Textbook Committee of the Paper Industry TAPPI, Atlanta 4, pp 213–243

Ibarra D, Romero J, Martínez MJ, Martínez AT, Camarero S (2006) Exploring the enzymatic parameters for optimal delignification of eucalypt pulp by Laccase-Mediator. Enzyme Microb Technol 39:1319–1327

Ibarra D, Köpcke V, Ek M (2010) Behaviour of different monocomponent endoglucanases on the accessibility and reactivity of dissolving-grade pulps for viscose process. Enzyme Microb Technol 47(7):355–362

Isogai A, Usuda M, Kato T, Uryu T, Atalla RH (1989) Solid-state CP/MAS 13C NMR study of cellulose polymorphs. Macromolecules 22:3168–3172

Jackson LS, Heitmann JA Jr, Joyce TW (1998) Production of dissolving pulp from recovered paper using enzymes. Tappi J 81:171

Jayme G, Schenck U (1949) Die Auswirkungverschieden eralkalischereredelungsbedingungen auf das Verhalten von Zellstoffenbei der Acetylierung. Das Papier 3:469–476

Kirk TK, Jeffries TW (1996) Roles for microbial enzymes in pulp and paper processing. In: Jeffries TW, Viikari L (eds) Enzymes for pulp and paper processing. American Chemical Society Symposium; 655, pp 2–14

Köpcke V (2010) Conversion of wood and non-wood paper-grade pulps to dissolving-grade pulps. Doctoral dissertation, KTH

Köpcke V, Ibarra D, Ek M (2008) Increasing accessibility and reactivity of paper grade pulp by enzymatic treatment for use as dissolving pulp. Nordic Pulp Pap Res J 23(4):363–368

Krässig HA (1993) Accessibility in intercrystalline reactions. In: Krässig HA (ed) Cellulose: structure, accessibility and reactivity, 1st edn. Gordon and Breach Science Publishers, Amsterdam, pp 187–214

Kvarnlöf N, Germgård U, Jönsson L, Söderlund CA (2005) Enzymatic treatment to increase the reactivity of a dissolving pulp for viscose preparation. Appita J 59(3):242–246

Kvarnlöf N, Germgård U, Jönsson L, Söderlund CA (2007) Optimization of the enzymatic activation of a dissolving pulp before viscose manufacture. Tappi J 6:6

Losonczi A, Csiszar E, Szakacs G, Bezur L (2005) Role of the EDTA chelating agent in bioscouring of cotton. Textile Res J 75:411–417

Lynd LR, Weimer PJ, van Zyl WH, Pretorius IS (2002) Microbial cellulose utilization: fundamentals and biotechnology. Microbiol Mol Biol R 66(3):506–577

Mansfield SD, Mooney C, Saddler JN (1999) Substrate and enzyme characteristics that limit cellulose hydrolysis. Biotechnol Prog 15(5):804–816

Mäntyranta H (2013) Dissolving pulp provides completely recyclable textiles. http://www.forestplatform.org/en/dissolving-pulp-provides-completely-recyclable-textiles. Accessed 12 Aug 2014

McGinnis G, Shafizadeh F (1979) Cellulose and hemicellulose. In: Casey JP (ed) Pulp and paper chemistry and chemical technology, vol 1, 3rd edn. Wiley, New York, pp 1–8

Miao Q, Chen L, Huang L, Tian C, Zheng L, Ni Y (2014) A process for enhancing the accessibility and reactivity of hardwood kraft-based dissolving pulp for viscose rayon production by cellulase treatment. Bioresour Technol 154:109–113

Mosai S, Wolfaardt JF, Prior BA, Christov LP (1999) Evaluation of selected lignin degrading fungi for biosulfite pulping. Bioresour Technol 68:89–93

Motta FL, Andrade CCP, Santana MHA (2013) A review of xylanase production by the fermentation of xylan: classification, characterization and applications. In: Chandel AK, Silva SS (eds) Biochemistry, genetics and molecular biology "sustainable degradation of lignocellulosic biomass – techniques, applications and commercialization". Intech, Rijeka, pp 251–283

Östberg L (2012) Some aspects on pulp pre-treatment prior to viscose preparation Licentiate thesis, Karlstad University Studies, Karlstad

Paës G, Berrin JG, Beaugrand J (2012) GH11 xylanases: structure/function/properties relationships and applications. Biotechnol Adv 30:564–592

Pathak P, Bhardwaj NK, Singh AK (2011) Optimization of chemical and enzymatic deinking of photocopier waste paper. BioResources 6(1):447–463

Prasad DY, Heitmann JA, Joyce TW (1993) Enzymatic deinking of colored offset newsprint. Nordic Pulp Pap Res J 8(2):28–286

Puls J, Janzon R, Saake B (2006) Comparative removal of hemicelluloses from paper pulps using nitren, cuen, NaOH, and KOH. Lenzinger Berichte 86:63–70

Quintana E, Valls C, Vidal T, Roncero MB (2013) An enzyme-catalysed bleaching treatment to meet dissolving pulp characteristics for cellulose derivatives applications. Bioresour Technol 148:1–8

Rabinovich ML, Melnick MS, Bolobova AV (2002a) Microbial cellulases (review). Appl Biochem Microbiol 38(4):305–321

Rabinovich ML, Melnick MS, Bolobova AV (2002b) The structure and mechanism of action of cellulolytic enzymes. Biochemistry 67(8):850–871

Rahkamo L, Siika-Aho M, Vehviläinen M, Dolk M, Viikari L, Nousiainen P, Buchert J (1996) Modification of hardwood dissolving pulp with purified *Trichoderma reesei* cellulases. Cellulose 3(3):153–163

Rahkamo L, Siika-Aho M, Viikari L, Leppänen T, Buchert J (1998) Effects of cellulases and hemicellulases on the alkaline solubility of dissolving pulps. Holzforschung 52(6):630–634

Rye CS, Withers SG (2000) Glycosidase mechanisms. Curr Opin Chem Biol 4:573–580

Scott GM, Akhtar M, Lentz M, Sykes M, Abubakr S (1996) Biosulfite pulping using *Ceriporiopsis subvermispora*. In: Srebotnik E, Messner K (eds) Biotechnology in the pulp and paper industry. Facultas-Universitätsverlag, Vienna, pp 217–220

Senior DJ, Hamilton J (1991) Use of xylanase to decrease the formation of AOX in kraft pulp bleaching. In: Proceedings of environment conference of the technical section, Canadian Pulp and Paper Association, 8–10 October, Quebec, pp 63–67

Senior DJ, Hamilton J (1992) Reduction in chlorine use during bleaching of kraft pulp following xylanase treatment. Tappi J 75(11):125–130

Shrinivasan MC (1992) Lignocellulose biotechnology, recent advances and technology prospects. In: Subba RNS, Balagopalan C, Ramakrishna SV (eds) New trends in biotechnology. Oxford and IBH Publishing Co. Pvt. Ltd., New Delhi

Sixta H (2006) Handbook of pulp, 1st edn. Wiley, Weinheim, p 2

Sixta H, Iakovlev M, Testova L, Roselli A, Hummel M, Borrega M, Heiningen A, Froschauer C, Schottenberger H (2013) Novel concepts of dissolving pulp production. Cellulose 20:1547–1561

Sunna A, Antranikian G (1997) Xylanolytic enzymes from fungi and bacteria. Crit Rev Biotechnol 7:425–430

Suurnäkki A, Tenkanen M, Siika-Aho M, Niku-Paavola M-L, Viikari L, And Buchert J (2000) *Trichoderma reesei* cellulases and their core domains in the hydrolysis and modification of chemical pulp. Cellulose 7:189–209

Teeri TT (1998) Fibre modification in nature: modes of action of cellulolytic enzymes. In: 7th international conference on biotechnology in the pulp and paper industry, Vancouver, pp A171–A174

Thakur VV, Jain RK, Mathur RM (2012) Studies on xylanase and laccase enzymatic prebleaching to reduce chlorine based chemicals during CEH and ECF bleaching. BioResources 7(2):2220–2235

Treiber E (1987) Quality demands for dissolving pulps for HWM fibers. Swedish Pulp and Paper Institute (STFI), Stockholm

Verma P, Bhardwaj NK, Singh SP (2013) Improvement in pulp dewatering through cellulases. Ippta J 25(3):105–108

Viikari L (2002) Trends in pulp and paper biotechnology. In: Viikari L, Lantto R (eds) Biotechnology in the pulp and paper industry: 8th ICBPPI meeting. Progress in biotechnology, vol 21. Elsevier, Amsterdam/New York, pp 1–5

Vitolo M (2009) Industrial uses of enzymes. In: Doelle HW, Rokem S (eds) Biotechnology VI, vol 6. Eolss, Paris, pp 135–138

Wallis AFA, Wearne RH (1990) Chemical cellulose from radiata pine kraft pulp. Appita J 43(5):355–357

Wang H, Pang B, Wu K, Kong F, Li B, Mu X (2014) Two stages of treatments for upgrading bleached softwood paper grade pulp to dissolving pulp for viscose production. Biochem Eng J 82:183–187

Wollboldt RP, Zuckerstätter G, Weber HK, Larsson PT, Sixta H (2010) Accessibility, reactivity and supramolecular structure of *E. globulus* pulps with reduced xylan content. Wood Sci Technol 44:533–546

Yang JL, Lou G, Eriksson K-EL (1992) The impact of xylanases on bleaching of kraft pulps. Tappi J 75(12):95–101

Zhan H, Yue B, Hu W, Huang W (2000) Kraft reed pulp TCF bleaching with enzyme treatment. Cellul Chem Technol 33(1–2):53–60

Zhang YHP, Lynd LR (2004) Toward an aggregated understanding of enzymatic hydrolysis of cellulose: noncomplexed cellulase system. Biotechnol Bioeng 88(7):797–824

Zhu S, Wu Y, Chen Q, Yu Z, Wang C, Jin S, Ding Y, Wu G (2006) Dissolution of cellulose with ionic liquids and its application: a mini-review. Green Chem 8:325

Functional Aspects of Xylanases Toward Industrial Applications

9

Vishal Kumar and Pratyoosh Shukla

Abstract

Microbial enzymes are biocatalysts applicable in a number of industries. Their ability to tolerate extreme pH and temperature makes them suitable for various industrial processes. Among a number of microbial enzymes, xylanases have a unique place. Xylanase production is reported in different types of microorganisms like actinomycetes, bacteria, fungi, and some yeast. Different types of xylanases have a conserved catalytic domain which establishes an evolutionary relationship between different xylanase producers. A fungus produces xylanases in a significant amount than bacteria and yeast. These enzymes have been used in several industries, viz., pulp and paper industry, biofuel industry, brewing and baking industry, food and feed industry, and pharmaceutical industry. Application of xylanases in several industries improves quality of different products, i.e., beer, juices, animal feed, baked products, paper, etc. However, a xylanase application particularly in pulp and paper industry is fascinating to the researchers. Thermostable and alkalistable cellulase-free xylanases play a very important role in pulp and paper industry.

Keywords

Xylanase • Xylan • Pulp and paper • Thermostable • Biobleaching • *Thermomyces lanuginosus*

9.1 Introduction

Enzymes are biological polymers that act as catalysts in the biochemical reactions that transform specific substrates to products. These are proteins with highly specific structure and bind to a substrate with high specificity. Enzyme lowers the activation energy to accelerate the rate of reaction without being consumed in the reaction. All enzymes are vital for the biochemical reactions; without these enzymes specific reaction cannot be possible for their application in several industrial processes (Haq et al. 2006). The functions of these enzymes in various bioprocesses have been acknowledged from ancient age. Existence of enzyme was related to the history of the past Greek where microbial enzymes were used in brewing and baking together with the

V. Kumar • P. Shukla (✉)
Enzyme Technology and Protein Bioinformatics Laboratory, Department of Microbiology, Maharshi Dayanand University, Rohtak 124001, Haryana, India
e-mail: pratyoosh.shukla@gmail.com

P. Shukla (ed.), *Frontier Discoveries and Innovations in Interdisciplinary Microbiology*,
DOI 10.1007/978-81-322-2610-9_9

production of yogurt, cheese, and alcohol. Recently, enzymes from microorganisms have occupied a significant place in the arena of biotechnology. The international market of industrial microbial enzymes was up to the billions of dollar in 1990, and now it has crossed \$2.0 billion mark in 2005 (Krishna 2005). It has been predicted that international market for industrial enzymes will attain \$7.1 billion mark in 2018.

Among the several hemicellulases xylanase plays a significant role in the decomposition of hemicellulose, a major constituent of lignocellulosic biomass. Xylan is composed of D-xylopyranose residues linked together by β-1, 4 glycosidic linkages in straight chain (Ebringerova et al. 2005). Actions of several enzymes are needed to degrade xylan; among them endo-1,4-β-xylanases act as a major enzyme. A number of diverse microorganisms have the capability to produce xylanase, including bacteria, yeasts (Bastwade et al. 1994; Poorma and Prema 2007; Haddar et al. 2012; Morias et al. 2013), actinomycetes (Nascimento et al. 2002; Ninawe et al. 2006), and fungi (Das et al. 2013; Facchini et al. 2012; Sorgatto et al. 2012; Shrivastava et al. 2013; Bakri et al. 2013; Kumar et al. 2014). Xylanolytic enzymes have a great prospective and physiological attribute for industrial applications. These enzymes have hydrolytic properties to hydrolyze xylan, the hemicellulosic biomolecule of plant cell wall. Nowadays, xylanases and the microorganisms either fungi or bacteria are used to decompose xylan to produce biofuels and chemicals, besides their use in pulp and paper industry, food and feed industry, and baking industry, where the enzymes support environment sustainability to reduce negative environmental impact (Collins et al. 2005). In addition to these applications, xylanases can be further applied to generate several biological products with high cumulative value. To fulfill the requirements of industries, more deliberation has been maintained on the stability of enzyme under harsh industrial conditions, such as high temperature, high pH, and inhibitors such as metal ions. It is known that there are several xylanases from microorganisms which possess the desired characteristics such as high thermostability and alkalistability, so an individual xylanase is incapable to meet all of such properties for any industrial applications. Therefore there is a need to isolate microorganisms from the extreme areas possessing extraordinary features which may be suitable for industrial processes.

9.2 Xylan, Xylanolytic Complex, and Xylanase

9.2.1 Xylan

Hemicelluloses, the second most abundant carbohydrate, including xylan, galactan, arabinan, and mannan are the major heteropolymers. These heteropolymers have been categorized in several types on the basis of sugar monomer present. The majority of the hemicelluloses are containing L-arabinose, D-galactose, D-mannose, and D-xylose as monomer unit. Xylan is a hemicellulose, found in the cell wall of plants as well as in some algae especially macrophytic siphonous genera. Xylan is a branched heteropolysaccharide which contains a straight chain of D-xylose units attached by β-1, 4-xylosidic bonds which is branched with acetyl, glucuronic, and arabinofuranosyl side chains connected to the xylose (an aldopentose) backbone (Yang et al. 2005, 2007). Xylans are present in each and every plant cell wall as well as cellulose found in plant cell walls. Xylans contain predominantly β-D-xylose units linked by glycosidic linkage as well as β-D-glucose in cellulose. Typically, the content of xylans varies from hardwood to softwood. Hardwoods contain greater amount of xylan 10–35 % of the hemicelluloses compared to softwoods 10–15 % of the hemicelluloses. The hardwood xylan contains O-acetyl-4-O-methylglucuronoxylan as a principal constituent. The straight chain of β-D-xylopyranosyl units which is connected by (1–4) glycosidic bonds is known as glucuronoxylans, as mentioned in Fig. 9.1. The 4-O-acetyl-α-D-glucuronopyranosyl units are connected to each other to form a linear strand. The D-glucuronosyl moieties of these strands acetylated at position 4 and connected to the location 2 or 3 of the β-D-xylopyranosyl. In softwood xylans the principal

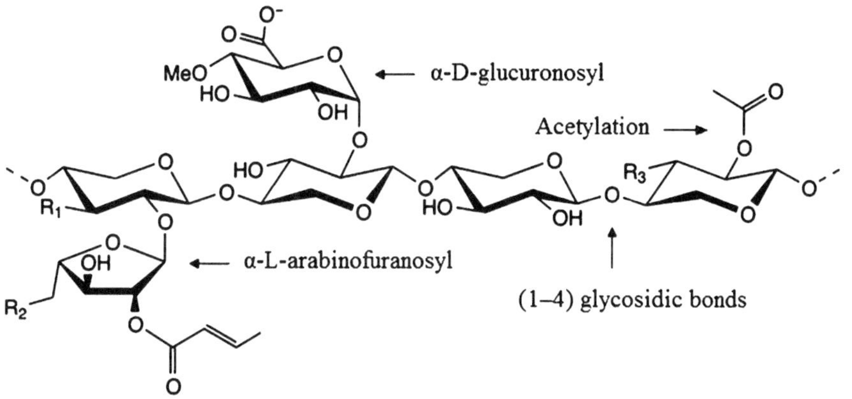

Fig. 9.1 Structure of xylan

component is arabino-4-O-methylglucuronoxylans, which are characteristically found in softwoods and also have the same xylan framework, but after an interval of ten β-D-xylopyranosyl moieties, these are substituted by α-L-arabinofuranosyl (Fig. 9.1). Softwoods have a high content of 4-O-methyl-α-D-glucuronopyranosyl units compared to hardwood xylan.

9.2.2 Xylanolytic Complex and Their Source

Xylanases accelerate the hydrolytic breakdown of xylan, a hemicellulose. Xylanases are widely distributed from eukaryotes to prokaryotes. Major part of xylanase production is produced by microorganisms such as bacteria, fungi, and actinomycetes, and these enzymes carry out hydrolysis of different components of the cell wall of plants, together with different other enzymes such as cellulases and hemicellulases that hydrolyze cellulose and hemicelluloses. Xylanases are also reported in arthropods, crustaceans, marine algae, mollusks, protozoans, and seeds of terrestrial plants (Bajpai 2014; Sunna and Antranikian 1997). Among the several microbial sources of xylanases, particularly filamentous fungi are more fascinating as they produce extracellular xylanase, and the amount of secretion is much greater than that found in yeasts and bacteria. As xylan has a heterogeneous structure, only one enzyme is not sufficient to complete hydrolysis of xylan, but it requires a xylanolytic complex. The different components of xylanolytic complex have been comprehensively studied; these are endoxylanase, 1,4-β-xylosidase, α-L-arabinofuranosidase, acetylxylan esterase, p-coumaric acid esterase, α-glucuronidase, and ferulic acid esterase.

9.2.3 Structure of Endo-1,4-β-Xylanases

Xylanases can be divided into two families on the basis study of homologous sequences and analysis of hydrophobic clusters; these are F and G families. F xylanases are somewhat larger than G xylanases, having a molecular mass of approximately 35 and 20 kDa, respectively. Family F and family G are corresponding to GH10 and GH11 families according to numerical categorization of glycosyl hydrolases. The three-dimensional structures of GH10 and GH11 endo-1,4-β-xylanases have been determined for a number of bacterial as well as fungal xylanases. The first xylanase enzyme which was structurally illustrated was Xyl-11 from *Bacillus pumilus* (Ko et al. 1992; Arase et al. 1993), but structural predetermination was not sufficient enough for accuracy, and three-dimensional structures were never submitted. Structures of xylanases from *Trichoderma harzianum* (Campbell et al. 1993) and BCX (Wakarchuk et al. 1994) were reported at the same time. There is a β-jelly-roll-like structure which comprised of two antiparallel β-sheets sculpting a deep and elongated cleft and acts as a catalytic domain of enzyme. The dominant force to stabilize framework is a number of hydrogen bonds between these two antiparallel β-strands. One of the β-sheets is partially perverted on itself, while another β-sheet is nearly planar to constructing a right angle. An exclusive α-helix is crammed beneath the β-sheet. The β-jelly roll which is the catalytic part of the xylanase is one of the super-fold structures that are more conserved compared to its sequence, and it possesses several functions (Paës et al. 2012). An evolutionary relationship probably may establish between all these enzymes which represent a jelly-roll super-fold structure. There are some significant differences in catalytic activities of endo-1,4-β-xylanases of GH11 and GH10 which may be credited to the variations in their tertiary structure. In general catalytic domain structure of the GH10 endo-1,4-β-xylanase is a barrel-like structure consisting of eight folds. There is a shallow furrow on the bottom of the "bowl"-like shape where substrate binds to the enzyme. The substrate binding sites of the GH10 endo-1,4-β-xylanases are apparently not such deep cleft as the substrate binding sites of GH11 endo-1,4-β-xylanases. The GH11 endo-1,4-β-xylanases are smaller and compact molecules with molecular infrastructure that primarily composed of β-pleated sheets as shown in Fig. 9.2. The catalytic groups present in the fissure are able to hold a string of 5–7 xylopyranosyl units. Thermostability is a significant characteristic due to their planned industrial applications. Thermostability and thermophilic characteristics of these enzymes can be described by a number of factors and structural parameters such as the significance of the disulfide bridges and aromatic sticky patches.

9.3 Industrial Applications of Xylanases

Xylanases have created great interest recently because of their potential application in several industrial processes. Nowadays, xylanases and xylan in biotechnological arena have been grown remarkably (Bhat 2000; Beg et al. 2001; Subramaniyan and Prema 2002; Techapun et al. 2003). The use of xylanases has been started since the 1980s; it begins with the preparation of fodder and animal feed and later in the biofuel industry, food industry, pulp and paper industry, and textile industry. The by-products of xylan decomposition by xylanases, which have enormous importance in industrial applications, are xylitol and furfural (Parajó et al. 1998; Kamat et al. 2014). Presently, cellulase, xylanase, and pectinases have taken financial credit of at least 20 % of the global enzyme trade (Polizeli et al. 2005; Sharma and Kumar 2013). A major part of commercial xylanases are produced by *Aspergillus*, *Thermomyces*, *Trichoderma*, *Penicillium*, *Aureobasidium*, and *Bacillus* sp.

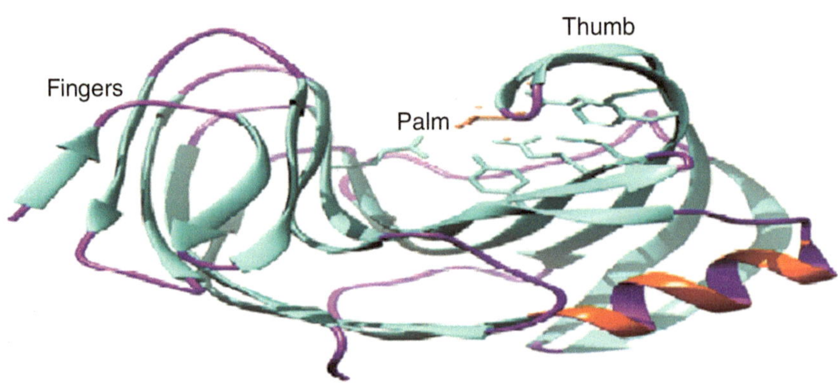

Fig. 9.2 Topological representation of catalytic domain of xylanase

Commercial xylanases are produced by industries in some countries, for example, in Canada, Denmark, Germany, Finland, Japan, Ireland, and the USA.

9.3.1 Paper and Pulp Industry

The traditional process of paper manufacturing at industrial scale required alkaline cooking of pulp at high temperature up to 170 °C followed by the removal of lignin using a chemical-based bleaching process. Chemical bleaching is generally performed by chlorine-based agents such as ClO_2, Cl_2, and hypochlorite which results in the creation of several organic derivatives of chlorine. Most of these are classified as the strong mutagens and suspected carcinogens, though an eco-friendly alternate is available for bleaching of pulps by using microbial enzymes such as xylanases and laccase. Application of biological agents such as enzymes or whole microorganism in bleaching process of pulp and paper industry is known as biobleaching.

9.3.1.1 Biobleaching

Endo-1,4-β-xylanases have been applied in paper and pulp industry to biobleaching processes. These enzymes remove xylan cross-linking layer of plant cell wall to facilitate the release of lignin and ultimately enhance the bleaching effect by chemicals (Birijlall et al. 2011; Ko et al. 2011). There are a number of reports available on biobleaching studies using endo-1,4-β-xylanases that are summarized in Table 9.1. In a few studies, cocktail of two enzymes xylanase and laccase is assessed in biobleaching process. Application of xylanases in biobleaching process significantly decreases the requirement of chlorine in bleaching process ranging from 10 to 42 % and thus helps to protect the environment from strong mutagens and suspected carcinogens. Biobleaching process reduces the overall cost of paper production and enhances the paper quality. For the development of xylanase as a suitable pre-bleaching agent, it is desirable that the xylanase should be thermostable and alkaline stable (Bajpai 1999).

Table 9.1 Biobleaching of different pulp samples using different xylanases

Sr. no.	Pulp	Microbial strain	Percentage decrease in chlorine consumption	References
1.	Kraft pulp	*Talaromyces thermophilus*	42 %	Maalej-Achouri et al. (2012)
2.	Eucalyptus kraft pulp	*Bacillus halodurans* FNP 135	20 %	Sharma et al. (2015)
3.	Wheat straw pulp	*Bacillus halodurans* C-125	10 %	Lin et al. (2013)
4.	Wheat straw pulp	*Bacillus stearothermophilus* SDX	20 %	Garg et al. (2011)
5.	Wheat straw pulp	*Streptomyces rameus* L2001	14.5 %	Li et al. (2010)
6.	Hardwood pulp	*Paenibacillus campinasensis* BL11	–	Ko et al. (2011)
7.	Wood kraft pulp	*Bacillus* sp. XTR-10	15 %	Saleem et al. (2009)
8.	Non-wood pulps	*Thermomyces lanuginosus* SSBP	25 %	Ziaie-Shirkolaee et al. (2008)
9.	Eucalyptus kraft pulp	*Bacillus pumilus* ASH	20 %	Battan et al. (2007)
10.	Wheat straw-rich soda pulp	*Streptomyces cyaneus* SN32	10 %	Ninawe and kuhad (2006)
11.	Wheat straw pulp and rice straw pulp	*Bacillus coagulans*	–	Chauhan et al. (2006)

9.3.1.2 Requirement of Endo-1,4-β-Xylanase Free of Cellulase

Despite the benefits of using xylanases, there are certain drawbacks also as these enzyme preparations have mixing of some extent of cellulases which adversely affects the paper quality because cellulases can easily carry out hydrolytic decomposition of cellulose, which is the major recovered product in pulp and paper industry. However, there are certain microorganisms which can produce higher levels of endo-1,4-β-xylanases with no cellulase activity or fragile cellulase activity which are most suitable to apply in paper and pulp industry to obtain a good quality of paper without loss of pulp viscosity. Biotechnological techniques further help to obtain cellulase-free xylanases by cloning techniques which are useful for the application of xylanases in paper and pulp industry.

9.3.2 Feed, Brewing, and Baking Industry

Xylanases have also been used in preparation of animal feed and fodder, food industry, and baking industry apart from paper and pulp industry. These enzymes are used in the baking industry as food additive to wheat flour to improve handling of dough and qualities of baked products. It is also used as food additives to poultry feed to improve digestibility of feed. Xylanases have also been used to the extraction of plant products such as coffee, oil, and starch (Harris and Ramalingam 2010). These enzymes have been found to improve the nutritional qualities of animal's grain feed and agricultural fodder (Café et al. 2002). Xylanases are also used in conjugation with pectinase and cellulase for clearing up of fruit juices (Biely 1985).

9.3.2.1 Xylanase in Brewing Industry

The wine and juice industries required a number of enzymes which make a considerable part of the enzyme market. These industries require effective methods of extraction, clearance, and stabilization to manufacture the fruit and vegetable juices. Endo-1,4-β-xylanase helps considerable increase in juice yield from fruits and also in the maceration process. Beside this, it also reduces the viscosity of the fruit juice and consequently improves its filterability (Biely 1985). Xylanases improve the extraction of fermentable sugar from barley and therefore are useful for making beer. Amylases and pectinases help to increase stability of the fruit pulp, liquefaction of fruit and vegetables, and decline viscosity which further leads to an increased yield of juice. These enzymes further help to enhance flavor of juices by improved recovery of aromas, essential oils, edible dyes, mineral salts, pigments and vitamins, etc. These enzymes carry out hydrolytic breakdown of components that hampers the chemical or physical clearance of the juices and wine that may create vagueness in the product (Polizeli et al. 2005). A genetically manipulated yeast strain was constructed for wine production with both the xln genes from *Aspergillus nidulans* for xylanase production to make wine having more prominent aroma (Ganga et al. 1999). Arabinoxylan is a long-chain polysaccharide which increases the viscosity of beer and gives muddy appearance to it which is not desirable; to short out this problem, xylanases have been used for hydrolytic decomposition of arabinoxylans to its smaller oligo-sugars which reduces the beer's viscosity and accordingly eliminating its grimy appearance (Dervilly et al. 2002).

9.3.2.2 Xylanase in Baking Industry

Xylanases have also been employed in the baking industry in addition with other enzymes like α-amylase, glucose oxidase, malting amylase, and proteases to improve the quality of baked product. The xylanases help in the evenly distribution of water in wheat flour by breaking down the hemicelluloses found in it. Enzymatic breakdown of non-starch carbohydrates leads to the improved rheological properties of baked food products (Martinez-Anaya and Jimenez 1997; Gray and BeMiller 2003). With the help of these enzymes, we can obtain breads with increased volume, better absorption quality, and enhanced fermentation resistance (Harbak and Thygesen 2002; Camacho and Aguilar 2003). Cocktail of xylanase and endoglucanase is used for the hydrolytic decomposition of arabinoxylan and starch, to separate out gluten from the starch in the wheat flour. The main beneficial characteristic for xylanases to apply in the food industry is its optimum activity at an acidic pH and its high stability.

9.3.2.3 Xylanase in Animal Feed and Fodder

Animal feed and fodder production using various enzymes is an important sector of agribusiness. Xylanases have been applied in animal feed and fodder in combination with different other enzymes such as cellulases, glucanases, amylases, pectinases, proteases, and lipases. The arabinoxylan is an anti-nutrient component of animal feed which is found in the cell walls of grains. These enzymes split arabinoxylans into the nutritive components of the feed, decreasing the viscosity of the raw material with enhanced nutritive value (Twomey et al. 2003). When these components are present in soluble form in animal feed, they may increase the viscosity of the ingested feed that renders absorption and mobility of ingested feed in the digestive tract of animals. If xylanase is applied to feed containing sorghum and maize, it may enhance the digestion capability of nutrients in the initial part of the digestive tract of animals, resulting in a better use of energy (Babalola et al. 2006). It has been reported that endo-1,4-β-xylanase together with cellulase treatment of forages produces a better quality silage that improves the subsequent rate of plant cell wall digestion by ruminants. There is a considerable amount of sugars sequestered in the xylan of plant biomass. As effect of endo-1,4-β-xylanase treatment, there is increased amount of nutritive sugar and that is useful for digestion in cow and other ruminants. It has been reported that endo-1,4-β-xylanase also produces some compounds which are the nutritive source for many ruminal microflora.

9.4 Conclusion

Xylanases are a significant industrial enzyme which hydrolytically decomposes the xylan, plant hemicelluloses. These are mostly reported in microorganisms and in flora but not in animals. Xylanases have been comprehensively studied from different microbial sources, viz., fungus, bacteria, yeast, and actinomycetes. Commercial xylanases have been mainly produced by fungus and bacteria. Thermostability and alkaline stability of xylanases make them a suitable agent for industrial applications, viz., pulp and paper industry, biofuel industry, brewing and baking industry, food and feed industry, and pharmaceutical industry. Its applications in biobleaching process in paper industries have been described which reduce chlorine consumption ultimately in reduction of environmental pollution. It acts as an additive in animal feedstock and bakery products to develop its nutritional value. There is much scope toward genetic engineering and protein engineering for microbial xylanases in order to take advantage of these enzymes more appropriately for industries. Nevertheless, it demands extremely specific and appropriate functionally stable enzymes toward their effective and demonstrated use in various industrial applications.

References

Arase A, Yomo T, Urabe I, Hata Y, Katsube Y, Okada H (1993) Stabilization of xylanase by random mutagenesis. FEBS Lett 316:123–127

Babalola TOO, Apata DF, Atteh JO (2006) Effect of β-xylanase supplementation of boiled castor seed meal-based diets on the performance, nutrient absorbability and some blood constituents of pullet chicks. Trop Sci 46(4):216–223

Bajpai P (1999) Application of enzymes in the pulp and paper industry. Biotechnol Prog 15:147–157

Bajpai P (2014) Sources, production, and classification of xylanases. In: Bajpai P (ed) Xylanolytic enzymes. Academic Press (Imprint of Elsevier), Tokyo, pp 43–52

Bakri Y, Jawhar M, Eddin Arabi MI (2013) Xylanase production by *Fusarium solani* in solid state fermentation. Res Biotechnol 4(1):31–37

Bastwade KB, Puntambekar US, Gokhale DV (1994) Optimization of cellulase-free xylanase production by a novel yeast strain. J Ind Microbiol 13:220–224

Battan B, Sharma J, Dhiman SS, Kuhad RC (2007) Enhanced production of cellulase-free thermostable xylanase by *Bacillus pumilus* ASH and its potential application in paper industry. Enzyme Microb Technol 41(6–7):733–739

Beg QK, Kapoor M, Mahajan L, Hoondal GS (2001) Microbial xylanases and their industrial applications: a review. Appl Microbiol Biotechnol 56:326–338

Bhat MK (2000) Cellulases and related enzymes in biotechnology. Biotechnol Adv 18:355–383

Biely P (1985) Microbial xylanolytic systems. Trends Biotechnol 3:286–290

Birijlall N, Manimaran A, Kumar KS, Permaul K, Singh S (2011) High level expression of a recombinant xylanase by *Pichia pastoris* NC 38 in a 5 L fermenter and its efficiency in biobleaching of bagasse pulp. Bioresour Technol 102:9723–9729

Café MB, Borges A, Fritts A, Waldroup W (2002) Avizyme improves performance of broilers fed corn-soybean meal based diets. J Appl Poult Res 11(1):29–33

Camacho NA, Aguilar OG (2003) Production, purification and characterization of a low molecular mass xylanase from *Aspergillus sp.* and its application in bakery. Appl Biochem Biotechnol 104:159–172

Campbell RL, Rose DR, Wakarchuk WW, To R, Sung WS, Yaguchi M (1993) A comparison of the structure of the 20 kD xylanases from *Trichoderma harzianum* and *Bacillus circulans*. In: Suominen P, Reinikainen T (eds) *Trichoderma reesei* cellulases and other hydrolases. Foundation for Biotechnical and Industrial Fermentation Research, Espoo, pp 63–72

Chauhan S, Choudhury B, Singh SN, Ghosh P (2006) Application of xylanase enzyme of *Bacillus coagulans* as a prebleaching agent on non-woody pulps. Process Biochem 41(1):226–231

Collins T, Gerday C, Feller G (2005) Xylanases, xylanase families and extremophilic xylanases. FEMS Microbiol Rev 29:3–23

Das A, Paul T, Halder SK, Jana A, Maity C, Mohapatra PKD, Pati BR, Mondal KC (2013) Production of cellulolytic enzymes by *Aspergillus fumigatus* ABK9 in wheat bran-rice straw mixed substrate and use of cocktail enzymes for deinking of waste office paper pulp. Bioresour Technol 128:290–296

Dervilly G, Leclercq C, Zimmerman D, Roue C, Thibault JF, Sauliner L (2002) Isolation and characterization of high molecular mass water-soluble arabinoxylans from barley malt. Carbohydr Polym 47:143–149

Ebringerova A, Hromadkova Z, Heinze T (2005) Hemicellulose. Adv Polym Sci 186:1–67

Facchini FD, Reis VR, Roth AP, Magalhaes KA, Peixoto-Nogueira SC, Casagrande DR, Reis RA, Polizeli MLTM (2012) Effects of *Aspergillus spp.* exogenous fibrolytic enzymes on in vitro fermentation of tropical forages. J Sci Food Agric 92:2569–2573

Ganga MA, Piñaga F, Vallés S, Ramón D, Querol A (1999) Aroma improving in microvinification processes by the use of a recombinant wine yeast strain expressing the *Aspergillus nidulans* xlnA gene. Int J Food Microbiol 47:171–178

Garg G, Dhiman SS, Mahajan R, Kaur A, Sharma J (2011) Bleach-boosting effect of crude xylanase from *Bacillus stearothermophilus* SDX on wheat straw pulp. New Biotechnol 28(1):58–64

Gray JA, Bemiller JN (2003) Bread staling: molecular basis and control. Compr Rev Food Sci F 2(1):1–21

Haddar A, Driss D, Frikha F, Ellouz-Chaabouni S, Nas M (2012) Alkaline xylanases from *Bacillus mojavensis* A21: production and generation of xylooligosaccharides. Int J Biol Macromol 51(4):647–656

Haq IU, Javed MH, Khan TM (2006) An innovative approach for hyperproduction of cellulolytic and hemicellulolytic enzymes by consortium of *Aspergillus niger* MSK-7 and *Trichoderma viride* MSK- 10. African J Biotech 5(8):609–614

Harbak L, Thygesen HV (2002) Safety evaluation of a xylanase expressed in *Bacillus subtilis*. Food Chem Toxicol 40:1–8

Harris AD, Ramalingam C (2010) Xylanases and its application in food industry: a review. J Exp Sci 1(7):01–11

Kamat S, Khot M, Zinjarde S, RaviKumar A, Gade WN (2014) Coupled production of single cell oil as biodiesel feedstock, xylitol and xylanase from sugarcane bagasse in a biorefinery concept using fungi from the tropical mangrove wetlands. Bioresour Technol 135:246–253

Ko EP, Akatsuka H, Moriyama H, Shinmyo A, Haya Y, Katsube Y et al (1992) Site-directed mutagenesis at aspartate and glutamate residues of xylanase from *Bacillus pumilus*. Biochem J 288:117–121

Ko CH, Tsai CH, Tu J, Yang BY, Hsieh DL, Jane WN et al (2011) Identification of *Paenibacillus* sp. 2S-6 and application of its xylanase on biobleaching. Int Biodeterior Biodegrad 65:334–339

Krishna C (2005) Solid-state fermentation systems-an overview. Crit Rev Biotechnol 25:1–30

Kumar V, Pandey P, Gupta S, Shukla P (2014) A reviving preliminary evoke on few xylanase producing fungal isolates from different ecological niche. Int J Curr Microbiol App Sci 3(4):501–506

Li X, She Y, Sun B, Song H, Zhu Y, Lv Y, Song H (2010) Purification and characterization of a cellulase-free, thermostable xylanase from *Streptomyces rameus* L2001 and its biobleaching effect on wheat straw pulp. Biochem Eng J 52(1):71–78

Lin X, Han S, Zhang N, Hu H, Zheng S, Ye Y, Lin Y (2013) Bleach boosting effect of xylanase A from *Bacillus halodurans* C-125 in ECF bleaching of wheat straw pulp. Enzyme Microb Technol 52(2):91–98

Maalej-Achouri I, Guerfali M, Romdhane IBB, Gargouri A, Belghith H (2012) The effect of *Talaromyces thermophilus* cellulase-free xylanase and commercial laccase on lignocellulosic components during the bleaching of kraft pulp. Int Biodeterior Biodegrad 75:43–48

Martinez-Anaya MA, Jimenez T (1997) Functionality of enzymes that hydrolyse starch and non-starch polysaccharide in breadmaking. Z Lebensm Unters Forsch 205:209–214

Morias CG, Uetanabaro APT, Rosa LH, Lachance MA, Rosa CA (2013) D-xylose-fermenting and xylanase producing yeast species from rotting wood of two Atlantic Rainforest habitats in Brazil. Fungal Genet Biol 60:19–28

Nascimento RP, Coelho RR, Marques Alver SL, Girio EPS, Amaral-Collago MT (2002) A novel strain of *Streptomyces malaysiensisi* from Brazilian soil produces high endo-1, 4 xylanase titres. Enzyme Microb Technol 31:549–555

Ninawe S, Kuhad RC (2006) Bleaching of wheat straw-rich soda pulp with xylanase from a thermoalkalophilic Streptomyces cyaneusSN32. Bioresour Technol 97(18):2291–2295

Ninawe S, Lal R, Kuhad RC (2006) Isolation of three xylanase producing strains of Actinomycetes and their identification using molecular methods. Curr Microbiol 53:178–182

Paës G, Berrin JG, Beaugrand J (2012) GH11 xylanases: structure/function/properties relationships and applications. Biotechnol Adv 30(3):564–592

Parajó JC, Domíngues H, Domíngues JM (1998) Biotechnological production of xylitol. Part 1, interest of xylitol and fundamentals of its biosynthesis. Bioresour Technol 65:191–201

Polizeli ML, Rizzatti AC, Monti R, Terenzi HF, Jorge JA, Amorim DS (2005) Xylanases from fungi: properties and industrial applications. Appl Microbiol Biotechnol 67:577–591

Poorma CA, Prema P (2007) Production of cellulose free endoxylanase from novel alkalophilic thermotolerant *Bacillus pumilus* by solid state fermentation & its application in waste paper recycling. Bioresour Technol 98:485–490

Saleem M, Tabassum MR, Yasmin R, Imran M (2009) Potential of xylanase from thermophilic *Bacillus sp.* XTR-10 in biobleaching of wood kraft pulp. Int Biodeterior Biodegrad 63(8):1119–1124

Sharma M, Kumar A (2013) Xylanases: an overview. Br Biotechnol J 3(1):1–28

Sharma P, Sood C, Singh G, Capalash N (2015) An eco-friendly process for biobleaching of eucalyptus kraft pulp with xylanase producing *Bacillus halodurans*. J Clean Prod 87:966–970

Shrivastava S, Shukla P, Deepalakshmi PD, Mukhopadhyay K (2013) Characterization, cloning and functional expression of novel xylanase from *Thermomyces lanuginosus* SS-8 isolated from self-heating plant wreckage material. World J Microbiol Biotechnol 12:2407–2415

Sorgatto M, Guimarães NCA, Zanoelo FF, Marques MR, Peixoto-Nogueira SC, Giannesi GC (2012) Purification and characterization of an extracellular xylanase produced by the endophytic fungus, *Aspergillus terreus*, grown in submerged fermentation. Afr J Biotechnol 11(32):8076–8084

Subramaniyan S, Prema P (2002) Biotechnology of microbial xylanases, enzymology, molecular biology, and application. Crit Rev Biotechnol 22:33–64

Sunna A, Antranikian G (1997) Xylanolytic enzymes from fungi and bacteria. Crit Rev Biotechnol 17:39–67

Techapun C, Poosaran N, Watanabe M, Sasaki K (2003) Thermostable and alkaline-tolerant microbial cellulose-free xylanases produced from agricultural wastes and the properties required for use in pulp bleaching bioprocesses: a review. Process Biochem 38:1327–1340

Twomey LN, Pluske JR, Rowe JB, Choct M, Brown W, McConnell MF, Pethick DW (2003) The effects of increasing levels of soluble non-starch polysaccharides and inclusion of feed enzymes in dog diets on faecal quality and digestibility. Anim Feed Sci Technol 108(1–4):71–82

Wakarchuk WW, Robert L, Campbell RL, Sung WL, Davoodi J, Yaguchi M (1994) Mutational and crystallographic analyses of the active site residues of the *Bacillus circulans* xylanase. Protein Sci 3:467–475

Yang R, Xu S, Wang Z, Yang W (2005) Aqueous extraction of corncob xylan and production of xylooligosaccharides. LWT- Food Sci Technol 38:677–682

Yang CH, Yang SF, Liu WH (2007) Production of xylooligosaccharides from xylans by extracellular xylanases from *Thermobifida fusca*. J Agric Food Chem 55:3955–3959

Ziaie-Shirkolaee Y, Talebizadeh A, Soltanali S (2008) Comparative study on application of *T. lanuginosus* SSBP xylanase and commercial xylanase on biobleaching of non wood pulps. Bioresour Technol 99(16):7433–7437

10 An Overview of Potential Health Hazards in Recreational Water Environments and Monitoring Programme in Porto Belo Bay, Brazil

Camila Burigo Marin, Katia Naomi Kuroshima, Lia Raquel de Campos Galuppo, Daniela de Souza Waltrick, Maria Letícia Biavatti, Giovana Faccio, Karoliny Fantim Fracalossi, Marcus Adonai Castro da Silva, Ana Lucia Zonta, Marco Antônio Bacellar Barreiros, and Diego Galiani

Abstract

The Porto Belo Bay region is formed by Itapema and Porto Belo cities, both considered important beaches from southern Brazil. During the summer, these cities welcome large numbers of tourists, fomenting their economies that are largely based on tourism activities. Given the poor sewage treatment in this region (60 % in Itapema City and none in Porto Belo City), a high population increase over the summer period can influence the water quality within Porto Belo Bay. The aim of this study was to evaluate the microbiological, chemical, and physical aspects within this area. Water samples were collected from 14 sites in the Perequê-Áçu River (five sampling stations) and in nine beach sampling stations from October 2000 to March 2014. The physical parameters (pH, salinity, DO, temperature), inorganic nutrients (NH^{4+}, PO_4^{3-}), organic compounds (POC, TOP, SPM), and biological and microbiological indicators (BOD_5, chlorophyll-*a*, total coliforms, and *E. coli*) indicated that areas close to the river discharge displayed lowest water quality, with a greater amount of inorganic and organic nutrients. On the other hand, better water quality was observed in two beach sites far from river or storm water discharges.

C.B. Marin (✉) • K.N. Kuroshima
L.R. de Campos Galuppo • D. de Souza Waltrick
M.L. Biavatti • G. Faccio • K.F. Fracalossi
M.A.C. da Silva • A.L. Zonta
Universidade do Vale do Itajaí – UNIVALI,
Rua Uruguai, n°458, Itajaí, SC, CEP: 88302202, Brazil
e-mail: milaoceano@gmail.com

M.A.B. Barreiros
Universidade Federal do Paraná – UFPR,
Palotina, Brazil

D. Galiani
Companhia Nacional de Saneamento – CONASA,
Londrina, Brazil

P. Shukla (ed.), *Frontier Discoveries and Innovations in Interdisciplinary Microbiology*,
DOI 10.1007/978-81-322-2610-9_10

Keywords
Water quality • Microbiology • Inorganic nutrients

Recreational coastal water and freshwater environments are defined as areas where any type of recreational activities is undertaken by an expressive number of users. The use of these areas is diverse; contact with water may be direct, thorugh the practice of sports such as swimming, surfing, and diving, or indirect, such as fishing, sailing, walking, and picnicking. However, activities with direct contact with water is of more concern in polluted waters due to the risk of ingestion (WHO 2000, 2003).

The social and economic importance of recreational uses of waters has increased over the last years (WHO 2000). The pursuit of leisure activities in contact with the natural environment, in order to counteract the urban way of life, is believed to be the main driver for this pattern.

Recreational swimming is a popular activity in many places, the European Union, the United States, Peru, China, Brazil, and others. Americans make an estimated 928 million trips to the beach each year (NOAA 2005).

Brazil has a vast coastline, composed of over 2000 beaches, along over 7000 km of coastline, making it a natural attraction for domestic and foreign tourists. In the south, beach tourism occurs in the summer (December–March) due to its subtropical climate. For its natural beauties, and services offered, the estate of Santa Catarina consolidates itself as the largest center of beach tourism in southern Brazil, attracting mainly Brazilian, Argentinean, Chilean, Uruguayan, and Paraguayan nationals (CUNHA 2010).

10.1 Coastal Waters

The marine coast is a fragile environment due to its intense dynamics. Coastal environments are subject to the action of tides, marine currents, and waves which together shape the coastline (Philippi Junior 2010). Coastal waters are used in various ways by societies including: leisure, transportation, food production, and sewage receiver of domestic and industrial waste. At times however, such uses of coastal waters may be incompatible with one another.

Recreational swimming in coastal waters is a popular activity in Brazil, supporting the tourist activity on the coast and moving economic resources both within states and between countries (WHO 2003; IBGE 2004). However, tourism activities have directly and indirectly caused serious environmental impacts. For instance, the construction of holiday homes and tourist facilities can disrupt the balance existing between the acting natural forces, increased water consumption, and production of sewage in large quantity (Vasconcelos and Coriolano 2008).

The release of untreated sewage to waterways may introduce pollutants to the environment, posing a risk to the health of local population and water users. Human sewage or other animal sources, for instance, can cause health problems due to the presence of contagious microorganisms (Boehm et al. 2009). Furthermore, pollution of coastal waters reaches estuarine environments such as mangroves, also affecting the fishery. Therefore, the monitoring of this indicator has implications on the health of the population, tourism, and sea fishing (IBGE 2004).

10.2 Factors Affecting the Quality of Recreational Coastal Waters

In densely populated areas, the water quality closely reflects human activity located not only along the beaches or rivers but also within its whole upstream watershed (Billen et al. 2001).

Water pollution occurs when substances or energy is directly or indirectly added to a water body, changing its natural characteristics.

Common pollutants are nutrients (eutrophication), pathogens, organic matter (nonbiodegradable and biodegradable), heavy metals, and suspended solids (Parsons and Takahashi 1998; Von Sperling 1996).

The general idea of pollution includes several processes of change in water quality, such as bacteriological, chemical contamination and eutrophication (Moraes and Jordão 2002). The resultant impact on human health by these contaminants originates mainly from the release of industrial and domestic wastewater, agricultural use of fertilizers, combustion of fossil fuels, and storm water flow into rivers, beaches, and lakes. Thus, the polluted water environment involves chemical, biological, and physical processes (WHO 2000; Segerson and Walker 2002).

Human activity significantly influences the cycling of nutrients, especially the movement of nutrients to estuaries and coastal waters (Billen et al. 2001).The polluting effect of effluent discharges in coastal waters offer a variable polluting effect that is dependent on the composition and quantity of the effluent as well as the water body depuration capacity. Thus, enclosed water systems with high residence time of water will be more readily affected by sewage discharges (Pilson 1985; Horita and Carvalho 1999; WHO 2000). The abundance of pathogenic microorganisms and nutrient concentration are strongly influenced by the environmental dispersion capacity through physical processes of vertical diffusion, overturning, ultraviolet radiation from sunlight, and current systems (Talley et al. 2011).

10.3 Water Quality Evaluation

Water quality can be evaluated in accordance with the components or substances contained therein, i.e., its composition. Parameters used to describe and classify water include: physicochemical parameters, such as water temperature and turbidity; chemical parameters, such as biochemical oxygen demand (BOD), dissolved oxygen (DO), nitrogen and phosphorus, and potential of hydrogen ion (pH); and biological parameters including the coliform group and nonpathogenic organisms, such as phytoplankton and zooplankton.

Most aquatic creatures such as fish, bacteria, and protozoa are aerobic and need oxygen to survive (Butkus and Manous 2005). Though dissolved oxygen will not have a direct effect on users, but it will directly influence aquatic life, chemical oxidation state of metals, and microbial activity (Baumgarten et al. 2010). Oxygen may enter the water through the roots of aquatic plants and algae, which undergo photosynthesis, and through the air-water interface. The water's ability to hold dissolved oxygen is directly dependent on the temperature, salinity, and pressure. However, aquatic environments under extreme conditions where oxygen concentrations are below 4 mg.L−1 are considered critical for aquatic organisms (Arana 1997; Esteves 1988).

A combination of bacteria and excess organic material in aquatic environments may lead to eutrophic conditions. As organic matter decays, bacteria can consume oxygen, thus depleting the available oxygen supply. The amount of dissolved oxygen consumed by microorganisms in water during the oxidation of organic matter is known as the biochemical oxygen demand (BOD_5). Sources of BOD_5 are biodegradable organic carbon and ammonia, common metabolic by-products of plant and animal wastes, and human activities. Some issues associated with the discharge of wastes containing high levels of BOD_5 are poor water quality, severe dissolved oxygen depletion, and fish kills (Butkus and Manous 2005).

Potential impacts to recreational users of water bodies may arise from extreme pH levels. Very acidic or very basic water conditions may cause skin, eye, hair, and mucous membrane irritation. However, the impact will also be dictated by the buffering capacity of the water, mainly saline water (WHO 2003).

Humans are not the only ones affected by pH. Extreme pH levels in water can also cause detrimental environmental effects by affecting the solubility and toxicity of chemicals and heavy metals, stress animal systems, reduce hatching and survival rates, and ultimately cause the death of aquatic organisms. For instance, a slight change in the pH of water can increase the solubility of phosphorus and other nutrients, making them more accessible for phytoplankton and increase growth of aquatic plant. As a consequence, algal blooms may occur and result in eutrophication of the water body (Baumgarten et al. 2010).

Natural occurring elements, such as phosphorous and nitrogen, are essential for the growth of plants and animals. Although these elements do not represent direct risks to human health, extreme concentrations (high or low) of these elements in the environment can lead to ecological disequilibrium (Carmouze 1994; Braga et al. 2000).

Extreme levels of phosphorus and nitrogen are usually result from anthropogenic input (i.e. domestic wastewater supply). While phosphorus is naturally released from decaying vegetation and soils, it is also present in industrialized products such as household detergents, soaps, and sewage (Carmouze 1994). Nitrogen naturally exists in the aquatic environment in many forms, including ammonia. Some potential exogenous sources of this Nitrogen to water bodies are: municipal effluent discharges, excretion from animals, nitrogen fixation process, air deposition, and runoff from agricultural lands and cities (Vitousek et al. 1997; Valigura et al. 2000).

Water quality can also be monitored using biological parameters.Indicator bacteria, including Escherichia coli and intestinal Enterococcus, are used for the monitoring of recreational waters. The coliformbacteria group are not pathogens, nontheless, large numbers of these organisms are found in the intestinal contents of warm-blooded animals and humans. Therefore, their use as an indicator of faecal contamination provides a good indicator of potential pathogens in water (Da Cunha 2006).

10.4 Health Hazards in Coastal Waters

Recreational coastal waters may offer a diverse range of hazards to human health. Health hazards may include physical hazards, such as contamination of beach sand, cuts and bruises from sharp objects, and exposure to microalgae, or chemical and physical agents which include presence of harmful aquatic organisms, poor water quality, as well as the presence of pathogenic microorganisms in recreational water (WHO 2003). Considering the risks posed to human health due to prolonged and direct exposure to hazards, it is required that recreational waters meet specific standards of water quality.

The link between water quality and diseases is very well studied. Contamination of natural waters represents one of the leading public health risks, particularly the increase in infant mortality (IBGE 2010). Basic sanitation is fundamental to life quality; it refers to the provision of facilities and services and the maintenance of hygienic conditions through services like sewerage system, water supply, garbage collection, and urban drainage (Philippi Jr. 2010).

In Brazil only 38.7 % of the total sewage generated is treated (SNIS 2014). Urbanization rapidly resulted in the informal settlement of low-income residents in neighborhoods without access to sanitation. Such arrangements generate concentrated pollution and serious drainage problems due to a number of factors (e.g., inadequate waste disposal, siltation, and consequently decrease in the speed of runoff), helping to disseminate diseases (Moraes and Jordão 2002).

The lack of basic sanitation and minimum hygiene conditions increase the exposure of the population to many types of illnesses. Some diseases associated with water pollution are cholera, gastrointestinal infections, diarrhoea, typhoid fever, amebiasis, schistosomiasis, hepatitis, and leptospirosis. The transmission of these infectious diseases occurs normally by contaminated water ingestion or direct contact with the upper respiratory tract, eyes, or skin (Da Motta et al. 1994; FUNASA 2010; Barsano et al. 2014).

In touristic cities there is a seasonal fluctuation of the population. A population rise during summer holidays and long weekends, for example, may overload the existing sewage systems, so the excess is released in streams many times, compromising bathing (CETESB 2010) and therefore increasing the number of pathogen load in the water. Furthermore, the higher water temperatures observed during this period may enhance the proliferation of certain types of these pathogens (Van Asperen et al. 1995).

Risks of microbial contamination in coastal waters are usually associated with the type of recreational activity involved, volume of water ingested, and periods of contact. Humans involuntarily swallow some water while practicing water sports. In recreational activities with limited water contact, such as fishing, kayaking, motor boating, canoeing, and rowing, the volume of water ingested is approximately 3–4 mL (Dorevitch et al. 2011). Surfing in contaminated water offers high risk for contracting gastrointestinal illness especially through ingestion; the mean exposure is 170 ml of water ingested per activity day (Stone et al. 2008). Recreational water users with whole-body contact such as a swimmer can expect to ingest 100–200 ml of water in one session, probably more than sailboard riders and water-skiers (WHO 2003).

Divers may run a higher risk of infection with waterborne pathogens than bathers due frequent and intense contact; in marine waters, they accidentally swallow around 9.4 ml per dive (Schijven and Husman 2006), considering the serious health hazards associated with aquatic pollution.

10.5 Monitoring Programs

Monitoring programs are designed to understand how the environment changes over time and, besides this, allow the evaluation and diagnosis of the environment in question. Long-term monitoring lets us evaluate information on conservation and environmental degradation of the region studied.

The concentration analysis of nutrient inputs to coastal waters and estuaries is necessary for the management of water quality. Often, nutrient inputs are estimated as part of some scientific research project and are published only in scientific literature. But the results must be used for management too (Valigura et al. 2000; Howart et al. 2002).

Many places of recreational waters suffer from population growth and pollution increase. Monitoring programs are used in few locations facing these problems worldwide, standing out as successful examples to follow. Some examples are: Marine Conservation Society, England; Beach Water Quality – Environmental Protection Department, Hong Kong, China; Calidad de agua de las playas de Montevideo, Uruguay; and Programa Integral de Playas Limpias, Mexico.

At Lake Geneva, located between France and Switzerland, we have another example. Tourism, trade, and wine growing constitute the main activities within the region, which are generally in conflict with the natural environment. After swimming was prohibited in many areas due to local pollution, integrated management between France and Switzerland was required. Domestic and industrial sewage systems have been installed resulting in significant improvement in water quality (WHO 2000).

The Ganga River is the largest and the most important river in India. The densely populated Ganga Basin is inhabited by 37 % of India's population; domestic and industrial wastes characterize the principal source of pollution. The Ganga Project started in June 1985 and was perceived as an investment providing demonstrable effects on river water quality. After finding pollution sources, mainly domestic and industrial wastes, the first step was the creation of facilities for interception, diversion, and treatment of the wastewater (WHO 1997). Results obtained during the Ganga monitoring program, 1995–2011, indicated that the bacterial contamination continued to be critical in the Ganga River. The municipal corporations at large are not able to treat the increasing load of municipal sewage. But the success of the program is visible through to the number of improvements commissioned (CPCB 2011).

The beach monitoring program in Lima, Peru, initiated in 1986, was essential to confirm the pollution sources and convince the authorities of the real necessity of the implementation of management actions (WHO 2000).

Hong Kong began a monitoring program in 1986 and currently uses a ranking to reflect the water quality of the beach. Under the dual rating system, beaches are categorized into "good," "fair," "poor," or "very poor." Those ranked "good" and "fair" have necessary conditions for bathing waters. In 1986 only 23 % of their beaches had good conditions, and 25 % were unsuitable to swimming; at the last report released in 2013, this number increased to 58.5 %, and no improper point for swimming was recorded (HKWQRC 2013).

The Australian Gold Coast has a water quality monitoring program; currently, approximately 230 sites are used for analyzing ocean beaches, rivers, creeks, estuaries, canals, freshwater, and tidal lakes. The water quality information is used for a range of purposes including information on the ecological health, identifying trends in water quality and assessing changes in water quality in association with landscape modification (HWDEHP 2013).

A comprehension about the dynamics of water systems, including chemical, physical, and biological interactions, the perceptions of the beach users, and economic and tourism interests, is essential for a successful management of coastal waters. Certainly, there are conflicts between these factors, but many of these conflicts can be resolved through effective communication and environmental education activities (WHO 2000; HKWQRC 2013).

10.6 Porto Belo Bay Program

Porto Belo Bay is a thriving touristic area, with a local population of nearly 70,000 people. The beaches are used for recreation as well as for and seafood extraction. The sea surrounding them is also the home of miscellaneous marine species, from microscopic phytoplankton to economic important fish species.

Porto Belo Bay is one of the most important places of summer tourism in southern Brazil. Population number in this area can more than quadruple in the summer months. The cities of Itapema and Porto Belo, located within Porto Belo Bay, presented a disordered growth both in population numbers and in real estate over the last decade. This has led to uncontrolled increase in sanitation problems due to the rise of domestic sewage and the lack of planning in the infrastructure to support such a discharge. As a consequence, raw sewage ends up being dumped on beaches and negatively influencing the quality of its waters.

Poor water quality as a result of increased wastewater discharge in the summer months has been reported by several authors in Porto Belo Bay (Vasques 2002; Marin 2007; Biavatti 2014). While Itapema currently has a wastewater treatment plant (WWTP) that serves up to 60 % of the population, Porto Belo City doesn't have a WWTP and therefore only uses septic tanks or releases its untreated effluents into water bodies. The population growth and tourism contribute to the city economical growth, however the absence of a WWTP causes negative effects to habitats as well as to public health. Consequently, local beaches have been hampered by the large amount of pathogenic organisms due to the inadequate disposal of municipal domestic wastewater.

Dividing the cities of Itapema and Porto Belo lays the Perequê-Áçu River, the largest river that flows into the Porto Belo Bay. While the Perequê-Áçu River is used to discharge treated effluent from the WWTP, untreated sewage is introduced to this water body through clandestine sewage connections from both Itapema and Porto Belo. Physicochemical and/or biological changes in estuarine and coastal waters undertake ecological balance and offer hazards to human health (Carmouze 1994; CETESB 1995; De Miranda et al. 2002; Pereira Filho et al. 2003).

There are conflicts about water uses in Perequê-Áçu River. Upstream, the river is used to provide drinking water to Porto Belo, Itapema, and the neighbouring City of Bombinhas (around 500,000 people in summer time). Downstream, as well as receiving untreated effluents from Itapema and Porto Belo, waters are used for irrigation in rice cultivation, navigation and fishing (Marin 2007).

During 2013, the WWTP received and treated by upflow anaerobic sludge blanket - UASB reactor equivalent to 2,060,897.08 m^3 of effluents. The average inlet and outlet BOD_5 was 216.91 mg/L and 36.77 mg/L, respectively (CONASA 2014), with a mean removal of 83 %.

Usually, the BOD_5 of wastewater varies from 100 to 300 mg/L, with posttreatment usually aiming to achieve a reduction of BOD_5 to a range of 20–30 mg/L (Jordão and Pessoa 1995). Thus, it is concluded that the WWTP is showing very good results in removal since the expected efficiency for UASB reactors (Teixeira et al. 2009).

The values quoted are removed from carbonaceous matter within the limits established by the Brazilian Law in CONAMA Resolution 430 that complements with 357 and provides for a minimum reduction of 60 % BOD_5.

At summer time in the years 2013 and 2014, the Perequê-Áçu River suffered a major drought which, together with the receipt of large organic load from the sewage, resulted in the deaths of dozens of species, including some tougher catfish and crabs, culminating in a negative image of the city and the interdiction of stretches of beach (Caldas and Felix 2013; Basso 2014). From this situation, population demanded actions of authorities and from the water company.

Thereby, the monitoring program in Porto Belo Bay emerged and put together university and water company in February, 2014, to start chemical, microbiological diagnoses of this environment. It serves the following purposes: (1) reveal water quality of marine and river waters and (2) provide a basis for the planning of pollution control strategies.

The monitoring program includes 14 water sampling stations, 9 in the bay and 5 along the Perequê-Áçu River (Fig. 10.1). Chemical and physical parameters, including temperature, salinity, pH, and dissolved oxygen, were measured in situ by a Yellow Springs 6600 V2 sonde.

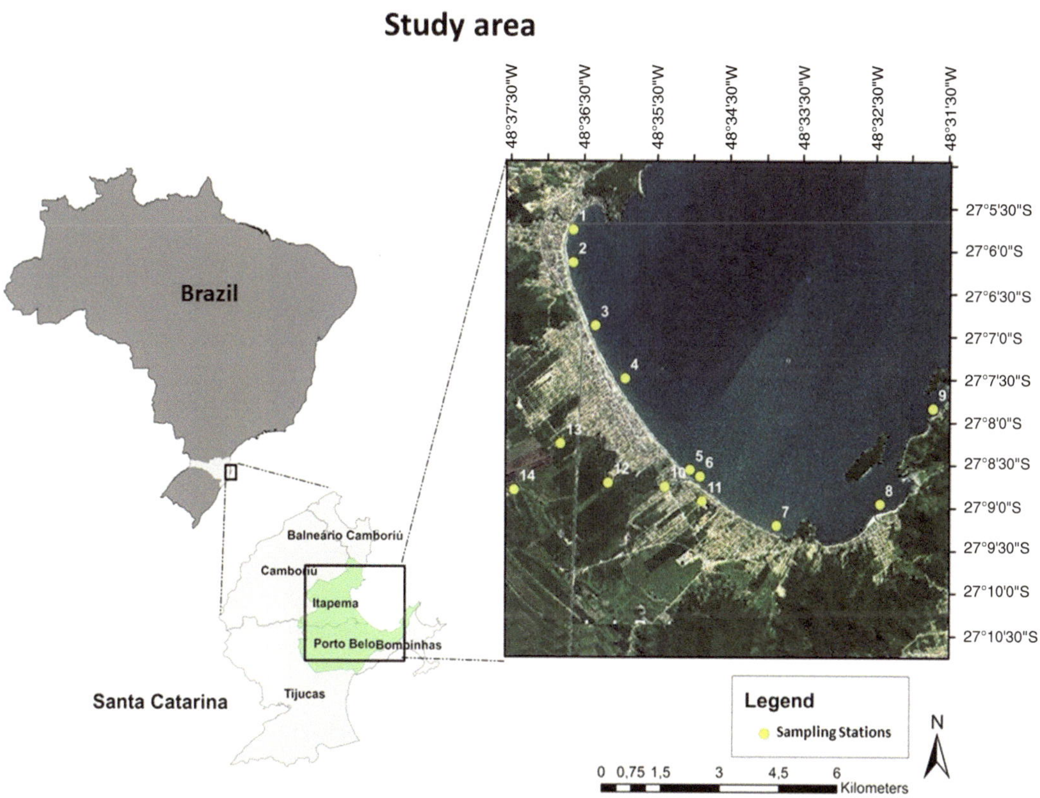

Fig. 10.1 Location of the 14 sampling sites at the Porto Belo Bay, SC, Brazil

Water samples were collected and sent to the university laboratories for the analysis of inorganic nutrients (NH_4^+, PO_4^{3-}), suspended particulate matter, particulate organic carbon, phosphorus organic, chlorophyll concentration, biochemical oxygen demand, and coliform bacteria. The methodologies for determining these parameters were followed, and the recommendations are described in American Public Health Association (APHA), the American Water Works Association (AWWA), and the Water Environment Federation (WEF) (1999). Water quality monitoring is generally conducted once a month and in summer time every week.

10.7 Data Collection

Prior to the commencement of the Porto Belo Bay monitoring program, other scientific research projects have estimated nutrient inputs; however, continuous data has never been collected in this area. Though the Porto Belo program is a continuing monitoring study, only one sampling month over the summer period (February–March 2014) and sampling points coinciding with previous studies were selected for the scope of this chapter. All existing data for this area was compiled and compared in Table 10.1.

The Porto Belo Bay monitoring program collected 99 samples throughout the year. However, in order to compare findings with previous studies, only the summer period will be presented here.

These projects (we used only the data that were collected at the same points of the current project) can be seen in Fig. 10.2, the equivalence of sample points.

Water samples were collected from 14 sites in the Porto Belo Bay (9 beach samples, 5 river samples), between October 2000 and March 2014 (Table 10.2). Each sample was then processed and their physical, chemical, and microbiological properties determined.

10.7.1 Physicochemical Parameters

The average water temperature during the sampled years was 24.9 °C for both areas, beach and river, with a maximum of 31.50 °C (March 2001) and minimum of 15.60 °C (July 2001).

The mean observed salinity for beach and river was 25.9 $g.L^{-1}$ and 25.4 $g.L^{-1}$, respectively. Ocean water normally has a mean salinity of 35. This observation is an indication of how dynamic this environment is, low values observed in Porto Belo Bay can be explained through entries of freshwater in some sampling stations. In the river, salinity was observed between 0 and 35.3, the low values being found upstream the river and high values downstream.

The mean neutral pH values observed in both river and beach are consistent with estuarine waters (Lau and Chu 1999). The lowest pH values observed on the beach always occurred during periods of low salinity, with dominance of fluvial environment or storm water discharges, possibly associated with excess nutrients (Table 10.2). Thus, the variation in pH levels observed in the beach sampling can be explained

Table 10.1 Summaries of the projects with their respective dates of beginning and end and the periodization and the points that were used for sampling

Project	Start date	Final date	Periodicity	N° sample stations	Total number of samples
Vasques 2002	October 2000	February 2002	Monthly	7	48
Univali 2003 (unpublished)	July 2001	February 2003	Monthly	2	82
Marin 2007	January 2007	March 2007	Weekly	8	48
Univali 2008 (unpublished)	January 2008	April 2008	Weekly	2	26
This program	February 2014	March 2014	Weekly	14	84
Total					288

Fig. 10.2 Summary of collection points and each previous project. *Blue* (Project 1-2000 ~ 2002), *pink* (Project 2-2001 ~ 2003), *green* (Project 3-2007), *yellow* (Project 4-2008), *dark red* with numbers (Project 5-2014)

by the presence of freshwater in the bay and consequently its weak buffer capacity.

For dissolved oxygen (DO), vital for the survival of aquatic organisms, the mean oxygen value for the entire study period was 7.1 $mg.L^{-1}$ for beach and 6.9 $mg.L^{-1}$ for river. This parameter is the one that best reflects the state of degradation of a water body. The annual mean surface water dissolved oxygen in beach ranged from 3.7 to 10.8 $mg.L^{-1}$ and in river from 0.8 to 8.2 $mg.L^{-1}$.

Dissolved oxygen can be produced during the photosynthetic processes or by input gas through the atmosphere-water interface, mainly during low temperature. The consumption can occur by various means such as the decomposition of organic matter, respiration of aquatic organisms, loss to the atmosphere-water interface, and oxidation of metallic ions (Esteves 1988). Low concentrations of dissolved oxygen, observed in the river, may be indicating a contribution of large amounts of organic matter in this environment, inducing its decomposition. It is important to note that the DO concentration of smaller than 4 $mg.L^{-1}$ is critical to aquatic organisms (Arana 1997). While mobile animals, like fish, can usually avoid hypoxic and anoxic areas, they sometimes become trapped against the shore and cannot escape. In some other situations, wind and tidal mixing may be so weak and respiration rates so high that even the surface waters can become hypoxic or anoxic (rarely) and cause fish kills (Nixon and Fulweiler 2009).

10.7.2 Inorganic Nutrients

The ammonium nutrient (NH_4^+) showed minimum and maximum values in the beach between 0,34 and 170 $\mu mol.L^{-1}$ and in the river between 0,57 and 2034 $\mu mol.L^{-1}$. The concentrations of orthophosphate phosphorus (PO_4^{2-}) in water in

Table 10.2 Chemical and microbiological parameters evaluated between October 2000 and March 2014 in Porto Belo Bay, Brazil. Number of samples (N), mean, standard deviation (SD), minimum (min), and maximum (max) are displayed for each determined variable: temperature (T °C), salinity, pH, dissolved oxygen (DO), ammonium (NH_4^+), phosphate (PO_4^{2-}), total organic phosphorus (TOP), particulate organic carbon (POC), suspended particulate matter (SPM), biochemical oxygen demand (BOD_5), chlorophyll-a concentration, total coliform, and fecal coliform

		Place	N	Mean	SD	Min	Max
Physicochemical parameters	T (°C)	Beach	238	24.9	3.1	15.6	31.5
		River	50	24.9	3	18.3	30.2
	Salinity ($g.L^{-1}$)	Beach	238	25.9	11.9	1.4	37.6
		River	50	25.4	12.2	0	35.3
	pH	Beach	238	7.6	0.7	4.1	8.9
		River	50	7.5	0.7	5.1	8.4
	DO ($mg.L^{-1}$)	Beach	238	7.1	1.2	3.7	10.8
		River	50	6.9	1.5	0.8	8.2
Inorganic nutrients	NH_4^+ ($\mu mol.L^{-1}$)	Beach	232	38.3	149.9	0.3	170.2
		River	50	56.7	207	0.57	2034.8
	PO_4^{2-} ($\mu mol.L^{-1}$)	Beach	222	0.7	1.3	0.05	2.53
		River	50	0.8	1.3	0.1	13.3
Organic compounds and SPM	TOP ($\mu mol.L^{-1}$)	Beach	158	3.8	4.9	0.1	28
		River	37	4	4.8	0.4	42.7
	POC ($mgC.L^{-1}$)	Beach	163	1	1.3	ND	13.3
		River	41	1.1	1.5	0.6	6.2
	SPM ($mg.L^{-1}$)	Beach	131	53.3	39.2	1	285
		River	29	53.1	39.1	0.67	76.7
Biological and microbiological indicators	BOD_5 ($mg\ O_2.L^{-1}$)	Beach	148	24.5	93	0.1	12.2
		River	42	30.7	48.2	0.3	594.9
	Chlorophyll-*a* ($\mu g.L^{-1}$)	Beach	206	1.2	1.7	0.001	10.2
		River	47	1.3	1.8	0.1	11.3
	Total coliform (MPN/100 ml)	Beach	124	10,140.3	19,214.5	10	68,670
		River	35	21,318.4	48,279.4	100	241,960
	Fecal coliform (MPN/100 ml)	Beach	54	1265.6	3255	0	24,196
		River	30	9773.5	21,634.6	73	120,330

Porto Belo Bay ranged from 0.05 to 2.53 $\mu mol.L^{-1}$ on the beach and 0.5 to 13.27 $\mu mol.L^{-1}$ on the river (Table 10.2).

High ammonium (NH_4^+) values observed in this study, in both environments, river and beach, could be related to the pollution of water through the input of organic matter and consequently bacterial decomposition. Thus, high levels of NH_4^+ may be indicating excess organic matter and pollution resultant from continental drainage. Phosphorus (PO_4^{2-}) is a nutrient of vital importance in aquatic systems, has origin of the dissolution of the compounds derived from the soil or decaying organic matter, but may also have anthropogenic origin, mainly due to domestic and industrial wastewaters, especially from detergents (Von Sperling 1996). These, together with the continual urban development stress around the estuary and beach, adversely affect the existence and survival of many biological species (Lau and Chu 1999).

The highest levels of inorganic nutrients (i.e., $NH_4^+$and PO_4^{2-}) were observed in the river sample from station number 13 during March 2014. These extreme levels can be explained by the high influx of tourists in Porto Belo Bay for the carnival holiday resulting in a greater production of effluent and significant increase in their supply to the environment. Similar results have been observed in the estuary Saco da Mangueira, where high levels of $NH_4^+$and PO_4^{2-} resulted from anthropogenic inputs of organic matter or nutrient-rich domestic and industrial effluents (Baumgarten et al.2001).

10.7.3 Organic Compounds and Suspended Particulate Matter

For total organic phosphorus (TOP) on the beach, values oscillated between 0.1–28 μmol.L^{-1} and 0.4–42.7 μmol.L^{-1} in the river.

Particulate organic carbon (POC) in the aquatic ecosystem can originate as fragments of dead plants and animals, suspended sediments, and water exchanges with adjacent systems or because of the contribution of effluents, particularly domestic (Carmouze 1994). Similar to inorganic nutrients, high levels of total organic phosphorus (TOP) and particulate organic carbon along the beach in Porto Belo Bay were associated with river discharges.

Suspended particulate matter (SPM) concerns the amount of suspended solids which interferes directly in water transparency and consequently to primary production. Physical factors such as wind regime, water circulation, and rainfall can cause the movement of the water column or increased river discharge and therefore influence the levels of solids suspended in the water column. For this parameter, a maximum of 285 mg.L^{-1} and a minimum of 0.67 mg.L^{-1} were found, considering both environments, river and beach.

10.7.4 Biological and Microbiological Indicators

Levels of biochemical oxygen demand (BOD_5) in Porto Belo Bay varied greatly throughout the sampling period (Table 10.2), reflecting the fluctuation in the presence of biodegradable material in the environment. The highest BOD_5 recorded (594.9 mg OD.L^{-1}) was observed in the river (station number 13) during the summer sampling of March 2014, after the carnival holiday, indicating harmful conditions for the environment, aquatic organisms, and humans that may have contact with water.

Chlorophyll concentrations showing primary produce levels, as measure of the biomass of phytoplankton. The mean water chlorophyll-a in beach ranged from 0.001 to 10.2 μg.L^{-1} and in river from 0.1 to 11.3 μg.L^{-1}. Similar values were found in Biscayne Bay, USA, which ranged from 0.001 to 9.18 μg.L^{-1} (Caccia and Boyer 2005).

Traditionally, bacteriological indicators are used to describe water quality. The fecal coliform in Porto Belo Bay displayed a gradual increase over the sampled years (Table 10.2). The highest values of coliforms in both beach and river environments were observed in the summer of 2014 (sampling station number 13), during which levels were larger (hundred times greater) than those determined in the Brazilian legislation (maximum of 2500 MPN/100 ml). As with previous parameter, such drastic increase may be associated with the increased population density in the region during the summer period. Results from beach stations varied widely, with low levels observed in sampling stations numbers 3 and 9, possibly because they are more distant areas from river and storm water discharges.

10.7.5 Final Analysis

Water quality assessment is usually made by indicator parameters such as nutrient levels and fecal indicator organisms. However, total coliforms and fecal coliforms do not necessarily correlate well with the presence of pathogenic organisms (Ferguson et al. 1996); the stability of the natural populations of fecal coliforms is affected and can be drastically inactivated in a few hours when exposed to sunlight (Fujioka et al. 1981). However, enhanced nutrient loading alone might also influence the abundance and survival of pathogens that are already resident in aquatic ecosystems (Smith et al. 1999).

Water quality is influenced by many natural factors interacting in the drainage basins (biological, meteorological, geological, topographical, and hydrological). Human influence on water quality is significant too and may be from several sources; the discharge of sewage and agricultural fertilizer runoff are the most common examples (Cood 2000).

Human activities have also profound impacts upon the biogeochemical cycles of carbon, phosphorus, and nitrogen (Vitousek et al. 1997; Smith et al. 1999). Phosphorus and nitrogen are essential nutrients for the plants and animals, but input increases in the marine environment represent a potential pollution problem (De Jong 2006).

There are many sources of phosphorus in aquatic systems, including wastewater, wastewater treatment plants, runoff from fertilized soil, and the widespread use of detergents, which have important contributions to the phosphorus loads (Carpenter et al. 1998; De Jong 2006; Baumgarten et al. 2010).

Nitrification is carried out in two steps. First, ammonium is converted to nitrite by ammonia-oxidizing bacteria. In the second step, nitrite-oxidizing bacteria convert nitrite to nitrate, resulting to the high oxygen demand for ammonium oxidation. High ammonium concentration can be found especially in municipal wastewater (Ruiz et al. 2003).

The water quality varies around the world. Some places have low nutrient levels; the high values are associated to high population density (Table 10.3).

The indicator ammonium is chosen for the comparison of coastal water quality around the world on the map (Fig. 10.3).

In Itapema beach, river and storm water discharges are the most important factors influencing its water quality, with chemical, physicochemical, and microbiological aspects.

The sample stations that showed better water quality were Praia Central (site 2) and Caixa d'Aço (site 9) located far away from river discharges, without its influences. The lowest water qualities were represented in the northern Itapema beach (site 1), Praia do Morretes (site 3), and Praia do Perequê (site 6) which have a great discrepancy in chemical and microbiological characteristics, both located much closer to discharges and with a high population density.

During all the periods monitored, the nutrients were very high, and this area was classified to eutrophic system, according to Lau and Chu (1999), Ingole and Kadam (2003), and Baumgarten et al. (2001).

The supervision of the outputs of rainwater on the beach is recommended, since they appear to contain illegal sewage. The illegal sewage also appears to be responsible for the poor water quality of the river.

Around Perequê-Áçu River over the last 10 years, many people without access to sanitation have set and dumped their sewage directly into the river. It is important to remember that Porto Belo City does not possess an effluent treatment plant.

We must make an environmental education work and awareness actions with the population, mainly scholar children, explaining the importance of treating sewage and pollution damages and its influences to human and environmental health.

Table 10.3 Water quality of some beaches around the world. The symbol * represents estuaries and indicator ammonium is chosen for the comparison of coastal water quality around the world on the map (A to I)

Local		Sal	NH_4^+ µmol.L^{-1}	PO_4^{3-} µmol.L^{-1}	References
Yucatán Bay, México A	Mean	37.45	4.45	0.51	Aranda-Cirerol et al. 2006
	Range	(24.0 ~ 39.7)	(0.40 ~ 42.26)	(0.03 ~ 1.79)	
Dzilam Lagoon, México	Mean ± SD	25.8 ± 1.4	2.8 ± 0.2	0.05 ± 0.03	Medina and Herrera 2003
Yucatán Bay, México	Range	36 ~ 38,3	4.2 ~ 5,2	0.46 ~ 0.62	Herrera-Silveira et al. 2004
Flórida, USA	Mean	34.25	0.474	0.015	Boyer et al. 2000
Biscayne Bay, USA B	Range	6.21 ~ 42,3	0 ~ 16.3	0 ~ 0.68	Caccia and Boyer 2005
Hawaii Beach, USA	Mean ± SD	34.5	0.8 ± 0.4	0.15 ± 0.07	Laws et al. 1999
Mangrove Mai Po, Hong Kong, China C*	Mean	8	485.7	37	Lau and Chu 1999
	Range	0 ~ 30	37.8 ~ 1450	1.9 ~ 83.9	
East of Hong Kong, China	Mean	30	1.6	0.3	Zhou et al. 2007
Dadar Beach of Mumbai. India D	Mean	27.9	135.7	25.8	Ingole and Kadam 2003
	Range	15.1 ~ 32.4	57 ~ 207	6.4 ~ 58	
Itacaré to Canavieiras, BA, Brazil	Mean ± SD	36.5 ± 0.51	0.63 ± 0.56	0.07 ± 0.03	Eça 2009
São Vicente Beach, SP, Brazil E	Range	32.23 ~ 35.56	0.82 ~ 4.94	0.22 ~ 4.42	Braga et al. 2000
São Sebastião, SP, Brazil	Mean ± SD	34.65 ± 0.95	0.71 ± 0.55	0.15 ± 0.11	Gianesella and Saldanha-Corrêa 2003
Conceição Lagoon, SC, Brazil	Mean ± SD	29 ± 1.1	6.0 ± 1.1	0.21 ± 0.15	Fonseca et al. 2002
Pântano do Sul, SC, Brazil	Mean ± SD	34.7 ± 0.82	1.3 ± 0.70	0.6 ± 0.18	Simonassi et al. 2010
North beaches of Santa Catarina, Brazil	Mean ± SD	30.9 ± 2.1	16.3 ± 16.1	0.68 ± 0.32	Kuroshima et al. 2006
Balneário Camboriú, SC, Brazil	Mean ± SD	29.4 ± 4.7	20.0 ± 36.4	0.71 ± 0.7	Kuroshima et al. 2007
Saco do Justino, RS, Brazil F*	Range	0 ~ 20	1.9 ~ 29.7	0 ~ 5.3	Baumgarten et al. 2005
Saco da Mangueira, RS, Brazil G*	Range	0 ~ 24	1.7 ~ 47.9	1.3 ~ 13.4	Baumgarten et al. 2001
Porto Belo Bay, this study H	Range	1.4 ~ 37.6	0.3 ~ 170.2	0 ~ 2.5	This study
Perequê River, this study I*	Range	0 ~ 35.3	0.6 ~ 2034.8	0.1 ~ 13.3	This study

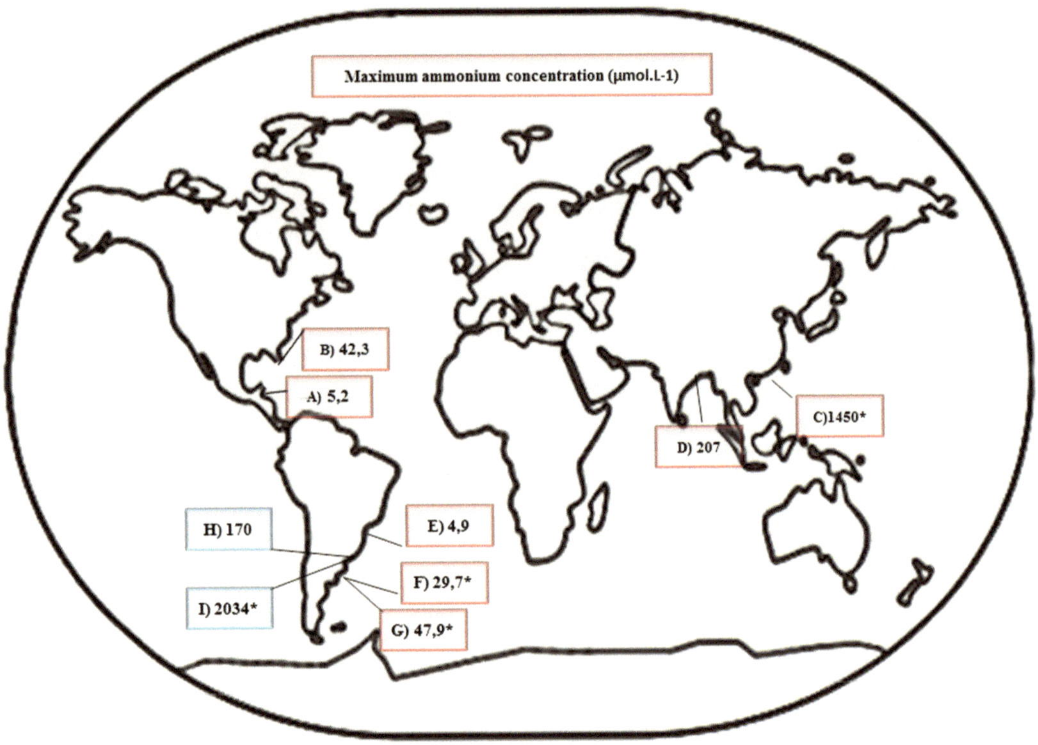

Fig. 10.3 Distribution of maximum values of ammonium. The symbol * represents estuary places. The letters represent places showed on Table 10.3. *Blue* represents this study

References

Arana LV (1997) Princípios químicos de qualidade da água em aquicultura. Universidade Federal de Santa Catarina, Florianópolis

Aranda-Cirerol NA, Herrera-Silveira JA, Comín FA (2006) Nutrient water quality in a tropical coastal zone with groundwater discharge, northwest Yucatán, Mexico. Estuar Coast Shelf Sci 68:445–454

APHA/AWWA/WEF – American Public Health Association – American Water Works Association – Water Environment Federation (1999) Standard methods for the examination of water and wastewater. CD – ROM

Barsano PR, Barbosa RP, Viana VJ (2014) Poluição ambiental e saúde pública. Erica, São Paulo

Basso M (2014) Peixes e siris aparecem mortos no rio Perequê, em Itapema. Jornal o Sol diário. http://osoldiario.clicrbs.com.br/sc/noticia/2014/01/peixes-e-siris-aparecem-mortos-no-rio-pereque-em-itapema-4379566.html. Accessed 8 Jul 2014

Baumgarten MGZ, Niencheski LFH, Veeck L (2001) Nutrientes na coluna d'água e na água intersticial de sedimentos de uma enseada rasa estuarina com aportes de origem antrópica, vol 23. Atlântica, Rio Grande, pp 101–116

Baumgarten MGZ, Niencheski LFH, Martins BAD (2005) Saco do Justino (RS-Brasil): amônio e fosfato na coluna d'água e na água intersticial de uma enseada não contaminada. Atlântica, Rio Grande 27(2):113–129

Baumgarten MGZ, Niencheski LFH, Wallner-Kersanach M (2010) Manual de análises em oceanografia química. Universidade Federal de Rio Grande, Centro

Biavatti ML (2014) Qualidade de água da enseada de Porto Belo: evolução de 2000 a 2014. Monography Universidade do Vale do Itajaí, Itajaí

Billen G, Garnier J, Ficht A, Cun C (2001) Modeling the response of water quality in the Seine River estuary to human activity in its watershed over the last 50 years. Estuar Coast Shelf Sci 24(6B):977–993

Boehm AB, Ashbolt NJ, Colford JM Jr et al (2009) A sea change ahead for recreational water quality criteria. J Water Health 07(1):9–20

Boyer JN, Sterling P, Jones RD (2000) Maximizing information from a water quality monitoring network through visualization techniques. Estuar Coast Shelf Sci 50:39–48

Braga ES, Bonetti CVDH, Burone L, Bonetti Filho J (2000) Eutrophication and bacterial pollution caused by industrial and domestic wastes at the baixada santista estuarine system – Brazil. Mar Pollut Bull 40(2):165–173

Butkus MA, Manous JD (2005) Biochemical oxygen demand. Water Encycl 1:639–642

Caccia VG, Boyer JN (2005) Spatial patterning of water quality in Biscayne Bay, Florida as a function of land use and water management. Mar Pollut Bull 50:1416–1429

Caldas J, Felix D (2013) Vazamento de esgoto irregular causa poluição e morte de peixes em SC. http://g1.globo.com/sc/santa-catarina/noticia/2013/01/vazamento-de-esgoto-irregular-causa-poluicao-e-morte-de-peixes-em-sc.html. Accessed 8 Jul 2014

Carmouze J-P (1994) O metabolismo dos ecossitemas aquáticos: fundamentos teóricos, métodos de estudo e análises químicas. Edgard Blucher: FAPESP, São Paulo

Carpenter S, Caraco N, Correll DL, Howarth RW, Sharpley AN, Smith VH (1998) Nonpoint pollution of surface waters with phosphorus and nitrogen. Ecology 8(3):559–568

CETESB – Companhia Tecnológica de Saneamento Ambiental (1995) Relatório de balneabilidade das praias paulistas. CETESB, São Paulo

CETESB – Companhia Tecnológica de Saneamento Ambiental (2010) Inventário de emissões de gases de efeito estufa no setor de resíduos e efluentes do estado de São Paulo 1990 a 2008. http://www.ambiente.sp.gov.br/proclima/files/2014/05/brasil_mcti_residuos.pdf. CETESB, São Paulo. Accessed 8 Apr 2014

CONASA – Companhia Nacional de Saneamento (2014) Dados do sistema comercial Itapema – SC. 2014

Cood GA (2000) Cyanobacterial toxins, the perception of water quality, and the prioritization of eutrophication control. Ecol Eng 16:51–60

CPCB – Central Pollution Control Board (2011) Environmental information system on GIS Platform. http://cpcb.nic.in/EIS_on_GIS.pdf. Accessed 14 Jul 2014

Cunha FC (2010) Indicadores de sustentabilidade socioambiental relacionados com as atividades do turismo em Itapema (SC), com ênfase na qualidade da água de recreação e na saúde da população. Monography. Universidade de Santa Catarina, Florianópolis

Da Cunha MA (2006) Métodos de detecção de microrganismos indicadores. Saúde & ambien em rev 1(1):9–13

Da Motta RS, Mendes AP, Mendes FE, Young CEF (1994) Perdas e serviços ambientais do recurso água para uso doméstico. Pes e Plan Econ 24(1):35–72

De Jong F (2006) Marine eutrophication in perspective: on the relevance of ecology for environmental policy. Springer, Heidelberg

De Miranda LB, De Castro BM, Kjerfve B (2002) Princípios de Oceanografia Física de Estuários. Edusp, São Paulo

Dorevitch S et al (2011) Water ingestion during water recreation. Water Res 45:2020–2028

Eça GF (2009). Biogeoquímica de nutrientes e clorofila-a na plataforma continental rasa entre Itacaré e Canasvieiras – Bahia. Dissertation, Universidade Estadual de Santa Cruz

Esteves FA (1988) Fundamentos de limnologia. Interciência, Rio de Janeiro

Ferguson CM, Coote BG, Ashbolt NJ, Stevenson IM (1996) Relationships between indicators, pathogens and water quality in an estuarine system. Water Res 30(9):2045–2054

Fonseca A, Braga ES, Eichler BB (2002) Distribuição espacial dos nutrientes inorgânicos dissolvidos e da biomassa fitoplanctônica no sistema pelágico da Lagoa da Conceição, Santa Catarina, Brasil (Setembro, 2000). Atlântica, Rio Grande 24(2):69–83

Fujioka RS, Hashimoto HH, Siwak EB, Young RH (1981) Effect of sunlight on survival of indicator bacteria in seawater. Appl Environ Microbiol 41(3):690–696

FUNASA – Fundação Nacional da Saúde (2010) Impactos na saúde e no sistema único de saúde decorrentes de agravos relacionados a um saneamento ambiental inadequado. Brasília, Brazil. http://www.funasa.gov.br/site/wp-content/files_mf/estudosPesquisas_ImpactosSaude.pdf. Accessed 25 Aug 2014

Gianesella SMF, Saldanha-Corrêa FMP (2003) Nutrientes, séston e biomassa fitoplanctônica na plataforma interna de São Sebastião – verão de 1994 e primavera de 1997. Boletim do Instituto da Pesca 29:161–172

Herrera-Silveira JA, Comin FA, Aranda-Cirerol N, Troccoli L, Capurro L (2004) Coastal water quality assessment in the Yucatán Peninsula: management implications. Ocean Coast Manage 47:625–639

HKWQRC (2013) Beach water quality in Hong Kong. www.beachwq.gov.hk/UpFile/en/bwq_report2013.pdf. Accessed 24 Sept 2014

Horita CO, Carvalho JLB (1999) Uma alternativa para a melhoria na qualidade d'água na enseada de Balneário Camboriú, SC: resultados preliminares. In: Abstracts of the XIII SBRH – Simpósio brasileiro de recursos hídricos, Belo Horizonte, 28-Sept – 2 Dec 1999

Howart R, Anderson D, Cloern J et al (2002) Nutrient pollution of coastal rivers, bays and seas. Issues Ecol 7:1–14

HWDEHP – Healthy Waters, Department of Environment and Heritage Protection (2013) Healthy Waters Management Plan Guideline: under the environmental protection (water) policy 2009. http://www.ehp.qld.gov.au/water/policy/pdf/healthy-water-mp-guideline.pdf. Accessed 27 Jul 2014

IBGE – Instituto Brasileiro de Geografia e Estatística (2004) Indicadores de desenvolvimento sustentável – Brasil 2004 Dimensão ambiental – oceanos, mares e áreas costeiras. http://www.ibge.gov.br/home/geociencias/recursosnaturais/ids/oceanos.pdf. Accessed 27 Aug 2014

IBGE – Instituto Brasileiro de Geografia e Estatística (2010) PNSB: Pesquisa nacional de saneamento básico 2008. http://www.ibge.gov.br/home/estatistica/populacao/condicaodevida/pnsb2008/PNSB_2008.pdf. Accessed 25 Sept 2014

Ingole SA, Kadam AN (2003) Pollution of some recreation beaches of Mubai, Maharashtra. J IAEM 30:172–175

Jordão EP, Pessoa CA (1995) Tratamento de esgotos domésticos. ABES, Rio de Janeiro

Kuroshima KN, Barreiros MAB, Costódio PFS, Bellaver PH, Lacava L (2006) Chemical and microbiological evaluation in bathing waters from Itajaí, Balneário Camboriú and Itapema cities (Brazil). J Coastal Res 39:1378–1382

Kuroshima KN, Barreiros MAB, Lacava LA, Costódio PFS (2007) The influence of dioxide chloride on water quality of Balneário Camboriú Beach, Brazil. Environ Monit Assess 129:179–187

Lau SSS, Chu LM (1999) Water quality degradation at the Mai Po Marshes Nature reserve (Hong Kong) with reference to nutrient enrichment. Hidrobiologia 403:195–203

Laws EA, Ziemann D, Schulman D (1999) Coastal water quality in Hawaii: the importance of buffer zones and dilution. Mar Environ Res 48:1–21

Marin CB (2007) Avaliação da qualidade de água no rio Perequê e enseada de Itapema – SC. Monography, Universidade do Vale do Itajaí, Itajaí

Medina-Gómez I, Herrera-Silveira JA (2003) Spatial characterization of water quality in a karstic coastal lagoon without anthropogenic disturbance: a multivariate approach. Estuar Coast Shelf Sci 58:455–465

Moraes DSL, Jordão BQ (2002) Water resources deterioration and its impact on human health. Revista Saúde Pública 36(3):370–374

Nixon SN, Fulweiler RW (2009) Nutrient pollution, eutrophication, and the degradation of coastal marine ecosystems. In: Duarte CM (ed) Global loss of coastal habitants rates, causes and consequences. Fundación BBVA, pp 23–60

NOAA – National Oceanographic Atmospheric Association (2005) Economic Statistics for NOAA. www.publicaffairs.noaa.gov/pdf/economic-statistics2005.pdf. Accessed 26 Sept 2014

Parsons TR, Takahashi M (1998) Biological oceanographic processes. Pergamon, Oxford

Pereira Filho J, Spillere LC, Schettini CAF (2003) Dinâmica de nutrientes na região portuária do estuário do rio Itajaí-Açu. Revista atlântica 25(1):11–20

Philippi Junior A (2010) Saneamento, saúde e ambiente: fundamentos para um desenvolvimento sustentável. Coleção ambiental 2. Manole, Barueri

Pilson MEQ (1985) On the residence time of water in Narragansett Bay. Estuar Coast Shelf Sci 8(1):2–14

Ruiz G, Jeisonb D, Chamya R (2003) Nitrification with high nitrite accumulation for the treatment of wastewater with high ammonia concentration. Water Res 37:1371–1377

Schijven J, Husman AMR (2006) A survey of diving behavior and accidental water ingestion among dutch occupational and sport divers to assess the risk of infection with waterborne pathogenic microorganisms. Environ Health Perspect 114(5):712–717

Segerson K, Walker D (2002) Nutrient pollution an economic perspective. Estuar Coast Shelf Sci 25(4b):797–808

Simonassi JC, Hennemann MC, Talgatti D, Marques Júnior AN (2010) Nutrient variations and coastal water quality of Santa Catarina Island, Brazil. Biotemas 23(1):211–223

Smith VH, Tilman GD, Nekola JC (1999) Eutrophication: impacts of excess nutrient inputs on freshwater, marine, and terrestrial ecosystems. Environ Pollut 100:179–196

SNIS – Sistema Nacional de Informações sobre Saneamento. Ministério das Cidades 2013. http://www.snis.gov.br/PaginaCarrega.php?EWRErterterTERTer=6. Accessed 20 Aug 2014

Stone DL et al (2008) Exposure assessment and risk of gastrointestinal illness among surfers. J Toxicol Environ Health 71(24):1603–1615

Talley LD, Pickard GL, Emery WJ, Swift JH (2011) Descriptive physical oceanography: an introduction. Elsevier, Boston

Teixeira AR, de Lemos Chernicharo CA, Tatiana Figueiredo da Silva S, De Paula FS (2009) Influência da alteração da distribuição do tamanho de partículas no desempenho de reator UASB tratando esgoto doméstico. Engenharia Sanitária e Ambiental. Dig J Mol Med. doi:10.1590/S1413-41522009000200003

Valigura RA, Alexander RB, Castro MS, Meyers TP, Paerl HW, Stacey PE, Turner RE (2000) Nitrogen loading in coastal water bodies: an atmospheric perspective, vol 57, Coastal and estuaries series. American Geophysical Union, Washington, DC

van Asperen IA, Rover CM, Schijven JF, Oetomo SB, Schellekens JFP, van Leeuwen NJ, Collé C, Havelaar AH, Kromhout D, Sprenger MWJ (1995) Risk of otitis externa after swimming in recreational freshwater lakes containing Pseudomonas aeruginosa. Br Med J 311:1407–1410

Vasconcelos FP, Coriolano LNMT (2008) Socio-environmental impacts in coastal environments: focus on tourism and integrated coastal zone management in Ceará State/Brazil. Revista de gestão costeira integrada 8(2):259–275

Vasques JR (2002) Caracterização química e microbiológica dos principais recursos hídricos do município de Porto Belo e sua utilização como proposta de educação ambiental para séries iniciais em uma escola municipal, Monography. Universidade do Vale do Itajaí, Itajaí

Vitousek PM, Aber J, Howarth RW, Likens GE, Matson PA, Schindler DW, Schlesinger WH, Tilman GD (1997) Human alteration of the global nitrogen cycle: causes and consequences. Ecol Appl 7(3):737–750

Von Sperling M (1996) Introdução à qualidade das águas e ao tratamento de esgotos. Universidade Federal de Minas Gerais, Belo Horizonte

WHO – World Health Organization (1997) Water pollution control: a guide to the use of water quality management principles. E & FN Spon, London

WHO – World Health Organization (2000) Monitoring bathing waters: a practical guide to the design and implementation of assessments and monitoring programmes. E & FN Spon, London

WHO – Word Health Organization (2003) Guidelines for safe recreational water environments: coastal and fresh waters, vol 1. WHO – Word Health Organization, Geneva

Zhou F, Huang GH, Guo H, Zhang W, Hao Z (2007) Spatio-temporal patterns and source apportionment of coastal water pollution in eastern Hong Kong. Water Res 41:3429–3439

Effects of Fullerene Nanocomposite in Marine and Estuarine Organisms

11

Samile Seber and Grasiela Lopes Leães Pinho

Abstract

The fullerene-C_{60} is a commercially attractive nanoparticle which increases the efficiency of drugs, cosmetics, and electronics. Like most of the contaminants when released into the environment, the final destination of fullerene-C_{60} might be the aquatic system, where it will be subject to interactions with the biota and other compounds. This study aimed to evaluate the possible effects of this nanoparticle on marine and estuarine biota. Thus, acute toxicity assays with *Acartia tonsa* and *Mysidopsis juniae* were conducted to evaluate the toxicity of aqueous fullerene suspension (nC_{60}). This suspension of C_{60} was prepared using 200 mg of fullerene-C_{60} dissolved in 1 L of Milli-Q water. Due to the low solubility of fullerene-C_{60} in water, the nominal concentration of suspension was 0.53 μg L^{-1}. The nominal concentrations used for the assays were 0.053 and 0.265 μg L^{-1}, which correspond to 10 % and 50 % of the concentration of nC_{60}, respectively. The results indicated that nC_{60} did not cause acute toxicity to *A. tonsa* and *M. juniae* at the tested concentrations.

Keywords

Fullerenes • Nanotoxicology • Toxicological assays

11.1 Nanoparticles and Nanotechnology

Over the past decades nanotechnology has raised the interest of the scientific community to the search for new compounds that act with high specificity in several areas such as electronics, medicine, physics, and chemistry (RCEP 2008). Nanomaterials are defined by the US National Nanotechnology Initiative as materials having at least one dimension between 1 and 100 nm (Oberdörster 2004), which enable numerous applications in several areas of knowledge. Nanosized materials are present on Earth for a millions of years (Nowack and Bucheli 2007), being found naturally in the environment, such as viral particles, the biogenic magnetite, and ferritin protein (Oberdörster 2004). Nanoparticles can

S. Seber (✉) • G.L.L. Pinho
Laboratório de Microcontaminantes Orgânicos e Ecotoxicologia Aquática, Universidade Federal do Rio Grande-FURG, Av. Itália, km 8, Rio Grande/RS, Brazil
e-mail: sseber@gmail.com

P. Shukla (ed.), *Frontier Discoveries and Innovations in Interdisciplinary Microbiology*, DOI 10.1007/978-81-322-2610-9_11

also be naturally originated from combustion processes, such as forest fires and volcanic activities (Oberdörster et al. 2005).

There are different types of nanoparticles currently synthesized, with many applications. However, all of them have at least one of the main characteristics of nanomaterials: large surface area. Examples of nanoparticles are carbon nanotubes, inorganic nanotubes, and nanoparticles of silver, aluminum, titanium dioxide, dendrimers, quantum dots, and fullerenes (C_{60}, C_{70}, C_{84}, C_{540}). The production of such manufactured nanomaterials, as well as other emerging compounds, has increased significantly in recent years, because they facilitate everyday life and increase the efficiency of engineering equipment.

The nanotechnology is a rapidly expanding field which results in a variety of commercially available products. The preference for the use of nano-sized materials is mainly due to their large surface/volume ratio (Colvin 2003), as previously mentioned. Thus, there are many applications of nanomaterials, including their use in paints, self-cleaning windows, and stain-resistant clothing (Kahru and Dubourguier 2010). Its features also favor its use in sunscreens, cosmetics, drug carriers, and catalysts in intracellular reactions (Colvin 2003). According to Moore (2006), nanotechnology promises huge improvements in health, industrial technologies, and environmental bioremediation and collaborates with economic growth.

Although nanoparticles always are present on the planet, the Industrial Revolution was responsible for increasing human exposure due to sources of thermodegradation, such as internal combustion engines (Oberdörster et al. 2005). Since then, the exhibition has grown increasingly.

11.2 Fullerenes

Among the elements used in nanotechnology, carbon (C) is one that has wider application, because in different molecular and crystal structures, it features a wide range of mechanical and electronic properties (Dresselhaus et al. 1996; Hulteen and Martin 1997). The carbon was highlighted in 1985, after the discovery of new allotropic form. This allotrope was called fullerene, because its shape is similar to the geodesic domes designed by American architect Richard Buckminster Fuller, which are architectural forms that follow the same principle of symmetry and stability of fullerenes (Dresselhaus et al. 1996; Hirsch 1997).

Fullerenes are molecules in the form of three-dimensional closed polyhedra formed by hexagons and pentagons, 42 to more than 1000 carbons. The most abundant contains 60 carbon atoms linked together, forming a structure similar to a soccer ball (Fig. 11.1). The molecule is composed of six subunits, called piracelênicas (Taylor and Walton 1993), and has the form of a non-regular icosahedron with 32 faces (20 hexagons and 12 pentagons) and 30 carbon-carbon bonds located at the vertices of fusions between pentagons and hexagons (Diederich and Thilgen 1996).

The fullerene molecules were discovered by Sir Harold W. Kroto, Robert F. Curl, and Richard E. Smalley during their experiments to reproduce the conditions that form the carbon clusters in the dust star present in the vicinity of celestial bodies called red giants (Kroto et al. 1985). In these experiments, a graphite plate was subjected to a high-frequency pulsed laser in a supersonic flow of helium gas, and the mixture containing carbon clusters generated in plasma was analyzed by mass spectrometry. The formation of molecules consisting exclusively of carbon atoms was observed, with formula Cn, where n=30–190, being C_{60} and C_{70} the most abundant.

The fullerene with 60 carbon atoms was separated from this mixture by Taylor et al. (1990) in amounts sufficient to start exploring its chemical and spectroscopic properties. However, the technique developed in 1985 did not permit large quantities of synthesis of fullerene. Then, Krätschmer et al. (1990) developed a method to synthesize it with a yield of 5 %. Since its first

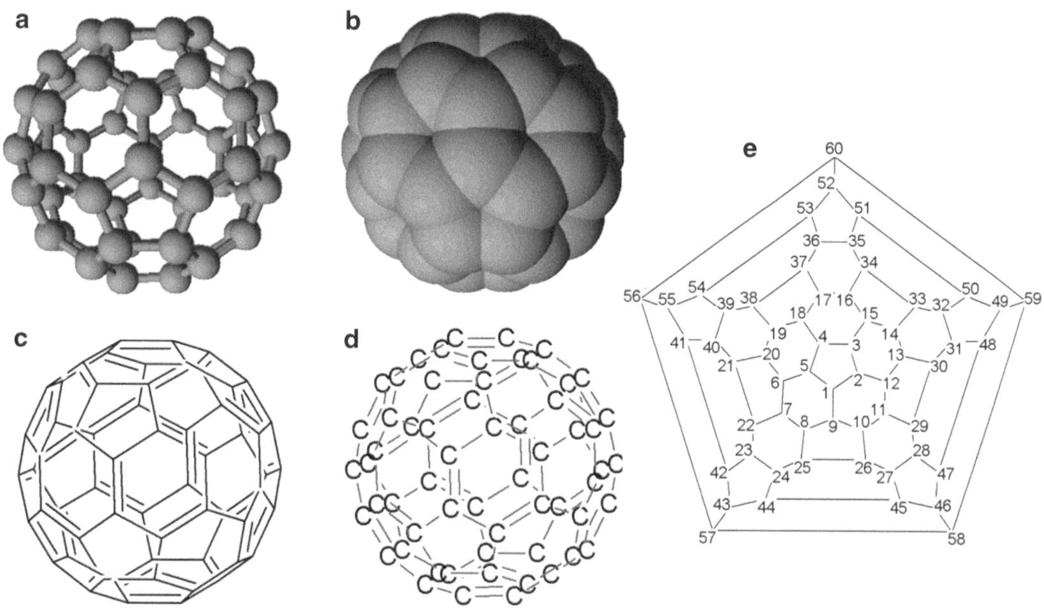

Fig. 11.1 Schematic representation of the fullerene-C_{60}. (**a**) Template spheres. (**b**) Space-filling model. (**c**) A linear condensed formula. (**d**) Complete structural formula. (**e**) Schlegel diagram showing the numbering of the carbon atoms

detection and mass production, fullerenes won a key role in the scientific field, reaching its peak in 1996, when the Chemistry Nobel Prize was awarded to Kroto, Curl, and Smalley for their discovery (Prato 1997).

The fullerene-C_{60} has hydrophobic properties with a log Kow 6.67 (Jafvert and Kulkarni 2008) and solubility 1.3×10^{-11} mg L^{-1} in water (Heymann 1996) and, when in an aqueous medium, can form stable agglomerates having diameters up to 500 nm (Fortner et al. 2005). Furthermore, a major problem for many of the possible applications of C_{60} is their low solubility in usual solvents (Ruoff et al. 1993). Since its discovery, the low solubility of C_{60} has been an obstacle to their practical application.

This nanoparticle effectively absorbs free radicals (Bianco and Ross 2007), is a reactive oxygen species generator for photodynamic therapy (Vileno et al. 2004), is a contrast agent for diagnosis conveyor (Bolskar et al. 2003), and is an inhibitor of enzymes and drug transporter (Markovic and Trajkovic 2008). It has also been used in the production of amphiphilic copolymers (Yu et al. 2005) and in the manufacture of photocells (Hasobe et al. 2005).

Despite the expansion of the use of nanomaterials, studies to evaluate whether they can cause undesired effects when administered for medical purposes or even during its manufacture are lacking (Oberdörster et al. 2005). Added to this, due to increased production of synthetic nanoparticles, it is likely that occupational and public exposure will dramatically increase in the coming years, as well as their possible release to the environment (Kahru and Dubourguier 2010). According to Shinohara et al. (2009), it is estimated that approximately 40 tons of fullerene-C60 are produced annually by industry. Thus, it is possible that this nanoparticle starts to be found in measurable concentrations (Oberdörster 2004) in various environments that have some association with its direct or indirect sources, including estuaries and coastal regions, acting potential risk to the biota of these sites.

Some studies presented data of toxicity of fullerene-C_{60} on aquatic species (Oberdörster 2004; Zhu et al. 2006; Tao et al. 2009), but other authors attributed this toxicity to organic solvents used to co-solubilize the nanoparticle (Fortner et al. 2005; Oberdörster et al. 2006). Thus, this nanoparticle in aqueous suspension could have its toxicity decreased considerably.

11.3 Toxicological Assays and Organisms

The toxicological assays can be used with a certain security to investigate adverse effects of nanocomposites on the natural environment. These assays have standardized methods, which assess the inherent ability of a substance or mixture to cause deleterious effects to living organisms, and efficient tools for evaluation of water and sediment (Zagatto 2006).

Aquatic environments can be the final destination of most contaminants, which can have deleterious effects, so it is extremely important to carry out studies to analyze the toxicity of fullerene-C60 on estuarine and marine species from different trophic levels.

The *Acartia tonsa* (Dana 1849) is one of the most studied copepod species in the world (Mauchline et al. 1998) and easily cultured in the laboratory (Kaminski 2004). It has broad geographic distribution, occurring in the Atlantic Ocean, from the coast of Canada to Argentina, and in the northern Pacific Ocean (Sabatini 1990). It supports a wide range of salinity and is adaptable to various environmental conditions. This species has proved sensitive to a variety of toxic chemicals (Medina and Barata 2004), which makes it interesting for use in toxicological assays (Bianchi et al. 2003). This toxicity test is standardized by the International Organization for Standardization (ISO 1999).

The *Mysidopsis juniae* (Silva 1979) is an epibentonic species and is part of the diet of several species of fish, so it is an important link between the branches of many marine food webs (Barnes 1996). The mysids are widely used as test organisms in toxicological studies due to favorable characteristics such as sensitivity to various toxic agents, ease of use and cultivation, direct development, and short life cycle (Badaró-Pedroso et al. 2002). Therefore, their sensitivity has been tested against various contaminants, either chemical compounds or complex mixtures. In addition, their use in toxicological assays was recommended and ruled for over a decade by the Company of Technology and Environmental Sanitation of the State of São Paulo (CETESB 1992) and later by the Brazilian Association of Technical Standards (ABNT 2005).

The present study aimed at investigating the role of aqueous fullerene suspension (nC_{60}) as contaminant carriers in aquatic systems. This was investigated in a series of acute toxicity assays with *A. tonsa* and *M. juniae*.

11.4 Aqueous Fullerene Suspension (nC_{60})

200 g of fullerene-C_{60} with 98 % purity, purchased from Sigma-Aldrich, was suspended in 1 L of Milli-Q water and stirred vigorously for 2 months during exposure to artificial light. This procedure was performed according to the method used by Baun et al. (2008). This process forms colloidal aggregates and, for this reason, represented as (nC_{60}). After this stage, the suspension was filtered through nylon filters of 0.1 μm (to ensure the size of the nanoparticles).

11.5 Acute Toxicity Assays with *Acartia tonsa*

The acute toxicity assays with *Acartia tonsa* were conducted following the methodology described in the standard published by the International Organization for Standardization (ISO 1999) in order to determine the mortality of organisms exposed to treatments. Adult organisms aged between 1 and 3 weeks were selected without distinction of sex. Ten organisms were randomly placed in each beaker of 60 ml capacity, containing 50 mL of the test solution. The nominal concentrations of nC_{60} tested were 0.053 and 0.265 μg L^{-1}.

For each treatment four replicates ($n=4$) with salinity 30 were prepared. The assays were maintained in a climatized room for a period of 48 h. After this time, the mortality rate was evaluated at each concentration, identified by the absence of animal movement, and observed under optical stereomicroscope.

11.6 Acute Toxicity Assays with *Mysidopsis juniae*

The acute toxicity assays with *Mysidopsis juniae* were conducted in accordance with the methodology described in ISO-15308 standard (ABNT 2005). Organisms were selected without distinction of sex, aged between 1 and 8 days. Ten organisms were randomly placed in each beaker of 400 mL capacity, containing 200 mL of the test solution. The nominal concentrations of nC_{60} tested were 0.053 and 0.265 μg L^{-1}.

For each treatment five replicates ($n=5$) with salinity 35 were prepared. The assays were maintained in a climatized room for a period of 96 h. The organisms were fed daily during the treatment period with 20–30 *Artemia* sp. for each *M. juniae*. After this time, the mortality rate was evaluated at each concentration, identified by the absence of animal movement after physical stimulation.

11.7 Determination of the Concentration Aqueous Fullerene Suspension (nC_{60})

The concentration was determinate by solid phase extraction (SPE). The procedure was optimized based on the methodology described by Chen et al. (2008). 100 μL of pre-concentrated extract was analyzed by liquid chromatography equipped with a high-efficiency UV detector (CLAE UV–vis Series 200, PerkinElmer).

11.8 Results and Discussion

There were no deleterious effects on *Acartia tonsa* in concentrations 0.053 and 0.265 μg L^{-1} (Fig. 11.2). The results showed no significant differences between treatment and control ($p>0.05$). The four experiments were pooled for the calculation of average survival of all replicates of each treatment ($n=16$).

The results of the assays with *Mysidopsis juniae* that were tested at concentrations of fullerene-nC_{60} 0.053 μg L^{-1} and 0.265 μg L^{-1} are shown in Fig. 11.3. At these concentrations the nanocompound showed no toxicity acute for the species. The average survival of organisms was greater than 90 % for both treatments being within the acceptable limits of survival for the groups of control on this species (ABNT 2005). Therefore, the median survival was not significantly different between treatments and controls ($p>0.05$).

Several studies have attributed toxicity for fullerene on studies with different organisms such as bacteria (Fortner et al. 2005), crustaceans (Lovern and Klaper 2006; Zhu et al. 2006), and fish embryos (Zhu et al. 2007). However, suspensions of fullerene used in these cases were obtained after dissolution of the nanocomposite in THF (tetrahydrofuran). According to Gharbi et al. (2005), conditions used for the removal of solvent (450 K under a pressure of 10–6 mmHg for at least 50 h) are insufficient for its complete removal. Thus, there is a high expectation that the presence of the solvent residues used in the preparation of aqueous solutions has been the cause of toxic effects observed in organisms (Gharbi et al. 2005; Henry et al. 2007). Thus, although the fullerene-C_{60} is slightly soluble in water, in the present study organic solvents were not used to dilute it.

According to Andrievsky et al. (1995), the aqueous fullerene suspension, which does not contain any organic solvent, has no toxicity for various biological systems and can still provide positive effects. On the other hand, some studies have reported sublethal effects of such suspensions for freshwater crustaceans (Lovern and Klaper 2006; Zhu et al. 2006; Oberdörster et al. 2006). However, it should be noted that these effects were caused by exposure to solutions with concentrations at least five thousand times the highest concentration used in this study.

Brant et al. (2005) studied the aggregation and deposition of fullerene-C_{60} in aqueous

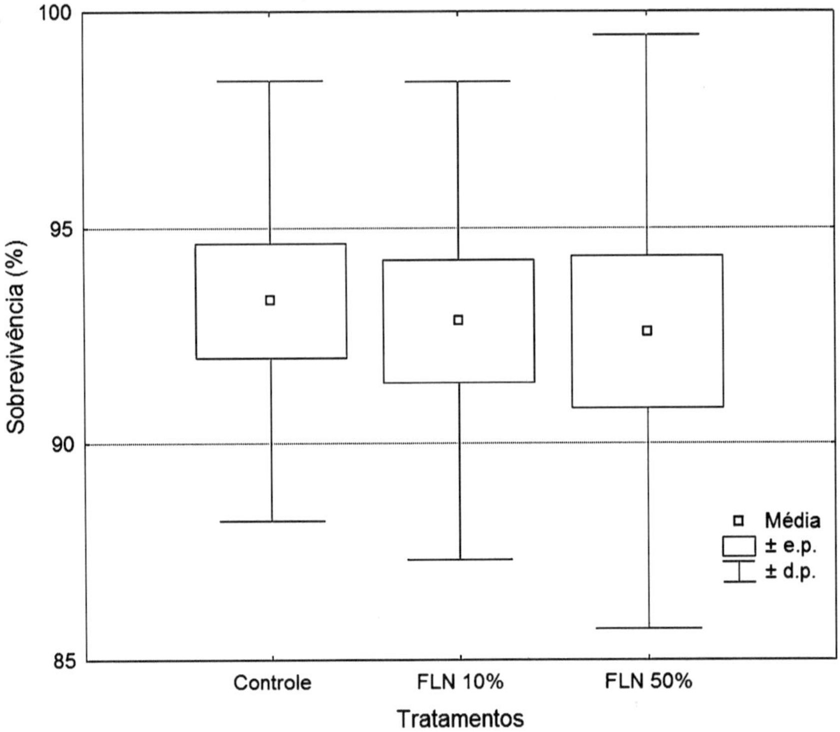

Fig. 11.2 Comparison of the mean values ($n = 16$) of survival of *Acartia tonsa* observed in the control and in the two concentrations of nC_{60} (FLN 10 % = 0.053 μg L^{-1} and FLN 50 % = 0.265 μg L^{-1})

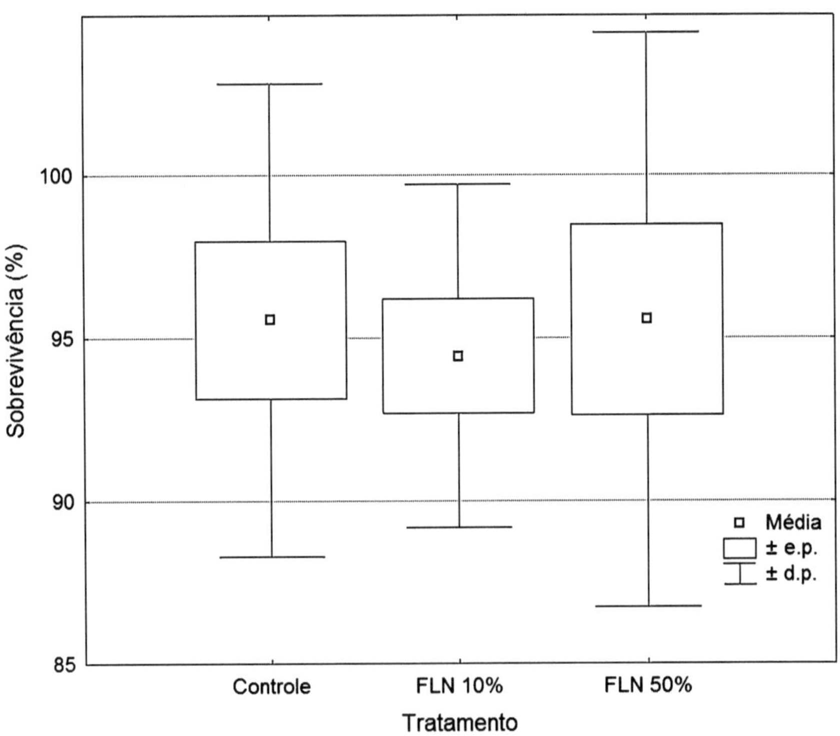

Fig. 11.3 Comparison of the mean values ($n = 16$) of survival of *Mysidopsis juniae* observed in the control and in the two concentrations of nC_{60} (FLN 10 % = 0.053 μg L^{-1} and FLN 50 % = 0.265 μg L^{-1})

suspension according to the variation of the ionic strength. According to these authors, electrolyte solutions tend to provide higher fullerene aggregates due to loss of stability in which they were. The stability of the n-C60 clusters in the absence of electrolyte and their corresponding aggregation at various salt concentrations strongly suggest that the origin of the n-C60's stability is due to electrostatic repulsion between colloids. In their experiments, they observed that in the absence of electrolytes, fullerene aggregates (nC_{60}) remained stable over time (the diameters did not vary), and aggregates from the same fullerene solution with addition of NaCl (final concentration 0.1 M) increased in diameter from 168 to 680 nm over a period of 2 months. The formation of large aggregates can lead to precipitation of the nanoparticles, which may contribute to the lack of its toxicity to microcrustaceans.

Despite the fullerene-nC_{60} has tendencies to form aggregates and suffers deposition in the presence of ions, it has been reported that other components in the environment can promote destabilization of the aggregates. Among them, the presence of organic matter and humic acids that are found in estuaries may have the potential to mobilize nC_{60} through changes in the charges on their surfaces. The unbundling process may result in the formation of smaller aggregates, which can be resuspended and again made available to the water column (see review of Christian et al. 2008).

References

ABNT (2005) Ecotoxicologia aquática – toxicidade aguda – "Método de ensaio com Misidáceos (Crustacea)'. Norma ABNT-NBR15308:2005, Associoção Brasileira de Normas Técnicas, Rio de Janeiro, p 17

Andrievsky GV, Kosevich MV, Volk OM, Shelkovsky VS, Vashchenko LA (1995) On the production of an aqueous colloidal solution of fullerenes. J Chem Soc 12, Chem Commun 1281–1282

Badaró-Pedroso C, Reynier MV, Prósperi VA (2002) Testes de Toxicidade Aguda com Misidáceos – Ênfase nas Espécies *Mysidopsis juniae e Mysidium gracile* (Crustacea: Mysidacea). In: Nascimento IA, Sousa ECPM, Nipper M (eds) Métodos em Ecotoxicologia Marinha: Aplicações no Brasil. Comp & Art, Salvador, pp 123–139

Barnes R (1996) Zoologia dos Invertebrados. Roca, São Paulo, p 1028

Baun A, Sorensen SN, Rasmussen RF, Hartmann NB, Koch CB (2008) Toxicity and bioaccumulation of xenobiotic organic compounds in the presence of aqueous suspensions of aggregates of nano-C60. Aquat Toxicol 86:379–387

Bianchi F, Acri F, Bernardi A, Berton A, Boldrin A, Cammati E, Cassin D, Comaschi A (2003) Can plankton communities be considered as bioindicators of water quality in the Lagoon of Venice? Mar Pollut Bull 46:964–971

Bianco A, Ros TD (2007) Biological applications of fullerenes. In: Langa F, Nierengarten JF (eds) Fullerenes. Principles and applications. The Royal Society of Chemistry, Cambridge, pp 301–328

Bolskar RD, Benedetto AF, Husebo LO, Price RE, Jackson EF, Wallace S, Wilson LJ, Alford JM (2003) First soluble M@C-60 derivatives provide enhanced access to metallofullerenes and permit in vivo evaluation of Gd@C-60[C(COOH)(2)](10) as a MRI contrast agent. J Am Chem Soc 125:5471–5478

Brant J, Lecoanet MR, Wiesner H (2005) Aggregation and deposition characteristics of fullerene nanoparticles in aqueous systems. J Nanoparticle Res 7:545–553

CETESB (1992) Água do mar – "Teste de toxicidade aguda com *Mysidopsis juniae*, Silva, 1979'. Norma Técnica L5.251:1992, Companhia de Tecnologia e Saneamento Ambiental, São Paulo, p 19

Chen Z, Westerhoff P, Herckes P (2008) Quantification of C60 fullerene concentrations in water. Environ Toxicol Chem 27:1852–1859

Christian P, Von der Kammer F, Baalousha M, Hofmann T (2008) Nanoparticles: structure, properties, preparation and behaviour in environmental media. Ecotoxicology 17:326–343

Colvin VL (2003) The potential environmental impact of engineered nanomaterials. Nat Biotechnol 21:1166–1170

Diederich F, Thilgen C (1996) Covalent fullerene chemistry. Science 271:317–324

Dresselhaus MS, Dresselhaus G, Eklund PC (1996) Structure of fullerenes. In: Dresselhaus MS, Dresselhaus G, Eklund PC (eds) Science of fullerenes and carbon nanotubes: their properties and applications. Academic, San Diego, pp 60–79

Fortner JD, Lyon DY, Sayes CM, Boyd AM, Falkner JC, Hotze EM, Alemany LB, Tao YJ, Guo W, Ausman KD, Colvin VL, Hughes JB (2005) C60 in water: nanocrystal formation and microbial response. Environ Sci Technol 39:4307–4316

Gharbi N, Pressac M, Hadchouel M, Szwarc H, Wilson SR, Moussa F (2005) [60]fullerene is a powerful antioxidant in vivo with no acute or subacute toxicity. Nano Lett 5:2578–2585

Hasobe T, Imahori H, Kamat PV, Ahn TK, Kim SK, Kim D, Fujimoto A, Hirakawa T, Fukuzumi S (2005) Photovoltaic cells using composite nanoclusters of porphyrins and fullerenes with gold nanoparticles. J Am Chem Soc 127:1216–1228

Henry TB, Menn FM, Fleming JT, Wilgus J, Compton RN, Sayler GS (2007) Attributing effects of aqueous C60 nano-aggregates to tetrahydrofuran decomposition products in larval zebrafish by assessment of gene expression. Environ Health Perspect 115:1059–1065

Heymann D (1996) Solubility of fullerenes C-60 and C-70 in seven normal alcohols and their deduced solubility in water. Fuller Sci Technol 4:509–515

Hirsch A (1997) Aspects of organic chemistry of fullerenes. J Phys Chem Solid 58:1729–1740

Hulteen C, Martin CR (1997) A general template-based method for the preparation of nanomaterials. J Mater Chem 7:1075–1087

ISO (1999) Water quality – "Determination of acute lethal toxicity to marine copepods (Copepoda, Crustacea)'. Standard 14669:1999, International Organization for Standardization, Geneva, p 16

Jafvert CT, Kulkarni PP (2008) Buckminsterfullerene's (C60) octanol-water partition coefficient (Kow) and aqueous solubility. Environ Sci Technol 42:5945–5950

Kahru A, Dubourguier HC (2010) From ecotoxicology to nanoecotoxicology. Toxicology 269:105–119

Kaminski SM (2004) Influência da alimentação sobre a reprodução e o desenvolvimento do copépodo calanoida Acartia tonsa DANA, 1984, em cultivo intensivo. Dissertação de Mestrado, Centro de Ciências Agrárias, Departamento de Aquicultura – Universidade Federal de Santa Catarina. Florianópolis. p 66

Krätschmer W, Lamb LD, Fostiropoulos K, Huffman DR (1990) Solid C60: a new form of carbon. Nature 347:354–358

Kroto HW, Heath JR, O'Brien SC, Curl RF, Smalley RE (1985) C60: buckminsterfullerene. Nature 318:162–163

Lovern SB, Klaper R (2006) Daphnia magna mortality when exposed to titanium dioxide and fullerene (C60) nanoparticles. Environ Toxicol Chem 25:1132–1137

Markovic Z, Trajkovic, V (2008) Biomedical potential of the reactive oxygen species generation and quenching by fullerenes (C60). Biomaterials 29: 3561–3573.

Mauchline J, Blaxter JHS, Southward AJ, Tyler PA (1998) Advances in marine biology – the biology of calanoid copepods – introduction. Academic Press LTD, London, p 124

Medina M, Barata C (2004) Static-renewal culture of acartia tonsa (copepoda: calanoida) for ecotoxicological testing. Aquaculture 229:203–213

Moore MN (2006) Do nanoparticles present ecotoxicological risks for the health of the aquatic environment? Environ Int 32:967–976

Nowack B, Bucheli TD (2007) Occurrence, behavior and effects of nanoparticles in the environment. Environ Pollut 150:5–22

Oberdörster E (2004) Manufactured nanomaterials (fullerene, C60) induce oxidative stress in the brain of juvenile largemouth bass. Environ Health Perspect 112:1058–1062

Oberdörster G, Oberdösrter E, Oberdörster J (2005) Nanotoxicology: an emerging discipline evolving from studies of ultrafine particles. Environ Health Perspect 113:823–839

Oberdörster E, Zhu S, Blickley TM, McClellan-Green P, Haasch ML (2006) Ecotoxicology of carbon-based engineered nanoparticles: Effects of fullerene (C60) on aquatic organisms. Carbon 44:1112–1120

Prato M (1997) [60]Fullerene chemistry for materials science applications. J Mater Chem 7:1097–1109

RCEP (2008) Novel materials in the environment: the case of nanotechnology. Twenty-seventh report, TSO (The Stationery Office), London, p 147

Ruoff RS, Tse DS, Malhotra R, Lorents DC (1993) Solubility of fullerene (C60) in a variety of solvents. J Phys Chem 97:3379–3383

Sabatini ME (1990) The developmental stages (copepodids I to VI) of Acartia tonsa Dana, 1849 (Copepoda, Calanoida). Crustaceana 59:53–61

Shinohara N, Matsumoto K, Endoh S, Maru J, Nakanishi J (2009) In vitro and in vivo genotoxicity tests on fullerene C60 nanoparticles. Toxicol Lett 191:289–296

Tao X, Fortner JD, Zhang B, He Y, Chen Y, Hughes JB (2009) Effects of aqueous stable fullerene nanocrystals (nC60) on daphnia magna: evaluation of sublethal reproductive responses and accumulation. Chemosphere 77:1482–1487

Taylor R, Walton DRM (1993) The chemistry of fullerenes. Nature 363:685–693

Taylor R, Hare JP, bdul-Sada AK, Kroto HW (1990) Isolation, separation and characterisation of the fullerenes C60 and C70: the third form of carbon. J Chem Soc Chem Commun 20:1423–1425

Vileno B, Sienkiewicz A, Lekka M, Kulik AJ, Forr L (2004) In vitro assay of singlet oxygen generation in the presence of water-soluble derivatives of C60. Carbon 42:1195–1198

Yu H, Gan LH, Hu X, Venkatraman SS, Tam KC, Gang YY (2005) A novel amphiphilic double-[60]fullerene-capped triblock copolymer. Macromolecules 38:9889–9893

Zagatto PA (2006) Ecotoxicologia. In: Zagatto PA, Bertoletti E (eds) Ecotoxicologia Aquática: Princípios e Aplicações. Rima, São Carlos, pp 1–13

Zhu S, Oberdorster E, Haasch ML (2006) Toxicity of an engineered nanoparticle (fullerene, C60) in two aquatic species, daphnia and fathead minnow. Mar Environ Res 62:S5–S9

Zhu XS, Zhu L, Li Y, Duan ZH, Chen W, Alvarez PJJ (2007) Developmental toxicity in zebrafish (danio rerio) embryos after exposure to manufactured nanomaterials: buckminsterfullerene aggregates (nC(60)) and fullerol. Environ Toxicol Chem 26:976–979

Comparative Analysis of Common and Unique Targets in Drug Resistant Strains of *Staphylococcus aureus*

12

Amandeep Kaur Kahlon, Mahendra P. Darokar, and Ashok Sharma

Abstract

The increasing number of drug-resistant strains of *Staphylococcus aureus* necessitated the identification of potential drug targets specific to the strain of interest. The chromosomal and plasmid genes of *S. aureus* strains N315, JH1, and JH9 were compared with Mu50 strain to find common (core) and unique (strain-specific) genes in the present study. In silico analysis of genomics data of *S. aureus* strains N315, JH1, and JH9 with vancomycin-resistant strain Mu50 showed total number of unique genes higher in Mu50 and least in N315. The presence of higher number of unique genes in Mu50 signifies their role in drug resistance mechanism. In *S. aureus* strain Mu50, the chromosomal genes showed tetracycline resistance protein and ABC transporters. Whereas, plasmid genes showed transposase, N-acetyltransferase, and truncated resolvase. In *S. aureus* strain N315, chromosomal genes showed more number of hypothetical proteins. Whereas, plasmid genes showed arsenical resistance proteins, arsenic efflux pump protein, CadD, CadX, blaR1, and penicillinase repressor. In *S. aureus* strains JH1 and JH9, chromosomal genes showed mainly phage regulatory proteins, and plasmid genes showed arsenic and cadmium resistance regulatory proteins. Furthermore, human proteome and gut flora non-pathogenic proteomes were compared with proteome of these strains to find target proteins specific to the strain of interest and not present in host. In conclusion, identification of unique genes in these strains provided insights on differences in drug resistance potential.

A.K. Kahlon, Ph.D (✉)
Recombinant Gene Products Group, ICGEB Campus, Aruna Asaf Ali Marg, New Delhi 110 067, India
e-mail: amangenomics@gmail.com

M.P. Darokar • A. Sharma
Biotechnology Division, CSIR-Central Institute of Medicinal and Aromatic Plants, P.O. CIMAP, Kukrail Picnic Spot Road, Lucknow 226015, Uttar Pradesh, India

P. Shukla (ed.), *Frontier Discoveries and Innovations in Interdisciplinary Microbiology*,
DOI 10.1007/978-81-322-2610-9_12

Keywords
Staphylococcus aureus • *Lactobacillus brevis* ATCC367 • *Lactobacillus reuteri* F275 • *Bifidobacterium longum* • Core genes • Unique genes • Drug resistance

Abbreviations

CGUG Core Genes Unique Genes
MRSA Methicillin-Resistant *Staphylococcus aureus*
MSSA Methicillin-Susceptible Staphylococcus *aureus*
VRSA Vancomycin-Resistant *Staphylococcus aureus*
VISA Vancomycin-Intermediate *Staphylococcus aureus*

12.1 Introduction

Staphylococcus aureus is an important cause of skin and soft tissue infections (SSTIs) such as endovascular infections, pneumonia, septic arthritis, endocarditis, osteomyelitis, and sepsis. Penicillin was found to be an effective antibiotic to treat *S. aureus* infections before the 1950s. When penicillin was introduced in 1944, over 94 % of *S. aureus* isolates were susceptible but later on by 1950 over 50 % were resistant (Livermore 2000). With the emergence of drug-resistant *S. aureus* strains, researchers posed serious concern to find effective antimicrobials worldwide. Resistance mechanisms include enzymatic inactivation of the antibiotic (penicillinase (Bondi and Dietz 1945) and aminoglycoside-modification enzymes), alteration of the target with decreased affinity for the antibiotic, trapping of the antibiotic (for vancomycin and possibly daptomycin), and efflux pumps (fluoroquinolones and tetracycline). Detection of the resistance mechanisms and their genetic basis is required for antibiotic susceptibility surveillance in *S. aureus* (Nickerson et al. 2009; Olson et al. 2007; Pantosti et al. 2007; Pantosti and Venditti 2009). *S. aureus* MGEs (mobile genetic elements) have provided detailed insight into the evolution of antimicrobial resistance mechanisms and virulence (Malachowa and Deleo 2010; Medini et al. 2005). *mecA* is part of a mobile genetic element called the staphylococcal cassette chromosome (SCC) *mec*. SCC*mec* is flanked by cassette chromosome recombinase genes (*ccrA*/*ccrB* or *ccrC*) that permit intra- and interspecies horizontal transmission of SCC*mec* (Chambers 1997; Gorden and Lowy 2008; Gustafson et al. 2006; Katayama et al. 2000). The transfer of mec sequences is horizontal, from coagulase-negative staphylococci to *S. aureus* (Archer et al. 1994). Community-associated MRSA (CA-MRSA) strains differ from the older, health care-associated MRSA strains. They infect a different group of patients, differ in antimicrobial susceptibility patterns, and spread rapidly among healthy people in the community (David and Daum 2010; DeLeo et al. 2009; Gorak et al. 1999). The differential distribution of large mobile elements carrying virulence and drug resistance determinants may be responsible for the clinically important phenotypic differences in these strains (Holden et al. 2004). The response made by society to reduce resistance involves surveillance, reduced usage, improved infection control, and the introduction of new antimicrobial agents (Hawkey 2008a). The Netherlands and Denmark followed strict isolation policies to prevent dissemination of MRSA. On the other hand, success of infection control measures was limited in the USA and England (Lowy 2003). The continued emergence of resistant strains of bacteria such as MRSA demands an urgent revival of the search for new antibiotics (Muzzi et al. 2007; Plata et al. 2009; Saxena and Gombar 2010; Srivastava et al. 2002; Hawkey 2008b; Tenover 2006).

The glycopeptide antibiotic vancomycin (molecular weight, 1,485.7) derived from

Nocardia orientalis (formerly known as *Streptomyces orientalis*) was clinically introduced in 1958 and was found to be effective against methicillin-resistant *S. aureus*. However, the first case of decreased susceptibility of *S. aureus* to vancomycin was reported from Japan in 1996, and similar reports came from the USA and other countries at the same time (Stryjewski and Chambers 2008; Chang et al. 2003). Experiments demonstrated that the thickened cell wall of vancomycin-intermediate *S. aureus* (VISA) could protect ongoing peptidoglycan biosynthesis in the cytoplasmic membrane from vancomycin inhibition, hence allowing the cells to continue producing nascent cell wall peptidoglycan to make the cells resistant to vancomycin (Cui et al. 2006; Avison et al. 2002). The reports on vancomycin-resistant *S. aureus* (VRSA) strains from the USA, France, Korea, South Africa, and Brazil have also confirmed the emergence of vancomycin resistance in *S. aureus* and decline of the clinical effectiveness of vancomycin in the hospitals (Hiramatsu 2001). The heterogeneous VISA strains could be involved in vancomycin therapeutic failure, especially in deep-seated or wound MRSA infections, given the weak tissue concentrations of vancomycin (Hiramatsu et al. 1997a). Point mutations in key regulatory genes contribute to the hVISA and VISA phenotypes (Hiramatsu et al. 1997b; Howden et al. 2010; Morell and Balkin 2010). Another report came from the northern part of India and found for the first time the emergence of VISA/VRSA in India. The major cause of this may be unawareness and indiscriminate use of broad-spectrum antibiotics. Use of antibacterial agents creates selective pressure for the emergence of resistant strains (Tiwari and Sen 2006). Minimizing the antibiotic pressure that favors the selection of resistant strains is very essential in order to control the emergence of these strains in the hospital and the community (Chambers 2001). Genomics could provide information regarding the overall conservation and hence therapeutic potential of antimicrobial targets (Chaudhuri et al. 2009; Enright et al. 2002; Lee et al. 2009; Moran and Mount 2003).

Due to differences in susceptibility patterns and antibiotic resistance in *S. aureus*, there is extensive need to find strain-specific genes responsible for such differences. In the present study, we performed in silico analysis to find the differences in drug-resistant strains of *S. aureus* responsible for varied drug response and susceptibility to antibiotics. Major concern was to know about the genetic differences in vancomycin-resistant strain Mu50 with other drug-resistant strains of *S. aureus*. First of all, we identified common genes present in all the strains under study and also found unique genes specific to strain of interest. Furthermore, *S. aureus* strains were compared with human proteome and gut flora organisms to find potential drug targets in each strain. This is the first report to find strain-specific unique genes to find differences in antibiotic resistance among drug-resistant *S. aureus* strains through in silico analysis.

12.2 Materials and Methods

12.2.1 Genomic Data Analysis

Chromosomal and plasmid genes of *S. aureus* strains Mu50 (accession(s), NC_002758 and NC_002774), N315 (accession(s), NC_002745 and NC_003140), JH1 (accession(s), NC_009632 and NC_009619), and JH9 (accession(s), NC_009487 and NC_009477) were retrieved using NCBI (http://www.ncbi.nlm.nih.gov/). Differences between drug-resistant strain(s) at the genome level were elucidated through in silico analysis. The vancomycin-resistant *S. aureus* strain Mu50 was compared with other drug-resistant strains N315, JH1, and JH9 for common and unique gene identification using CGUG 3.1. CGUG is available at http://binf.gmu.edu/geneorder.html (Mahadevan et al. 2009; Turner et al. 2013), and it performs iterative BLASTp analyses using a reference genome and up to four query genomes to provide a table of genes common to these genomes. Genomic sequences were compared using CGUG 3.1 with BLASTp threshold set at 75.

12.2.2 In Silico Drug Target Prediction

Protein-coding genes of *S. aureus* strains were compared against human and gut flora non-pathogenic microbes using UniDrug-Target tool (Chanumolu et al. 2012). This server is freely available at http://117.211.115.67/UDT/main.html and stand-alone application source codes are also available at http://www.bioinformatics.org/ftp/pub/bioinfojuit/UDT.rar. Host proteome and gut flora microbes, *Lactobacillus brevis* ATCC367, *Lactobacillus reuteri* F275, and *Bifidobacterium longum* were compared against each *S. aureus* strain to identify strain-specific potential drug targets.

12.2.3 In Silico Essential and Virulence Factor Analysis

Database of essential genes (http://www.essentialgene.org/ DEG 10.6) and the virulence factor database (http://www.mgc.ac.cn/VFs/ VFDB) were used in the study (Chen et al. 2012; Zhang and Lin 2009). BLASTp analysis against human proteome was performed to find essential genes and virulence factor nonhuman homologues. Threshold value was considered 10^{-3}, cutoff values for bit score were considered >100, % identity at amino acid level ≥35 %, and query coverage ≥50 % to identify the human homologues.

12.3 Results

12.3.1 In Silico Identification of Common and Strain-Specific Genes in Drug-Resistant Strains of *Staphylococcus aureus*

In *S. aureus* strain Mu50, the chromosomal genes showed unique genes, hexose phosphate transport protein, NADH dehydrogenase subunit, glycerol-3-phosphate transporter, tetracycline resistance protein, NLP/P60 family lipoprotein, ATP-/GTP-binding protein, transcriptional regulator, FtsK/SpoIIIE family protein, transposase, bacteriophage terminase small subunit, ferric hydroxamate receptor 1, integrase, excisionase, ssDNA-binding protein, 2-oxoglutarate/malate translocator, phi ETA ORF proteins, cell wall hydrolase, tail fiber, holin, amidase, peptide chain release factor 3, ABC transporter, Mrp protein, dUTP pyrophosphatase, and phi PVL ORF proteins, whereas plasmid genes showed transposase, N-acetyltransferase, and truncated resolvase.

In *S. aureus* strain N315, chromosomal genes showed truncated amidase, and hypothetical proteins and plasmid genes showed arsenical resistance operon repressor, truncated Pre protein, arsenic efflux pump protein, CadD, CadX, bla regulator protein blaR1, penicillinase repressor, and arsenate reductase as unique genes.

In *S. aureus* strain JH1, chromosomal genes showed phi PV83 orf 10-like protein, phiSLT ORF 87-like protein, DNA-binding protein, phi PV83 orf 10-like protein, Rha family phage regulatory protein, SLT orf 71-like protein-related protein, PVL orf 37-like protein, VRR-NUC domain-containing protein, phage terminase small subunit P27, phage protein, phi13 family phage major tail protein, Ig domain-containing protein, phi77 ORF100-like protein, and PV83 orf 4-like protein-related protein. In *S. aureus* strain JH9, chromosomal genes showed XRE family transcriptional regulator, phage head-tail adaptor, HK97 family phage protein, SPP1 family holing, Rha family phage regulatory protein, VRR-NUC, phage terminase, P27 phage protein DNA-packaging protein, phi13 family phage major tail protein, Ig domain-containing protein, and SH3 type 3 domain-containing protein, whereas plasmid genes in both JH1 and JH9 showed orotidine 5′-phosphate decarboxylase, sugar isomerase (SIS), HxlR family transcriptional regulator, integrase catalytic subunit, MarR family transcriptional regulator, quinone oxidoreductase putative (YhdH/YhfP), regulatory protein (ArsR), cadmium resistance transporter CadD, 6-phosphogluconate dehydrogenase, and glucose-6-phosphate isomerase as unique proteins compared with Mu50 plasmid genes. Plasmid genes of JH9 showed the presence of CopY family transcriptional regulator but it was absent in JH1 plasmid genes. Comparative analysis data is shown in Table 12.1 and

Table 12.1 List of strain-specific/unique genes in chromosomal and plasmid genes of *S. aureus* strains Mu50, N315, JH1, and JH9

Name of the strain	Strain-specific genes in chromosome and their function		Strain-specific genes in plasmid and their function	
	Accession: NC_002758		Accession: NC_002774	
S. aureus strain Mu50	GI:15923212	Hexose phosphate transport protein	GI:14141837	Transposase
	GI:15923262	NADH dehydrogenase subunit	GI:14141846	Transposase
	GI:15923326	Glycerol-3-phosphate transporter	GI:14141847	Transposase
	GI:15923382	Integrase-like protein	GI:14141848	Putative transposase
	GI:15923388	Tetracycline resistance protein	GI:14141849	N-acetyltransferase
	GI:15923390	NLP/P60 family lipoprotein	GI:14141851	Putative transposase
	GI:15923392	ATP/GTP-binding protein	GI:14141852	Transposase
	GI:15923398	Transcriptional regulator	GI:14141853	Truncated resolvase
	GI:15923399	FtsK/SpoIIIE family protein		
	GI: 15923405	Transposase		
	GI: 15923773	Integrase		
	GI: 15923790	Bacteriophage terminase small subunit		
	GI:15923793	Ferric hydroxamate receptor 1		
	GI:15923837	Integrase		
	GI:15923838	Excisionase		
	GI:15923845	Anti-repressor		
	GI:15923854	ssDNA-binding protein		
	GI:15923875	Small terminase		
	GI:15923876	Large terminase		
	GI:15923892	phi ETA orf 54-like protein		
	GI:15923893	phi ETA orf 55-like protein		
	GI:15923894	phi ETA orf 56-like protein		
	GI:15923895	phiETA ORF57-like protein		
	GI:15923896	phiETA ORF58-like protein		
	GI:15923897	phiETA ORF59-like protein		
	GI:15923898	phiETA ORF60-like protein		
	GI:15923899	Cell wall hydrolase		
	GI:15923900	Tail fiber		
	GI:15923901	phi ETA orf 63-like protein		
	GI:15923902	Holin		
	GI:15923903	Amidase		
	GI:57634622	Peptide chain release factor 3		
	GI:57634629	ABC transporter		
	GI:57634638	Mrp protein		
	GI:15924775	Transposase		
	GI:15924776	Transposase		
	GI:15924777	Transposase		
	GI:15924794	Transposase		
	GI:15924795	Transposase		
	GI:15924964	dUTP pyrophosphatase		
	GI:15924966	phi PVL ORF 52-like protein		
	GI:15924967	phi PV83 orf 27-like protein		
	GI:15924968	PVL orf 51-like protein		
	GI:15924969	phi PVL ORF 50-like protein		
	GI:15924985	phi PVL orf 33-like protein		
	GI:15924986	phi PVL orf 32-like protein		
	GI:57634648	Multi-drug transporter		
	GI:15925685	2-oxoglutarate/malate translocator		

(continued)

Table 12.1 (continued)

Name of the strain	Strain-specific genes in chromosome and their function		Strain-specific genes in plasmid and their function	
	Accession: NC_002745		Accession: NC_003140	
S. aureus strain N315	GI:15927518	Truncated amidase	GI:16119205	CadD
	GI:15927519	Truncated amidase	GI:16119206	CadX
			GI:16119211	Bla regulator protein blaR1
			GI:16119212	Penicillinase repressor
			GI:16119216	Arsenical resistance operon repressor
			GI:16119217	Arsenic efflux pump protein
			GI:16119218	Arsenate reductase
			GI:16119223	Truncated Pre protein
			GI:16119224	Truncated Pre protein
			GI:16119225	Truncated Pre protein
	Accession: NC_009632		Accession: NC_009619	
S. aureus strain JH1	GI:150392796	Phi PV83 orf 10-like protein	GI:150375694	Regulatory protein ArsR
	GI:150392797	Phi SLT ORF 87-like protein, DNA-binding	GI:150375695	Cadmium resistance transporter CadD
	GI:150392833	Phage head-tail adaptor	GI:150375706	6-Phosphogluconate dehydrogenase
	GI:150392834	HK97 family phage protein	GI:150375707	Glucose-6-phosphate isomerase
	GI:150393350	Phi PV83 orf 10-like protein	GI:150375708	Orotidine 5′-phosphate decarboxylase
	GI:150393406	SPP1 family holing	GI:150375709	Sugar isomerase (SIS)
	GI:150393517	Rha family phage regulatory protein	GI:150375710	Helix-turn-helix HxlR type
	GI:150393520	SLT orf 71-like protein	GI:150375717	Potential ATP-binding protein
	GI:150393522	PVL orf 37-like protein	GI:150375718	Integrase catalytic region
	GI:150393546	VRR-NUC domain-containing protein	GI:150375720	Penicillinase repressor
	GI:150393551	Phage terminase small subunit P27	GI:150375721	Beta-lactamase
	GI:150393555	Phage protein	GI:150375722	Bla regulator protein blaR1
	GI:150393556	Phi13 family phage major tail protein	GI:150375724	Regulatory protein MarR
	GI:150393557	Ig domain-containing protein	GI:150375725	Quinone oxidoreductase putative YhdH/YhfP
	GI:150394491	Phi77 ORF 100-like protein		
	GI:150394558	PV83 orf 4-like protein-related protein		

(continued)

Table 12.1 (continued)

Name of the strain	Strain-specific genes in chromosome and their function		Strain-specific genes in plasmid and their function	
	Accession: NC_009487		Accession: NC_009477	
S. aureus strain JH9	GI:148266750	XRE family transcriptional regulator	GI:148244140	Orotidine 5′-phosphate decarboxylase
	GI:148266788	Phage head-tail adaptor	GI:148244141	Sugar isomerase (SIS)
	GI:148266789	HK97 family phage protein	GI:148244142	HxlR family transcriptional regulator
	GI:148267346	SPP1 family holin	GI:148244149	Integrase catalytic subunit
	GI:148267461	Rha family phage regulatory protein	GI:148244151	CopY family transcriptional regulator
	GI:148267488	VRR-NUC	GI:148244154	MarR family transcriptional regulator
	GI:148267492	Phage terminase, P27 small subunit	GI:148244155	Quinone oxidoreductase putative, YhdH/YhfP
	GI:148267497	Phage DNA-packaging protein	GI:148244156	Regulatory protein, ArsR
	GI:148267501	Phi13 family phage major tail protein	GI:148244157	Cadmium resistance transporter CadD
	GI:148267502	Ig domain-containing protein	GI:148244167	6-Phosphogluconate dehydrogenase
	GI:148268418	SH3 type 3 domain-containing protein	GI:148244168	Glucose-6-phosphate isomerase

Figs. 12.1 and 12.2. It was also observed that hypothetical proteins were more in number among identified unique genes in each strain under study. Total number of hypothetical proteins was 112 in *S. aureus* strain Mu50, 19 in *S. aureus* strain N315, 114 in *S. aureus* strain JH1, and 83 in *S. aureus* strain JH9.

12.3.2 Comparative Genomics Analysis of *S. aureus* Genomes with Human and Gut Flora Non-pathogenic Microbes

UniDrug target tool analysis of *S. aureus* strain JH9 genome showed high number of unique proteins (with respect to human and gut flora non-pathogenic proteomes) and was least in *S. aureus* strain N315 (Table 12.2). DEG 10.6 database searches showed 302 essential genes in *S. aureus* strain N315. BLASTp analysis of these essential genes against human proteome showed 212 essential genes as nonhuman homologues. These genes were also found in other strains of *S. aureus* taken in the study. Virulence factor database analysis and text mining showed 84–90 genes in the strains under study (Table 12.3).

12.4 Discussion

Drug resistance development of *S. aureus* strains to available antibiotics is an important point of concern among health practitioners. According to NCBI genome database information (http://www.ncbi.nlm.nih.gov/genome/genomes/154), the total number of genome sequences of *S. aureus* strains assembled and

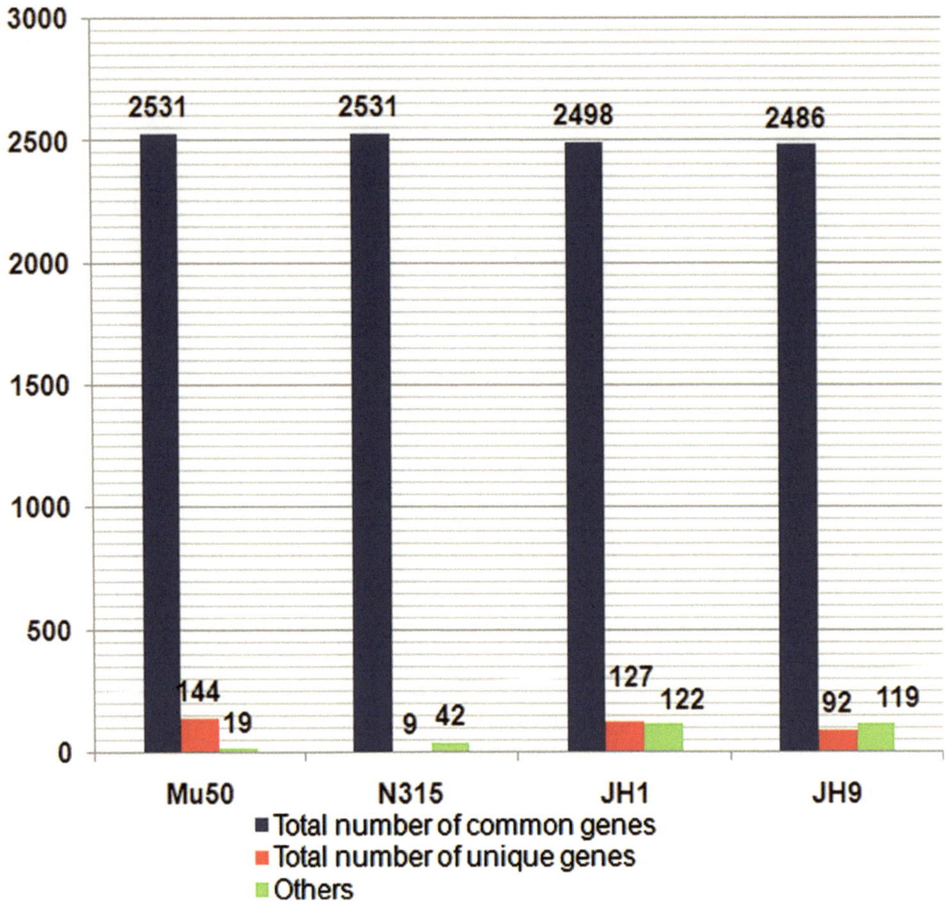

Fig. 12.1 Chromosomal genes data of *S. aureus* strain Mu50 compared through in silico analysis with N315, JH1, and JH9 strains

annotated is 4193 and the number of genome groups involved and reported worldwide is 35. Genome database showed 143 plasmid annotation data among the genome sequences of *S. aureus* strains annotated till date. *S. aureus* subsp. aureus strain Mu50 was isolated in 1997 from the pus of a Japanese male baby with a surgical wound infection that did not respond to vancomycin (Kuroda et al. 2001). This strain was taken as a reference strain to compare with other drug-resistant strains N315, JH1, and JH9 taken in the study. *S. aureus* strain N315 (submitted by Juntendo Univ.) is a methicillin-resistant (MRSA) strain and was isolated from a pharyngeal smear of a Japanese patient in 1982. *Staphylococcus aureus* subsp. aureus strain JH9 was submitted by the US DOE Joint Genome Institute. The chronologically earliest strain (JH1) was vancomycin sensitive, while the latest strain (JH9) exhibited increased vancomycin resistance (VISA-type vancomycin resistance). Understanding the differences between *S. aureus* genomes and the controls that govern these changes could help to predict the evolution of superbugs (Lindsay and Holden 2004; Mills 2003). The present study identified the differences in terms of presence and absence of drug-resistant genes in the drug-resistant *S. aureus* strains. Tetracycline resistance protein, ABC transporter and cell division proteins, and FtsK/Spo IIIE family protein were mainly found as strain-specific genes in *S. aureus* strain

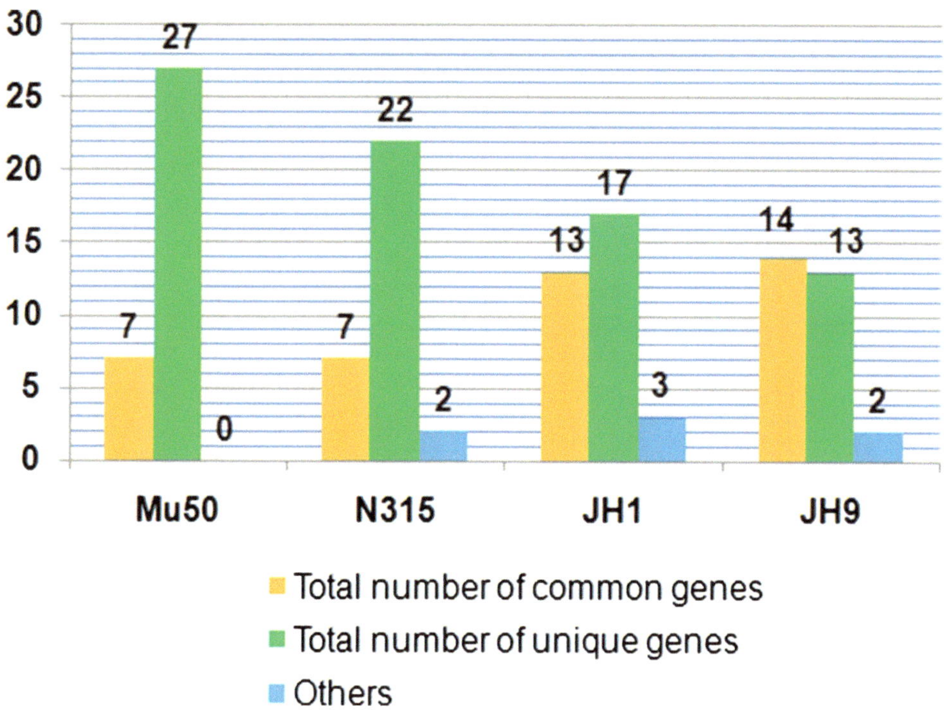

Fig. 12.2 Plasmid genes data of *S. aureus* strain Mu50 compared through in silico analysis with N315, JH1, and JH9 strains

Table 12.2 In silico prediction of drug targets in *Staphylococcus aureus* with respect to human proteome and gut flora non-pathogenic proteome

S. no.	Name of the strain	Unique proteins with respect to human and gut flora non-pathogenic proteome(s)	Highest matching similarity score in unique proteins with non-pathogenic proteome(s)
1.	*Staphylococcus aureus* strain N315	364	0–30.083 %
2.	*Staphylococcus aureus* strain Mu50	398	0–29.825 %
3.	*Staphylococcus aureus* strain JH1	415	0–29.825 %
4.	*Staphylococcus aureus* strain JH9	417	0–39.9 %

Table 12.3 Comparative genomics data analysis of virulence factors in *S. aureus* strains using VFDB database

Virulence factors	Related genes	*S. aureus* Mu50 chromosome NC_002758 (2878529 bp)	*S. aureus* N315 chromosome NC_002745 (2814816 bp)	*S. aureus* JH1 chromosome NC_009632 (2906507 bp)	*S. aureus* JH9 chromosome NC_009487 (2906700 bp)
Autolysin	Atl	SAV1052	SA0905	SaurJH1_1135	SaurJH9_1112
Cell wall-associated fibronectin-binding protein	Ebh	SAV1435, SAV1434	SA1268, SA1267	SaurJH1_1524	SaurJH9_1495
Clumping factor A	clfA		SA0742	SaurJH1_0828	SaurJH9_0812
Clumping factor B	clfB	SAV2630	SA2423	SaurJH1_2710	SaurJH9_2654
Elastin-binding protein	Ebp	SAV1481	SA1312	SaurJH1_1568	SaurJH9_1537

(continued)

Table 12.3 (continued)

Virulence factors	Related genes	*S. aureus* Mu50 chromosome NC_002758 (2878529 bp)	*S. aureus* N315 chromosome NC_002745 (2814816 bp)	*S. aureus* JH1 chromosome NC_009632 (2906507 bp)	*S. aureus* JH9 chromosome NC_009487 (2906700 bp)
Extracellular adherence protein/MHC analogous protein	eap/map	SAV1938		SaurJH1_0999	SaurJH9_1996
Fibrinogen-binding protein	Efb	SAV1158	SA1003	SaurJH1_1239	SaurJH9_1215
Fibronectin-binding proteins	fnbA	SAV2503	SA2291	SaurJH1_2578	SaurJH9_2526
	fnbB	SAV2502	SA2290	SaurJH1_2577	SaurJH9_2525
Intercellular adhesion	icaR	SAV2665	SA2458	SaurJH1_2746	SaurJH9_2689
	icaA	SAV2666	SA2459	SaurJH1_2747	SaurJH9_2690
	icaD	SAV2667	SA2460	SaurJH1_2748	SaurJH9_2691
	icaB	SAV2668	SA2461	SaurJH1_2749	SaurJH9_2692
	icaC	SAV2669	SA2462	SaurJH1_2750	SaurJH9_2693
Surface-anchoring and processing protein[a]	**SrtA**[a]	SA2316	SAV2528	SaurJH1_2603	SaurJH9_2551
	SrtB[a]	SA0982	SAV1135	SaurJH1_1216	SaurJH9_2551
Ser-Asp rich fibrinogen-binding proteins	sdrC	SAV0561	SA0519	SaurJH1_0598	SaurJH9_0584
	sdrD	SAV0562	SA0520	SaurJH1_0599	SaurJH9_0585
	sdrE	SAV0563	SA0521	SaurJH1_0600	SaurJH9_0586
Staphylococcal protein A	Spa	SAV0111	SA0107	SaurJH1_0102	SaurJH9_0098
Staphyloxanthin pigment synthesis[a]	**CrtM**[a]	SA2349	SAV2562	SaurJH1_2638	SaurJH9_2584
Cysteine protease	SspB	SAV1047	SA0900	SaurJH1_1129	SaurJH9_1106
	SspC	SAV1046	SA0899	SaurJH1_1128	SaurJH9_1105
Hyaluronate lyase	HysA	SAV2202	SA2003	SaurJH1_2271	SaurJH9_2232
Lipase	Lip	SAV2671	SA2463	SaurJH1_2751	SaurJH9_2694
	Geh	SAV0320	SA0309	SaurJH1_0310	SaurJH9_0303
Serine protease	SplA	SAV1813	SA1631	SaurJH1_1899	SaurJH9_1864
	SplB	SAV1812	SA1630	SaurJH1_1898	SaurJH9_1863
	SplC	SAV1811	SA1629	SaurJH1_1897	SaurJH9_1862
	SplD	SAV1810	SA1628	SaurJH1_1896	SaurJH9_1861
	SplF	SAV1809	SA1627		
Serine V8 protease	SspA	SAV1048	SA0901	SaurJH1_1130	SaurJH9_1107
Staphylocoagulase	Coa	SAV0230	SA0222	SaurJH1_0220	SaurJH9_0214
Staphylokinase	Sak	SAV1944	SA1758	SaurJH1_2037	SaurJH9_2002
Thermonuclease	Nuc	SAV0815	SA1160	SaurJH1_0832	SaurJH9_0816
Capsule	–	SAV0153, SAV0152, SAV0151, SAV0150, SAV0149, SAV0154, SAV0155, SAV0156, SAV0157, SAV0158, SAV0159, SAV0160, SAV0161, SAV0162, SAV0163, SAV0164	SA0147, SA0146, SA0145, SA0144, SA0148, SA0149, SA0150, SA0151, SA0152, SA0153, SA0154, SA0155, SA0156, SA0157, SA0158, SA0159	SaurJH1_0140, SaurJH1_0141, SaurJH1_0142, SaurJH1_0143, SaurJH1_0144, SaurJH1_0145, SaurJH1_0146, SaurJH1_0147, SaurJH1_0148, SaurJH1_0149, SaurJH1_0150, SaurJH1_0151, SaurJH1_0152, SaurJH1_0153, SaurJH1_0154, SaurJH1_0155	SaurJH9_0135, SaurJH9_0136, SaurJH9_0137, SaurJH9_0138, SaurJH9_0139, SaurJH9_0140, SaurJH9_0141, SaurJH9_0142, SaurJH9_0143, SaurJH9_0144, SaurJH9_0145, SaurJH9_0146, SaurJH9_0147, SaurJH9_0148, SaurJH9_0149, SaurJH9_0150

(continued)

Table 12.3 (continued)

Virulence factors	Related genes	*S. aureus* Mu50 chromosome NC_002758 (2878529 bp)	*S. aureus* N315 chromosome NC_002745 (2814816 bp)	*S. aureus* JH1 chromosome NC_009632 (2906507 bp)	*S. aureus* JH9 chromosome NC_009487 (2906700 bp)
Type VII secretion system	EsxA	SAV0282	SA0271	SaurJH1_0273	SaurJH9_0266
	EsaA	SAV0283	SA0272	SaurJH1_0274	SaurJH9_0267
	EssA	SAV0284	SA0273	SaurJH1_0275	SaurJH9_0268
	EsaB	SAV0285	SA0274	SaurJH1_0276	SaurJH9_0269
	EssB	SAV0286	SA0275	SaurJH1_0277	SaurJH9_0270
	EssC	SAV0287	SA0276	SaurJH1_0278	SaurJH9_0271
	EsaC	SAV0289	SA0277	SaurJH1_0279	SaurJH9_0272
	EsxB	SAV0290	SA0278	SaurJH1_0280	SaurJH9_0273
Alpha hemolysin	hly/hla	SAV1163	SA1007	SaurJH1_1246	SaurJH9_1222
Beta hemolysin	Hlb	SAV2003	SA1811	SaurJH1_2030[a]	SaurJH9_2061
Delta hemolysin	Hld	SAV2035	SAS065	SaurJH1_2109	SaurJH9_2072[a]
Enterotoxin A	Sea	SAV1948			
Enterotoxin C	Sec	SAV2009	SA1817		
Enterotoxin G	Seg	SAV1824	SA1642	SaurJH1_1912	SaurJH9_1878
Enterotoxin I	Sei	SAV1828	SA1646	SaurJH1_1915	SaurJH9_1881
Enterotoxin Yent1	yent1	SAV1827	SA1645		
Enterotoxin Yent2	yent2	SAV1826	SA1644		
Enterotoxin-like L	Sell	SAV2008	SA1816		
Enterotoxin-like M	Selm	SAV1829	SA1647	SaurJH1_1916	SaurJH9_1882
Enterotoxin-like N	Seln	SAV1825	SA1643	SaurJH1_1913	SaurJH9_1879
Enterotoxin-like O	Selo	SAV1830	SA1648	SaurJH1_1917	SaurJH9_1883
Enterotoxin-like P	Selp		SA1761		
Exfoliative toxin type A	Eta	SAV1173	SA1016	SaurJH1_1256	
Exotoxin	set6	SAV0422	SA0382	SaurJH1_0457	SaurJH9_0445
	set7	SAV0423	SA0383	SaurJH1_0458	SaurJH9_0446
	set8	SAV0424	SA0384	SaurJH1_0459	SaurJH9_0447
	set9		SA0385	SaurJH1_0460	SaurJH9_0448
	set10	SAV0425	SA0386	SaurJH1_0461	SaurJH9_0449
	set11	SAV0426	SA0387	SaurJH1_0462	SaurJH9_0450
	set12	SAV0427	SA0388	SaurJH1_0463	SaurJH9_0451
	set13	SAV0428	SA0389	SaurJH1_0464	SaurJH9_0452
	set14	SAV0429	SA0390	SaurJH1_0465	SaurJH9_0453
	set15	SAV0433	SA0393	SaurJH1_0468	SaurJH9_0456
Gamma hemolysin	HlgA	SAV2419	SA2207	SaurJH1_2493	SaurJH9_2445
	HlgC	SAV2420	SA2208	SaurJH1_2494	SaurJH9_2446
	HlgB	SAV2421	SA2209	SaurJH1_2495	SaurJH9_2447
Leukotoxin D	LukD	SAV1819	SA1637	SaurJH1_1904	SaurJH9_1870
Leukotoxin E	LukE	SAV1820	SA1638	SaurJH1_1905	SaurJH9_1871
Toxic shock syndrome toxin	Tsst	SAV2011	SA1819		

[a]Additional genes identified through in silico analysis are shown in bold letters

Mu50. Arsenic efflux pump proteins and cadmium resistance transporter (CadD) were mainly found in *S. aureus* strains N315, JH1, and JH9. Further in silico analysis identified the potential drug targets.

12.5 Conclusion

The present study identified the gene specific differences in these strains, which were responsible for variation in drug resistance to antibiotics, viz., methicillin and vancomycin. This study also identified the potential unique target proteins in these strains, which are non-homologues to human and gut flora organism proteomes (*Lactobacillus brevis* ATCC367, *Lactobacillus reuteri* F275, and *Bifidobacterium longum*). We found more numbers of hypothetical proteins in *S. aureus* strains, suggesting their functional role in the organism. Functional characterization of these hypothetical proteins could help in determining their importance in drug resistance differences in these *S. aureus* strains in the future.

Acknowledgment The authors acknowledge the Department of Biotechnology (DBT), New Delhi, India, for financial support under the BTISnet program.

References

Archer GL, Niemeyer DM, Thanassi JA, Pucci MJ (1994) Dissemination among *Staphylococci* of DNA sequences associated with methicillin resistance. Antimicrob Agents Chemother 38:447–454

Avison MB, Bennett PM, Howe RA, Walsh TR (2002) Preliminary analysis of the genetic basis for vancomycin resistance in *Staphylococcus aureus* strain mu50. J Antimicrob Chemother 49:255–260

Bondi JA, Dietz CC (1945) Penicillin resistant staphylococci. Proc Soc Exp Biol Med 60:55–58

Chambers HF (1997) Methicillin resistance in *Staphylococci*: molecular and biochemical basis and clinical implications. Clin Microbiol Rev 10:781–791

Chambers HF (2001) The changing epidemiology of *Staphylococcus aureus*? Emerg Infect Dis 7:178–182

Chang S, Sievert DM, Haqeman JC, Boulton ML, Tenover FC, Downes FP, Shah S, Rudrik JT, Pupp GR, Brown WJ, Cardo D, Fridkin SK (2003) Infection with vancomycin-resistant *Staphylococcus aureus* containing the vanA resistance gene. N Engl J Med 348:1342–1347

Chanumolu SK, Rout C, Chauhan RS (2012) UniDrug-Target: A computational tool to identify unique drug targets in pathogenic bacteria. PLoS One 7(3), e32833

Chaudhuri RR, Allen AG, Owen PJ, Shalom G, Stone K, Harrison M, Burgis TA, Lockyer M, Lara JG, Foster SJ, Pleasance SJ, Peters SE, Maskell DJ, Charles IG (2009) Comprehensive identification of essential *Staphylococcus aureus* genes using transposon-mediated differential hybridisation (TMDH). BMC Genomics 10:291

Chen L, Xiong Z, Sun L, Yang J, Jin Q (2012) VFDB 2012 update: toward the genetic diversity and molecular evolution of bacterial virulence factors. Nucleic Acids Res 40(Database issue):D641–D645

Cui L, Iwamoto A, Lian JQ, Neoh H, Maruyama T, Horikawa Y, Hiramatsu K (2006) Novel mechanism of antibiotic resistance originating in vancomycin-intermediate *Staphylococcus aureus*. Antimicrob Agents Chemother 50:428–438

David MZ, Daum RS (2010) Community-associated methicillin-resistant *Staphylococcus aureus*: epidemiology and clinical consequences of an emerging epidemic. Clin Microbiol Rev 23:616–687

DeLeo FR, Diep BA, Otto M (2009) Host defense and pathogenesis in *Staphylococcus aureus* infections. Infect Dis Clin North Am 23:17–34

Enright MC, Robinson DA, Randle G, Fell EJ, Grundmann H, Spratt BG (2002) The evolutionary history of methicillin-resistant *Staphylococcus aureus* (MRSA). Proc Natl Acad Sci U S A 99:7687–7692

Gorak EJ, Yamada SM, Brown JD (1999) Community-acquired methicillin-resistant *Staphylococcus aureus* in hospitalized adults and children without known risk factors. Clin Infect Dis 29:797–800

Gorden RJ, Lowy FD (2008) Pathogenesis of methicillin-resistant *Staphylococcus* infection. Clin Infect Dis 46:S350–S359

Gustafson AM, Snitkin ES, Parker SCJ, Delisi C, Kasif S (2006) Towards the identification of essential genes using targeted genome sequencing and comparative analysis. BMC Genomics 7:265

Hawkey PM (2008a) The growing burden of antimicrobial resistance. J Antimicrob Chemother 62:i1–i9

Hawkey PM (2008b) Molecular epidemiology of clinically significant antibiotic resistance genes. Br J Pharmacol 153:S406–S413

Hiramatsu K (2001) Vancomycin-resistant *Staphylococcus aureus*: a new model of antibiotic resistance. Lancet Infect Dis 1:147–155

Hiramatsu K, Aritaka N, Hanaki H, Kawasaki S, Hosoda Y, Hori S, Fukuchi Y, Kabayashi I (1997a) Dissemination in Japanese hospitals of strains of *Staphylococcus aureus* heterogeneously resistant to vancomycin. Lancet 350:1670–1673

Hiramatsu K, Hanaki H, Ino T, Yabuta K, Oquri T, Tenover FC (1997b) Methicillin-resistant *Staphylococcus aureus* clinical strain with reduced vancomycin susceptibility. J Antimicrob Chemother 40:135–136
Holden MTG, Feil EJ, Lindsay JA, Peacock SJ et al (2004) Complete genomes of two clinical *Staphylococcus aureus* strains: evidence for the rapid evolution of virulence and drug resistance. Proc Natl Acad Sci U S A 101:9786–9791
Howden BP, Davies JK, Johnson PDR, Stinear TP, Grayson ML (2010) Reduced vancomycin susceptibility in *Staphylococcus aureus*, including vancomycin-intermediate and heterogeneous vancomycin-intermediate strains: resistance mechanisms, laboratory detection and clinical implications. Clin Microbiol Rev 23:99–139
Katayama Y, Ito T, Hiramatsu K (2000) A new class of genetic element, staphylococcus cassette chromosome mec, encodes methicillin resistance in *Staphylococcus aureus*. Antimicrob Agents Chemother 44:1549–1555
Kuroda M, Ohta T, Uchiyama I, Baba T, Yuzawa H et al (2001) Whole genome sequencing of methicillin resistant *Staphylococcus aureus*. Lancet 357:1225–1240
Lee DS, Burd H, Liu J, Almaas E, Wiest O, Barabasi AL, Oltvai ZN, Kapatral V (2009) Comparative genome-scale metabolic reconstruction and flux balance analysis of multiple *Staphylococcus aureus* genomes identify novel antimicrobial drug targets. J Bacteriol 191:4015–4024
Lindsay JA, Holden MTG (2004) *Staphylococcus aureus*: superbug, super genome? Trends Microbiol 12: 378–385
Livermore DM (2000) Antibiotic resistance in *Staphylococci*. Int J Antimicrob Agents 16:S3–S10
Lowy FD (2003) Antimicrobial resistance: the example of *Staphylococcus aureus*. J Clin Invest 111:1265–1273
Mahadevan P, King JF, Seto D (2009) CGUG: *in silico* proteome and genome parsing tool for the determination of "core" and unique in the analysis of genomes up to ca. 1.9 Mb. BMC Res Notes 2:168
Malachowa N, Deleo FR (2010) Mobile genetic elements of *Staphylococcus aureus*. Cell Mol Life Sci 67:3057–3071
Medini D, Donati C, Tettelin H, Masignani V, Rappuoli R (2005) The microbial pan-genome. Curr Opin Genet Dev 15:589–594
Mills SD (2003) The role of genomics in antimicrobial discovery. J Antimicrob Chemother 51:749–752
Moran GJ, Mount J (2003) Update on emerging infections: news from the centers for disease control and prevention. Ann Emerg Med 41:148–151
Morell EA, Balkin DM (2010) Methicillin resistant *Staphylococcus aureus*: a pervasive pathogen highlights the need for new antimicrobial development. Yale J Biol Med 83:223–233
Muzzi A, Masignani V, Rappouli R (2007) The pan-genome: towards a knowledge-based discovery of novel targets for vaccines and antibacterials. Drug Discov Today 12:429–439
Nickerson EK, Hongsuwan M, Limmathurotsakul D, Wuthiekanun V, Shah KR, Srisomang P et al (2009) *Staphylococcus aureus* bacteraemia in a tropical setting: patient outcome and impact of antibiotic resistance. PLoS One 4, e4308
Olson LH, Fitzgerald JR, Musser JM, Kapur V (2007) Molecular correlates of host specialization in *Staphylococcus aureus*. PLoS One 2, e1120
Pantosti A, Venditti M (2009) What is MRSA? Eur Respir J 34:1190–1196
Pantosti A, Sanchini A, Monaco M (2007) Mechanisms of antibiotic resistance in *Staphylococcus aureus*. Future Microbiol 2:323–334
Plata K, Rosato AE, Wegrzyn G (2009) *Staphylococcus aureus* is an infectious agent: overview of biochemistry and molecular genetics of its pathogenicity. Acta Biochim Pol 56:597–612
Saxena S, Gombar C (2010) Surmounting antimicrobial resistance in the millennium superbug: *Staphylococcus aureus*. Cent Eur J Med 5:12–29
Srivastava A, Dick JD, Perl TM (2002) Vancomycin resistance in *Staphylococci*. Clin Microbiol Rev 15:430–438
Stryjewski ME, Chambers HF (2008) Skin and soft tissue infections caused by community acquired methicillin resistant *Staphylococcus aureus*. Clin Infect Dis 46:S368–S377
Tenover FC (2006) Mechanisms of antimicrobial resistance in bacteria. Am J Infect Control 34:S4–S10
Tiwari HK, Sen MR (2006) Emergence of vancomycin resistant *Staphylococcus aureus* (VRSA) from a tertiary care hospital from northern part of India. BMC Infect Dis 6:156
Turner D, Reynolds D, Seto D, Mahadevan P (2013) CoreGenes3.5: a webserver for the determination of core genes from sets of viral and small bacterial genomes. BMC Res Notes 6:140
Zhang R, Lin Y (2009) DEG 5.0, a database of essential genes in both prokaryotes and eukaryotes. Nucleic Acids Res 37:D455–D458

13 Enzyme Technology, Functional Proteomics, and Systems Biology Toward Unraveling Molecular Basis for Functionality and Interactions in Biotechnological Processes

Mehak Baweja, Puneet Kumar Singh, and Pratyoosh Shukla

Abstract

Metagenomics provides the opportunity to uncover many unculturable extremophilic microorganisms, which represent majority of the planet's biological diversity and are being utilized in industries to further furnish industrial process and products. This strategy has resulted in the isolation of novel biocatalysts and bioactive molecules. Here in this chapter, we will review various strategies for manipulating enzyme attributes like its activity, stability, inhibition, designing novel substrates, and substrate-specific binding. We have also recapitulated the idea of functional proteomics and systems biology approach in protein engineering.

Keywords

Enzyme technology • Proteomics • Systems biology • Inulinase • Protease • Biotechnological process

13.1 Introduction

Biotechnology is the use of biological systems or organisms, cells, or cellular components to develop or modify products intended to improve the quality of human life. Biotechnology is quite a broad term which has been divided into several branches to categorize the area of concern. White biotechnology is the branch of biotechnology (also known as industrial biotechnology) that utilizes microbes and enzymes to make effectual products in the field of nutrition, pharmaceuticals, textiles, chemicals, paper, etc. White biotechnology holds many promises in achieving sustainable development bringing the low cost and eco-friendly substrates in the service with minimum waste generation. White biotechnology has been considered to surmount the burden

M. Baweja • P.K. Singh • P. Shukla (✉)
Enzyme Technology and Protein Bioinformatics Laboratory, Department of Microbiology, Maharshi Dayanand University, Rohtak 124001, Haryana, India
e-mail: pratyoosh.shukla@gmail.com

P. Shukla (ed.), *Frontier Discoveries and Innovations in Interdisciplinary Microbiology*, DOI 10.1007/978-81-322-2610-9_13

in petroleum-dependent industrialized nations worldwide. Since white biotechnology flaunts on enzymes, it is of prior requirement to increase the collection of enzymes with enhanced catalytic power, stability, specificity, etc. that specializes variety of enzyme from others.

13.2 Enzyme Improvement: Recent Strategies

Enzymes are proteinaceous biomolecules that function as specialized catalysts in various chemical reactions. They often have higher caliber to conventional chemical catalysts; on this rationale, they are being explored increasingly in today's high-technological society. Currently, researchers are familiar with approximately 4,000 enzymes, and of these, almost 200 wild microbial types are well known for their utilization for commercial purpose. However, on an industrial scale, only 20 enzymes are truly produced (Li et al. 2012). With improvement in technology and knowledge, various developments are being made to design methods that may improve the production, recovery, and application of respective enzyme in various areas. Systems biology is another branch that helps to understand the microbial interaction that may facilitate to control the intermediates or end product of the reaction (Singh and Shukla 2014).

The application of enzymes as successful industrial biocatalysts requires certain properties to be fulfilled, viz., good activity, specificity, and stability under process conditions. However, native enzymes are often not capable of encompassing all these properties, and the growing industrial demand could be fulfilled using natural enzymes. Recently, with increasing trend of protein engineering, there has been tremendous progress in the improvement of marketable enzymes into superior industrial catalysts. The increasing trend of bioinformatics and its various tools plays a big role in reducing time and labor and provides the direction to the experiment. There are various tools available that help to model the enzyme and select suitable substrate for it. In recent studies, Karthik et al. (2012b) modeled keratinase enzyme using PROVE software with RMS Z score of 1.432. In another study by Singh and Shukla (2012), inulinase modeling was completed using Swiss model and further docked with two substrates, ketose and fructose 6-phosphate, where fructose 6-phosphate presented better interaction compared to ketose. Furthermore, the interactions between the enzyme and its ligand were studied by Karthik et al. (2012a), and they firstly modeled an important therapeutic enzyme, i.e., 3-hydroxy-3-methylglutaryl coenzyme A reductase (HMGCR), and two of its isoforms were docked with several ligands, in which atorvastatin was found to be the most potent inhibitor against both isoforms.

Protein engineering is the technique that edits amino acid in a protein through changes in nucleotide of DNA that has enabled to generate engineered proteins or "biocatalysts." During the process, libraries are generated that are screened and often have better and improved characters than the wild enzyme. It employs two fundamental strategies – rational design and directed evolution. These techniques are not mutually exclusive.

13.2.1 Rational Design

Rational design is the technique for already known biocatalyst with full knowledge about its 3-D structure, viz., active sites, conserved sequences in molecules, and its motifs and mechanism of the reaction (Adrio and Demain 2014). There are various tools and techniques available to deduce the structural information, viz., Evolutionary algorithm (Vijayvargiya and Shukla 2012; Deb et al. 2002) were identified using Pareto genetic algorithm, etc. The rational design strategy is based on structural and functional relationship. Knowing the importance of each amino acid in the enzyme, it is comparatively easy to judge the substitution of particular amino acid. It includes site-directed mutagenesis where the mutations are made at specific known site. From most of studies, it is evident that changes are mainly made in active site to broaden the substrate specificity and to alter the catalytic residues to modify the function of enzyme. However, many researchers have observed that substitutions made far from active site had significant effect on the function of the protein. The advancement of new techniques like metagenomics have uncovered the microbes and their enzymes from the extreme niches. It has increased the database information and nowadays it is comparatively easier to find the related protein sequences and their structure. These techniques are helpful towards making the effectual rational design for variety of enzymes.

13.2.2 Directed Evolution

Directed evolution is an elementary technique which can proceed without much knowledge about the structure of the enzyme. This strategy follows the natural process of variant generation under specified conditions and includes entire range of molecular techniques known for creating genetic diversity. It involves random mutagenesis instead of site-directed mutagenesis of protein-encoding genes. The various techniques involved are error-prone polymerase chain reaction (PCR), repeated oligonucleotide-directed mutagenesis, or chemical mutagenesis. These techniques help to generate diversity of variants giving the bigger library to screen out the best possible output. Novozymes introduced two types of variants from rational protein design: Lipolase Ultra and LipoPrime. They were effective in cold washing (Table 13.1).

13.3 Enzymes of Industrial Importance

Modern era industries are totally enzyme dependent; therefore, there is a high need to improve the enzyme status. Here, progress relies on isolation of a large number of microorganisms from heterogenous niches of the earth and exploration of their peculiar properties and employment of their unusual characteristics of enzymes commercially. The most profound use of the enzyme is in detergent industry in which major contributors are protease and lipase. Proteases (EC. 3.4.21-24, 99) are hydrolytic enzymes that catalyze the cleavage of peptide bonds that tends to remove protein-based stains. Subtilisin Carlsberg, subtilisin BPN', and Savinase are some of the commercial products of proteases. However, with expansion in biotechnology, few newer preparations, such as Durazym, Maxapem, and Purafect, are now also available in the market which are developed using techniques of site-directed mutagenesis and/or random mutagenesis (Gupta et al. 2002). Protein engineering is also employed to determine the functional aspect of the amino acid residues. With the aim to learn structural and functional effects, Teplyakov et al. (1995) mutated the serine protease subtilisins from the alkalophilic *Bacillus* strain PB92. Substitution of Val 102 by Trp resulted in a 12-fold increase in activity that inferred the importance of substitution of single amino acid. The main goal behind protein engineering is to enhance the use of the enzyme in the specific industrial or biotechnological process so that it gets better fit into the process. Miyazaki et al. (2000) contributed to enhance the thermostability of the psychrophilic enzyme without losing its activity at low temperature through directed evolution. In addition to proteases, lipases also share a great contribution in detergent industry particularly in removing lipid-based stains. Apart from their key function of hydrolyzing carboxylic ester bonds, lipases can act as catalyst in esterification, interesterification, and transesterification reactions in non-aqueous media (Houde et al. 2004). Such versatility makes lipases the preferred enzyme for most of the other industrial application. There are various databases available for different enzymes that enable one to understand structure and functional relationships, related information on their sequences, active sites, etc. For lipases, the Lipase Engineering Database (LED), the Data Warehouse System (DWARF), and the alpha/beta-Hydrolase Fold 3DM Database (ABHDB)

Table 13.1 Comparison of features of rational redesign and directed evolution

Features	Rational redesign	Directed evolution
Preknowledge of structure	Required	Not essential
Number of variants	Less	More
Screening of variants or selection	Easy	High-throughput screening is essential
Cost effective	Yes	No

have been designed and maintained by various institutes. They empower the public databases like GenBank, PDB, and others by adding sequence, structure, and annotation information available, with the aim to recognize functionally significant residues and for the development of variants with valuable properties (Table 13.2).

13.4 Systems Biology Approach for Gene Regulatory Network

The gene regulatory networks (GRNs) which were used previously through computational approaches using gene expression data are not proficient in illustrating dynamic regulations among genes of small-scale networks. Figure 13.1 displays the construction of GRNs. Information theory models and basically correlation networks are examples of GRNs (Steuer et al. 2002; Stuart et al. 2003). These models are elementary, simple to compute, and static models that could not coregulate a target gene with multiple genes. Continuous gene expressions' nature is important in GRNs, and it cannot be described by Boolean networks because it is a discrete dynamic network, and in it binary variables are used for the representation of gene that is either on or off (Akutsu et al. 2000; Shmulevich et al. 2002; Bornholdt 2008). Bayesian networks (BNs) take hold of Bayesian rule that provides adaptable frame for combining divergent data and prior knowledge (Zou and Conzen 2005; Needham et al. 2007). The vector autoregressive (VAR) and state space models (SSMs) are discrete dynamic models which generally demand equally spaced and intensive time series data in order to get reliable inference results for model parameters. Differential equation models help to compute the change rate (derivative) of gene expression in quantified manner in a system and as function of expression levels for all other genes (Hirose et al. 2008; Shimamura et al. 2009). Differential equation models' dynamic feature is that it quantifies the expression and it is a

Table 13.2 An overview of various techniques used for enzyme activity improvement

Enzyme	Organism	Improved property	Technique	References
Lipase	*P. aeruginosa*	Twofold increase in amidase activity	Random mutagenesis and screening	Fujii et al. (2005)
Protease BYA	*Bacillus* sp. Y	Specific activity 1.5-fold higher	Site-directed mutagenesis	Tobe et al. (2006)
Endo-1,4-β-xylanase II	*Trichoderma reesei*	Increased alkali stability	Site-directed mutagenesis	Fenel et al. (2006)
α-Amylase	*Bacillus* sp. TS-25	10 °C improvement in thermostability	Directed evolution	Jones et al. (2008)
Laccase	*Bacillus* HR03	Threefold improved k_{cat} and thermostability	Directed mutagenesis	Mollania et al. (2011)
Xylanase		Improved thermostability	Directed evolution	Dumon et al. (2008)
Xylanase	Termite gut isolated metagenomically	Improved reducing sugar	Directed evolution	

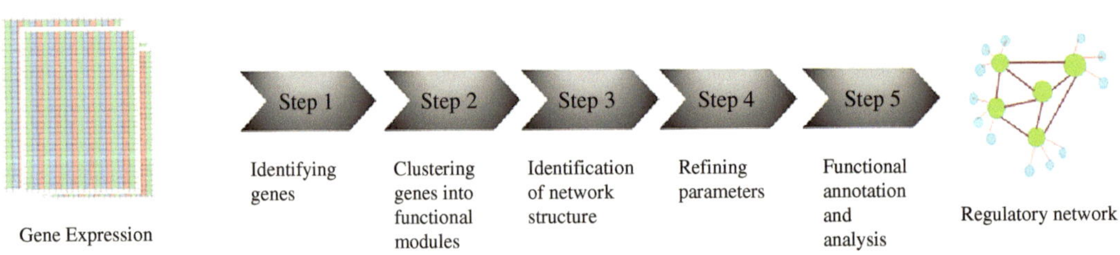

Fig. 13.1 A schematic diagram for reconstructing genome-wide dynamic GRNs

directed network graph model. Identifying model structure and parameter estimation with efficiency is a big challenge in using differential equation model for reconstructing GRNs.

13.5 ODE (Ordinary Differential Equation) for Genes of Industrially Important Enzymes

Many biological laws and relations can be mathematically represented as differential equation, and they are of basic importance in molecular biology. With the help of systems biology, dynamics of biochemical networks can be modeled, and the interactions between nodes (molecules) and edges (molecular interactions) are deliberated.

A system is defined by two aspects: one is a set of variables and how the variable values change in time. The variable could be deterministic or stochastic; deterministic value calculates the state of system at the next time based on the previous time and variable states. In stochastic, on the other hand, time and variable state defines the probability of the states at different points of time. Ordinary differential equations (ODE) are often applied in the perspective of systems biology. Dynamical models consist of Ordinary differential equations (ODE) which make easy investigation to the mechanism of action of a system in a systematic manner. For example, the cellular response to perturbations can be examined in molecular reactions (Raue et al. 2013). Table 13.1 lists the simulation language, software, and stochastic model simulator.

13.6 Modeling of the System

While modeling a system of interest, the following points should be kept in mind:

- Isolate your system of interest.
- Identifying the problem and entities which are required in the system where enzymatic reaction is occurring.
- Quantities which will be output and that can be analyzed.
- Entities which are to be required for the enzymatic reaction in the system.
- Characterize the environment of the system.

The model is composed of independent variables, dependent variables, parameters, and constraints. Independent variable, for example, time (t) in a system, implies the rate of addition of substrate in a system, product extraction, and growth. Dependent variables depend on the concentration of independent variable, for example, concentration of substrate and product at time (t). Parameters do not depend on independent variables and can be varied under different conditions. Lastly, constraints are fixed, e.g., Avogadro constant and gravitational constant (Table 13.3).

Table 13.3 Math-based simulation language, software, and stochastic model simulator list with their URL

Math-based simulation language and software	URL
Mathcad	http://www.ptc.com/product/mathcad
Matlab	http://in.mathworks.com/products/matlab/
Simulink	http://in.mathworks.com/products/simulink/
VisSim	http://www.vissim.com/
SBML	http://sbml.org/
Stochastic model simulator	
Dizzy	http://magnet.systemsbiology.net/software/Dizzy/
COPASI	http://www.copasi.org/
SimBiology	http://in.mathworks.com/products/simbiology/
ODEion	http://129.16.225.150/identification/ODEion/intro.html

Acknowledgments The authors acknowledge the support from SERB, Department of Science and Technology (DST), Government of India (DST Fast Track Grant. No. SR/FT/LS-31/2012), and University Grants Commission (UGC), New Delhi, India (Grant No. 42-457/2013(SR)). Mehak Baweja duly acknowledges the University Research Scholarship (URS) from M.D. University, Rohtak, India.

References

Adrio JL, Demain AL (2014) Microbial enzymes: tools for biotechnological processes. Biomolecules 4:117–139

Akutsu T, Miyano S, Kuhara S (2000) Inferring qualitative relations in genetic networks and metabolic pathways. Bioinformatics 16:727–734

Bornholdt S (2008) Boolean network models of cellular regulation: prospects and limitations. J R Soc Interface 5:S85–S94

Deb K, Pratap A, Agarwal S, Meyarivan T (2002) A fast and elitist multiobjective genetic algorithm: NSGA-II. IEEE Tran Evol Comput 6(2):182–197

Dumon C, Varvak A, Wall MA, Flint JE, Lewis RJ, Lakcy JH, Morland C, Luginbühl P, Healey S, Todaro T, DeSantis G, Sun M, Gessert LP, Tan X, Weiner DP, Gilbert HJ (2008) Engineering hyperthermostability into a GH11 Xylanase is mediated by subtle changes to protein structure. J Biol Chem 283:22557–22564

Fujii R, Nakagawa Y, Hiratake J, Sogabe A, Sakata K (2005) Directed evolution of *Pseudomonas aeruginosa* lipase for improved amide-hydrolyzing activity. Protein Eng Des Sel 18:93–101

Gupta R, Beg QK, Lorenz P (2002) Bacterial alkaline proteases: molecular approaches and industrial applications. Appl Microbiol Biotechnol 59(1):15–32

Hirose O, Yoshida R, Imoto S, Yamaguchi R, Higuchi T et al (2008) Statistical inference of transcriptional module-based gene networks from time course gene expression profiles by using state space models. Bioinformatics 24:932–942

Houde A, Kademi A, Leblanc D (2004) Lipases and their industrial applications: an overview. Appl Biochem Biotechnol 118(1–3):155–170

Jones A, Lamsa M, Frandsen TP, Spendler T, Harris P, Sloma A, Xu F, Nielsen JB, Cherry JR (2008) Directed evolution of a maltogenic α-amylase from *Bacillus* sp. TS-25. J Biotechnol 134:325–333

Karthik MV, Satya Deepak MV, Shukla P (2012a) Explication of interactions between HMGCR isoform 2 and various statins through in silico modeling and docking. Comput Biol Med 42(2):156–163. doi:10.1016/j

Karthik MVK, Syed H, Goswami T, Shukla P (2012b) Model and molecular docking substrate stabilization of *Microsporum canis* keratinase. Online J Bioinform 13(1):33–40

Li S, Yang X, Yang S, Zhu M, Wang X (2012) Technology prospecting on enzymes: application, marketing and engineering. Comput Struct Biotechnol J 2(3):1–11

Miyazaki K, Wintrode PL, Grayling RA, Rubingh DN, Arnold FH (2000) Directed evolution study of temperature adaptation in a psychrophilic enzyme. J Mol Biol 297:1015–1026

Mollania N, Khajeh K, Ranjbar B, Hosseinkhani S (2011) Enhancement of a bacterial laccase thermostability through directed mutagenesis of a surface loop. Enzyme Microb Technol 49:446–452

Needham C, Bradford J, Bulpitt AJ, Westhead D (2007) A primer on learning in Bayesian networks for computational biology. PLoS Comput Biol 3, e129

Raue A, Schilling M, Bachmann J, Matteson A, Schelke M, Kaschek D, Hug S, Kreutz C, Harms BD, Theis FJ, Klingmüller U, Timmer J (2013) Lessons learned from quantitative dynamical modeling in systems biology. PLoS One. doi:10.1371/journal.pone.0074335

Shimamura T, Imoto S, Yamaguchi R, Fujita A, Nagasaki M et al (2009) Recursive regularization for inferring gene networks from time-course gene expression profiles. BMC Syst Biol 3:41–54

Shmulevich I, Dougherty ER, Kim S, Zhang W (2002) Probabilistic Boolean networks: a rule-based uncertainty model for gene regulatory networks. Bioinformatics 18:261–274

Singh PK, Shukla P (2012) Molecular modeling and docking of microbial inulinases towards perceptive enzyme-substrate interactions. Indian J Microbiol 52(3):373–380

Singh PK, Shukla P (2014) Systems biology as an approach for deciphering microbial interactions. Brief Funct Genomics. doi:10.1093/bfgp/elu023

Steuer R, Kurths J, Daub CO, Weise J, Selbig J (2002) The mutual information: detecting and evaluating dependencies between variables. Bioinformatics 18:S231–S240

Stuart JM, Segal E, Koller D, Kim SK (2003) A gene-coexpression network for global discovery of conserved genetic modules. Science 302:249–255

Teplyakov AV, van der Laan JM, Lammers AA, Kelders H, Kalk KH, Misset O, Mulleners LJ, Dijkstra BW (1995) Protein engineering of the high-alkaline serine protease PB92 from Bacillus alcalophilus: functional and structural consequences of mutation at the S4 substrate binding pocket. Protein Eng 5(5):413–420

Tobe S, Shimogaki H, Ohdera M, Asai Y, Oba K, Iwama M, Irie M (2006) Expression of *Bacillus* protease (Protease BYA) from *Bacillus* sp. Y in *Bacillus subtilis* and enhancement of its specific activity by site-directed mutagenesis-improvement in productivity of detergent enzyme. Biol Pharm Bull 29:26–33

Vijayvargiya S, Shukla P (2012) A niched Pareto genetic algorithm for finding variable length regulatory motifs in DNA sequences. 3Biotech (Springer) 2(2):141–148

Zou M, Conzen S (2005) A new dynamic Bayesian network approach for identifying gene regulatory networks from time course microarray data. Bioinformatics 21:71–79

14 Archaeology vis-à-vis Microbiology: Discovering the Vistas of Interdisciplinary Research

Pranjal Garg and Pratyoosh Shukla

Abstract

The analysis of archaeological samples is very vital toward documentation and significance of archaeological remains and its interdisciplinary understanding. However, due to lack of integration between the two disciplines, i.e., archaeology and microbiology, it becomes very difficult to carry out a wholesome and scientifically supported research. Analysis of various samples like mummies, bones, organic archaeological remains, grave goods, etc., found at various archaeological sites has witnessed a new direction of analysis that has led to drastic changes in the formulation of understanding and theories. There are adequate records of such observations obtained from the analysis of samples from excavations and other archaeological remains and artifacts from different parts of the world. Our efforts should now concentrate on doing microbiological analysis of such rare archaeological samples and try to rationalize the need for such integration toward a better scientific and comprehensive approach.

Keywords

Archaeological remains • Microbiology • Next-generation sequencing • Multidisciplinary studies

P. Garg
Department of Archaeology, JRN Rajasthan Vidyapeeth (Deemed University), Udaipur 313001, Rajasthan, India

P. Shukla (✉)
Enzyme Technology and Protein Bioinformatics Laboratory, Department of Microbiology, Maharshi Dayanand University, Rohtak 124001, Haryana, India
e-mail: pratyoosh.shukla@gmail.com

14.1 Introduction

The coming together of archaeology and microbiology has revolutionized several aspects of analysis and understanding of archaeological remains, which, in the want of an interdisciplinary research, had remained untouched. Several

P. Shukla (ed.), *Frontier Discoveries and Innovations in Interdisciplinary Microbiology*,
DOI 10.1007/978-81-322-2610-9_14

new breakthrough discoveries in collaboration with microbiologists have advanced in the field of archaeology. Analysis of several samples from mummies, bones, organic archaeological remains, grave goods, marine archaeological samples, any material that contains bacterial or fungal growth, etc., from various archaeological sites has witnessed a new direction of analysis that has led to drastic changes in the formulation of understanding and theories (Suzuki et al. 2010; Child 1995; Reinhard and Bryant 1992). Ceramic investigation and other archaeological studies of the artifacts from the excavation make up the greater part of most archaeological reports, but microbiological and chemical analysis is an important and singular contribution to the field. Numerous facets of an archaeological site remain untouched and enigmatic without the microbial analysis (Schug et al. 2013). Microbiology has not just been adding to the data of archaeological importance but rectifying and asserting chemical analysis that had sometimes given alarmingly faulty results (The Mission Magazine, Spring 1996). If mummy or bones in a burial environment in examinable conditions are found, the bacterial and fungal isolates could tell details of genetic and dietary habits and help us understand the decomposing bones (Child 1995; Cavka et al. 2010; Antoine et al. 1992 for 1991).

An archaeological artifact cannot be restored to its original, but microbiology can definitely tell the archaeologist that if the artifact is not rescued soon, microbes are likely to eat it up completely. While this discovery was made, microbiologists also discovered a new species of a microbe of a previously known genus (Sánchez-Porro et al. 2010). So, often the process turns out to have mutual benefits and helps both fields grow in their research.

14.2 Microbiological Work in Concurrence to Archaeology

Microbiology has a significant relevance, contribution, and importance to the field of archaeology. *Mycobacterium leprae* was for the first time detected using whole-genome amplification and polymerase chain reaction of a sample from archaeological skeletal remains in the Far East of Japan (Suzuki et al. 2010). The Vedic literature has the earliest literary mention (c. mid second millennium BC) of the leprosy, while the more accepted evidence of the first millennium from the Indian surgery science *Sushruta Samhita* and *Arthashastra* (a treatise of statecraft, economic policy, and military strategy) clearly states the details about leprosy. Microbiologists took samples from the Chalcolithic archaeological site of Balathal (Robbins et al. 2007), located in the south of Rajasthan, India, to understand better the timing of first infection as well as the pattern of transmission of the disease and its geographic origin, which is believed to have originated from Africa. Evidence from India during 2000 BC indicates the presence of lepromatous leprosy. It is the oldest documented skeletal substantiation that supports the earliest Indian literary evidence of leprosy (Robbins et al. 2009).

The decline and disappearance of the Indus Valley Civilization still suffer inconclusive theories that don't apply unanimously to the entire extent of the civilization homogeneously. Nevertheless microbiology has joined hands with archaeology to confirm and assert the localized causes of decline of the habitation and effects of various infectious diseases and sociobiological methods by analyzing the samples from the mortuary remains (Schug et al. 2013). A sample from the late Indus valley period at Harappa was studied for the paleopathology to determine the outcomes of the climate change and the social as well as economic disturbances that likely led to the decline of the civilization (Fig. 14.1).

The analysis of a bioarchaeological evidence demonstrates a continuous prevalence and a gradual increase in infectious disease which is seen at a higher risk of infectious diseases spread unevenly among the burial communities. Analogous mortuary disparities indicate a greater vulnerability in the framework of uncertain climatic conditions for the socially and economically marginalized communities at Harappa. Along with the previous evidence for escalating levels of aggression within the society during that

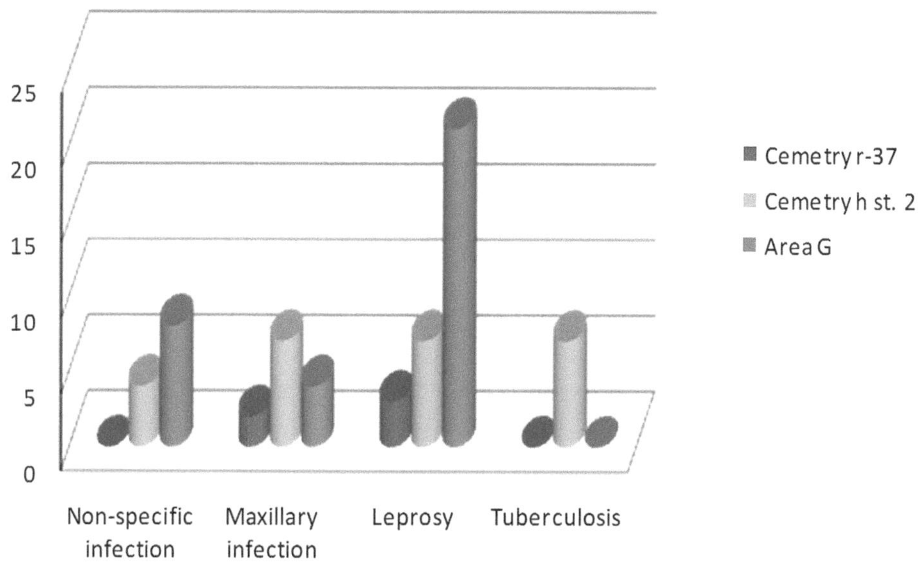

Fig. 14.1 Prevalence of infection and disease in three mortuary assemblages from Harappa (Schug et al. 2013)

time, the statistics sustains an increasing fight for attaining supremacy at the site of Harappa after 2000 B.C. (Schug et al. 2013). Scientists could go as far as to detail the socioeconomic conditions of the settlement from the pathological research, and analysis is indeed a creditable help to the field of archaeology. In fact, some sampling from other sites around or close to Harappa and a multidisciplinary analysis and approach could further strengthen and improve the theory stated. These observations on the relation between the climatic change and the social processes in the protohistoric urban settlements display decisive lessons in terms of susceptibility, diffidence, and the enduring consequences of temporary strategies that could cope with the change in climate (Schug et al. 2013).

A team of multidisciplinary scientists from microbiology, paleopathology, and paleoradiology in collaboration with an anthropologist worked on the samples of a mummy that was kept in the collection of the Zagreb Archaeological Museum, Croatia (Cavka et al. 2010). They documented a report of the outcome of the analysis of the samples of a few genera that belong to the groups of fungi and bacteria and could be considered crucial for carrying out further investigations. These samples were taken specifically from the oral and the abdominal cavity and the bandages tied around the mummy. Analysis conducted at the laboratory in Zagreb indicated that all of the found organisms belonged principally to the following groups of saprophytic fungi: *Monilia* spp., *Penicillium* spp., *Alternaria* spp., *Aspergillus fumigatus* spp., *Aspergillus nidulans* spp., *Rhizopus* spp., *Chrysosporium* spp., and a few genera of saprophytic bacteria, *Bacillus* spp., etc. (Cavka et al. 2010).

The results of the microorganisms isolated for microbiological analysis clearly demonstrate that they pose no threat to human health. They are not harmful, rarely associated with diseases, and ubiquitous in nature. With this, we cannot only understand the dietary habits of the dead but also its health at the time of the death and sometimes even the cause of the death.

Analysis of few samples from 21 archaeological sites from central Germany provided interesting diachronic insight into the epidemiology of carious defects in teeth of the people settled there over a period of 4,000 years (Nicklisch et al. 2015). The dietary habits and socioeconomic structures of the period were to be studied with the data obtained from a total of 494 individuals with preserved teeth. The results indicated that with the development of the society, meat and dairy products became more and more important, and the prevalence of caries significantly decreased between two cultural sub-eras. In addition to this, some sex-specific observations were made, which showed that a greater number of women were affected by caries than men. Undoubtedly, dental health reflects both biological and sociocultural and economic structures of the society.

Microbiologists have also tried to understand the decomposition by microorganisms from buried bone samples in archaeological sites (Antoine et al. 1992). A survey of literature suggests that a dead cell releases enzymes in the form of proteases and DNases. This causes an increase in autolytic destruction of the tissues resulting in microbial decomposition at a greater speed. As a result oxic and anoxic environments are created within the rotting flesh and "bone taint" appears within 6 h after the death (Roberts and Mead 1986). Post death, the gut flora affects the abdominal and thorax bones to a greater degree of putrefaction than the long bones and the skull. Firstly the gut flora putrefies the esophagus and diaphragm and then spreads into the thoracic region. Hagihara et al. (1988) have reported that causal organisms for dental caries present in the mouth and the gut feature both demineralization and collagenolytic characteristics, e.g., the yeast *Candida albicans*.

It has been suggested that the first phase of decomposition is carried out by microbial flora of the interred corpse, while in the second phase soil microorganisms are also involved. The soft tissues of the abdomen are mainly decomposed by anaerobic oxidation, which produces organic acids. These organic acids formed demineralize the surrounding bone. Some observed this to be the case; however, other studies indicate the reverse (Child 1995).

Analysis of microbes contained in the soil from archaeological sites in Italy and Belize was conducted by picking up the soil samples from the pottery of both sites. The samples were cultured for aerobic and anaerobic bacteria, acid-fast bacilli, fungi, and actinomycetes. This study was a challenge for the clinical microbiology. The organisms that were isolated were quite similar to those that might be found in contemporary soil specimens from equivalent geographical areas, though a more diverse and/or exotic microbial flora was expected owing to the presumed ages of the soil samples. The microscopy of the microbial flora and the results of the cultures of soil from the two archaeological sites did not differ much, except that the flora from the Lubaantun site, Belize, displayed more complex structure (Southern 2008).

In another report the archaeologists explored the great legendary "unsinkable ship" Titanic. They isolated the bacterium called *Halomonas titanicae* from the samples of the rusticles collected during the Akademik Keldysh (1991) expedition. Ballard (1989) and MacInnis (1992) have described "rusticles" as bioconcretious structures made up of rust, which look like icicles. These pieces of rusticles were sampled from the hull with the help of the articulated arm of the Mir-2 submersible. In this study, *Halomonas titanicae* strain BH1T was isolated from the rusticles using streak plate method on Bacto marine agar 2216 medium (Difco). The pure culture was subsequently preserved in lyophilized form till gene-sequencing program was initiated (Kaur 2004; Wells and Mann 1997). Microbiologists carried out a polyphasic taxonomic study and found the characteristics of the strain in detail. Analysis of the sample revealed the discovery of a new strain of the genus *Halomonas*. Overall, these studies prove that the abovementioned strain BH1T is showing phylogenetical similarity to various different species of the genus *Halomonas*, though they have essential differences in their phenotypic and chemotaxonomic features. However, DNA–DNA hybridization studies established it as a distinct species. Backed by this data, it was observed by scientists and microbiologists that a novel strain BH1T represents a distinctively separate and a new species of the genus *Halomonas*, named *Halomonas titanicae* sp. *nov*. (Sánchez-Porro et al. 2010). They realized that this isolated bacterium discovered from the rust samples of the RMS Titanic appears to accelerate the disintegration of the wreck of Titanic.

14.3 The Avenues and Challenges for Cross-Disciplinary Research

Another major problem stood in front of scientists when they carbon-date certain artifacts, and one of the most astonishing ones was in the case of the Shroud of Turin, precisely the burial cloth of Jesus of Nazareth. The radiocarbon dates showed its antiquity to a maximum of thirteenth to fourteenth century CE when it was realized that what had been radiocarbon-dated was in fact a jumble of linen and bacteria and fungi and bioplastic coating that had the occasion to grow on the fibers for centuries. The linen itself had not yet been dated. Often the conventional methods of cleansing a sample fail to remove bacteria and fungi. For example, biogenic varnishes were discovered on an ancient Mayan carved jade called "Itzamna Tun." The scientists had to decontaminate the sample of the plastic-like coating that is a by-product of bacteria and fungi (The Mission Magazine, Spring 1996 issue).

Archaeological remains from the burial environment have often yielded good samples for understanding dietary habits of people. In the nineteenth century, it was discovered that coprolites are the most noteworthy and remarkable dietary remains recoverable from archaeological contexts. Harshberger (1896) discovered for the first time the potential importance of human coprolite and concluded that bones and seeds found in ancient feces could very well offer clues to ancient diet. Later, Netolitsky (1911, 1912) studied the coprolites from Egyptian mummies and identified a number of important dietary components in it. Either dead bacteria can directly be identified in coprolites or else live bacteria have to be cultured for analytical examination. Two types of bacteria species are present in coprolite, viz., cystic and noncystic anaerobic bacteria. Endospore-containing bacteria, for instance, *Clostridium*, can be isolated and characterized from coprolites. However, noncystic disease-causing bacteria, viz., *Salmonella* and *Shigella*, are vulnerable to decay in the postdepositional surroundings and are difficult to be characterized (Reinhard and Bryant 1992). In addition, to these bacteria, there are reports of well-preserved hyphal segments and fungal spores in coprolites. These evidences help to understand the coprolite preservation strategies and also offer dietary data. Interpreting these remains is obstructed by the lack of taxonomical identity for specific class of fungi (Reinhard 1985a, b, c). Majority of research activities on coprolite have enlarged to include the study of many different disciplines along with microbiology. Nevertheless, its results and interpretations still remain a challenge for the scientists.

14.4 The Success Stories and Applying Next-Generation Sequencing

Next-generation sequencing (NGS) allows us quite easily to see if we are dealing with authentic ancient human DNA. The technique of NGS is metagenomic advancement in which, before sequencing, the DNA fragments in a sample are amplified, together with little samples of ancient DNA that are not possible by traditional DNA isolation methods. This offers an extensive overview of the genetic material of sample under study and recommends an extra precise data of the accurate ratio of prehistoric DNA to modern DNA. The advantage of the speed of next-generation platforms enables us to sequence a sample repeatedly, making it easier to verify the patterns of damage that distinguish ancient DNA from recent DNA (Cossins 2013).

The new NGS findings have not gone unchallenged, however. Some researchers have questioned Pusch's suggestion that the proportion of human DNA in samples from the Egyptian heads is comparable to that from permafrost-preserved specimens—such as a 4000-year-old Greenlandic mummy known as the Saqqaq man—subjected to NGS (Khairat et al. 2013).

This technique is specially being applied to the samples of mummified bodies as the mummification process protects DNA by rapidly desiccating the tissue, removing the water required for hydrolytic depurination and deamination, two of the primary mechanisms by which DNA is broken down.

The comparison to the ice-preserved specimens is misleading because the vast majority of the sequences obtained from the Egyptian samples were unidentifiable. For instance, in one of the Egyptian samples put forward as comparable, the researchers could not identify 97.5 % of the sequences, so they used the 2.5 % they could, of which 96 % were human. For comparison, 75 % of the sequences for Saqqaq man could be identified, 99.7 % of which were human (Khairat et al. 2013).

Even so, the latest study is a first step in the right direction and that whole-genome sequencing of ancient Egyptian mummies will likely become commonplace in the near future—that is, so long as sufficient funds are available and researchers can get their hands on specimens.

It could also bring about a new era of "molecular archaeology" in which mysteries concerning the ancestral origins of a certain ancient race, and how they lived and died, can finally be solved.

14.5 Microbes Are the Natural Data Bank of Archaeologists

Microbiologists have tried to determine paleoclimatic conditions through viable paleosol microorganisms, which could also help date the archaeological stratas.

The paleosol microorganisms were cultured by scientists from a complete chronosequence of the archaeological site of Hell Gap, Wyoming, USA (Grund et al. 2014), on an experimental basis to see if the colonies contain information regarding the climatic conditions of that region. They even went to the extent to do a relative dating of the archaeological site by looking at the change in the density of the microorganisms present in the soil sample. Studies of pathogens in archaeological contexts and microorganisms in ceramic residues are widespread, though not very frequent. But this is altogether a revolutionary idea, which could supplement the paleobotanic soil and phytolith analysis results. The results of the samples from Hell Gap didn't meet greatly the expectations, but the thorough study of the results helped understand the other external factors involved that could make the analysis' results go wayward (Grund et al. 2014). Those factors could be kept in mind while undertaking such a study again elsewhere. This experimental study has opened up new horizons for bridging collaborative efforts of microbiology and archaeology to get a comprehensive scientific study of archaeological data.

14.6 Microbes Determine Settlement Pattern

The analysis of soils and cultural layers at an archaeological site in Russia showed that repeated addition of various organic materials increases a microbial biomass and enzymatic activity (Chernysheva et al. 2015). This occurs in particular with urease activity, which increases in soil because of continuous input of compost, manure, and other organic residues and income of ureolytic bacteria along with organic fertilizers. The urease activity was observed in the cultural layers of the site in that area with ruined wall remains. The activity was almost twice higher than the area without wall remains. These results showed that the structures in the settlement were used as cattle pens. So it was inferred that comparison of urease activity, scattered potsherds, and phosphorus content in soil could help demarcate the boundaries of ancient manured lands. These determining parameters could be useful in revealing the sites of cattle keeping, areas of ancient arable lands, and other clues to the settlement pattern of a site.

14.7 Conclusion

Archaeology is a field of multidisciplinary contributions, which has much scope and potential for research. There's much contribution asked of microbiology in the field of archaeology. Archaeologists depend mostly on manual analysis of the samples of the excavations; instead, even the fundamental analysis of the soil, bone, and other samples in microbiological labs would surface manifold inferences that can be drawn to get a more integral understanding of an archaeological site. Our efforts should now concentrate on doing microbiological analysis of such rare archaeological samples and try to rationalize the need for such integration toward a better scientific and comprehensive approach.

Acknowledgment We are grateful and privileged to be inspired and encouraged by the Association of Microbiologists of India (AMI) to go ahead with this interdisciplinary project so that more research can be conducted in the field of interdisciplinary archaeology.

References

Antoine SE, Child AM, Nicholson RA et al (1992 for 1991) The biochemistry and microbiology of buried human bone, in relation to dietary reconstruction. Circaea 9(2):65–79

Ballard RD (1989) The discovery of the titanic. Penguin/ Madison Press Books, Toronto

Barrett J (1996) Microbiology meets archaeology in a renewed quest for answers. The Mission Magazine, spring issue. (http://uthscsa.edu/mission/spring96/ shroud.htm)

Cavka M et al (2010) Microbiological analysis of a Mummy from the Archaeological Museum in Zagreb. Coll Antropol 34(3):803–805

Chernysheva EV, Korobov DS, Khomutova TE et al (2015) Urease activity in cultural layers at archaeological sites. J Archaeol Sci 57:24–31

Child AM (1995) Towards an understanding of the microbial decomposition of archaeological bone in the Burial environment. J Archaeol Sci 22:165–174

Cossins D (2013) The mummy code. The Scientist 27(8). http://www.the-scientist.com/?articles.view/articleNo/36667/title/The-Mummy-Code/#

Grund BS, Williams SE, Surovell TA (2014) Viable paleosol microorganisms, paleoclimatic reconstruction, and relative dating in archaeology: a test case from Hell Gap, Wyoming, USA. J Archaeol Sci 46:217–228

Hagihara Y, Kaminishi H, Cho T et al (1988) Degradation of human dentine collagen by an enzyme produced by the yeast *Candida albicans*. Arch Oral Biol 33(8):617–619. http://dx.doi.org/10.1016/0003-9969(88)90138-0

Harshberger JW (1896) The purposes of ethnobotany. Am Antiq 17:73–81

Kaur B (2004) Investigation of the identity and biocorrosive ability of a novel deep-sea bacterium, belonging to the genus Halomonas from the Titanic. PhD thesis, Dalhousie University, Halifax, Nova Scotia

Khairat R, Ball M, Chang CH et al (2013) First insights into the metagenome of Egyptian mummies using next generation sequencing. J Appl Genet 54(3):309–325. doi:10.1007/s13353-013-0145-1

MacInnis JB (1992) Unlocking the secrets. In: Shulman S snd Charlottesville VA (ed) Titanic in a new light, Thomasson-Grant, USA, pp. 89–94

Netolitsky F (1911) Nahrungs und Heilmittel der Uragypter. Die Umschau 46:953–56

Netolitsky F (1912) Hirse und Cyperus aus dem Prehistorischen Agypten. Beihefte zum Botanischen Centralblatt 29:1–11

Nicklisch N, Ganslmeiera R, Siebert A et al (2015) Holes in teeth–Dental caries in Neolithic and Early Bronze Age populations in Central Germany. Ann Anat (article in press). http://dx.doi.org/10.1016/j.aanat.2015.02.001

Reinhard KJ (1985a) Recovery of helminths from prehistoric feces: The cultural ecology of ancient parasitism. MS thesis, Department of Biology, Northern Arizona University

Reinhard KJ (1985b) Parasitism at Antelope House: a Puebloan village in Canyon de Chelly, Arizona. In: Merbs CF and Miller RJ (eds) Health and disease in the Prehistoric Southwest, Tempe: Arizona State University anthropological research papers, no 34, pp 220–233

Reinhard KJ (1985c) Strongyloides stercoralis in the Prehistoric Southwest. In: Merbs CF and Miller RJ (eds) Health and disease in the Prehistoric Southwest, Tempe: Arizona State University anthropological research papers, no 34, pp 234–242

Reinhard KJ, Bryant VM Jr (1992) Coprolite analysis: a biological perspective on archaeology. J Archaeol Method Th 4:245–288

Robbins G, Tripathy VM, Misra VN et al (2007) Report on the human remains at Balathal. Man Environ 31:50–65

Robbins G, Tripathy VM, Misra VN et al (2009) Ancient skeletal evidence for leprosy in India (2000 B.C.). PLoS One 4(5), e5669. doi:10.1371/ journal.pone.0005669

Roberts TA, Mead GC (1986) Involvement of intestinal anaerobes in the spoilage of red meats, poultry and fish. In: Barnes EM, Mead GC (eds) Anaerobic bacteria in habitats other than Man. Blackwell Scientific, Oxford, pp 333–349

Sánchez-Porro C, Kaur B, Man H et al (2010) *Halomonas titanicae* sp nov a halophilic bacterium isolated from the RMS Titanic. Int J Syst Evol Microbiol 60:2768–2774. doi:10.1099/ijs.0.020628-0

Schug GR, Blevins KE, Cox B et al (2013) Infection, disease, and biosocial processes at the end of the Indus Civilization. PLoS One 8(12), e84814. doi:10.1371/ journal.pone.0084814

Southern PM (2008) Archaeology meets clinical microbiology: analysis of the microbial content of soil from archaeological sites in Italy and Belize. Lab Med 39:601–602

Suzuki K, Takigawa W, Tanigawa K et al (2010) Detection of *Mycobacterium leprae* DNA from Archaeological skeletal remains in Japan using whole genome amplification and polymerase chain reaction. PLoS One 5(8), e12422. doi:10.1371/journal.pone.0012422

Wells W, Mann H (1997) Microbiology and formation of rusticles from the RMS Titanic. Resour Environ Biotechnol 1:271–281